EUROVAL – An European Initiative on Validation of CFD Codes

Edited by
Werner Haase
Frans Brandsma
Eberhard Elsholz
Michael Leschziner
Dieter Schwamborn

Notes on Numerical Fluid Mechanics (NNFM)   Volume 42

Series Editors: Ernst Heinrich Hirschel, München
Kozo Fujii, Tokyo
Bram van Leer, Ann Arbor
Keith William Morton, Oxford
Maurizio Pandolfi, Torino
Arthur Rizzi, Stockholm
Bernard Roux, Marseille

Volume 26 Numerical Solution of Compressible Euler Flows (A. Dervieux / B. van Leer / J. Periaux / A. Rizzi, Eds.)
Volume 27 Numerical Simulation of Oscillatory Convection in Low-Pr Fluids (B. Roux, Ed.)
Volume 28 Vortical Solution of the Conical Euler Equations (K. G. Powell)
Volume 29 Proceedings of the Eighth GAMM-Conference on Numerical Methods in Fluid Mechanics (P. Wesseling, Ed.)
Volume 30 Numerical Treatment of the Navier-Stokes Equations (W. Hackbusch / R. Rannacher, Eds.)
Volume 31 Parallel Algorithms for Partial Differential Equations (W. Hackbusch, Ed.)
Volume 32 Adaptive Finite Element Solution Algorithm for the Euler Equations (R. A. Shapiro)
Volume 33 Numerical Techniques for Boundary Element Methods (W. Hackbusch, Ed.)
Volume 34 Numerical Solutions of the Euler Equations for Steady Flow Problems (A. Eberle / A. Rizzi / H. E. Hirschel)
Volume 35 Proceedings of the Ninth GAMM-Conference on Numerical Methods in Fluid Mechanics (J. B. Vos / A. Rizzi / I. L. Ryhming, Eds.)
Volume 36 Numerical Simulation of 3-D Incompressible Unsteady Viscous Laminar Flows (M. Deville / T.-H. Lê / Y. Morchoisne, Eds.)
Volume 37 Supercomputers and Their Performance in Computational Fluid Mechanics (K. Fujii, Ed.)
Volume 38 Flow Simulation on High-Performance Computers I (E. H. Hirschel, Ed.)
Volume 39 3-D Computation of Incompressible Internal Flows (G. Sottas / I. L. Ryhming, Eds.)
Volume 40 Physics of Separated Flow – Numerical, Experimental, and Theoretical Aspects (K. Gersten, Ed.)
Volume 41 Incomplete Decompositions (ILU) – Algorithms, Theory and Applications (W. Hackbusch / G. Wittum, Eds.)

Volumes 1 to 25 are out of print.
The addresses of the Editors and further titles of the series are listed at the end of the book.

EUROVAL –
An European Initiative on Validation of CFD Codes

Results of the EC/BRITE-EURAM Project
EUROVAL, 1990–1992

Edited by
Werner Haase
Frans Brandsma
Eberhard Elsholz
Michael Leschziner
Dieter Schwamborn

This publication is supported by the VALUE Programme of the Commission of the European Communities, which promotes the dissemination and exploitation of information from Community research and technological development programmes.

All rights reserved
© Friedr. Vieweg & Sohn Verlagsgesellschaft mbH, Braunschweig/Wiesbaden, 1993

Vieweg ist a subsidiary company of the Bertelsmann Publishing Group International.

No part of this publication may be reproduced, stored in a retrieval system or transmitted, mechanical, photocopying or otherwise, without prior permission of the copyright holder.

Produced by W. Langelüddecke, Braunschweig
Printed on acid-free paper
Printed in Germany

ISSN 0179-9614
ISBN 3-531-07642-9

Preface

Early in 1986, the Commission of the European Community observed a number of developments in world aviation, which were seen to threaten continuing success of the European aeronautic industry. In particular, a remarkable increase in research and development activities was noticed worldwide. These observations prompted the Commission to approach nine major European aircraft companies with the suggestion that action should be taken at Community level. It was decided that all European aircraft companies should collaborate, and a consortium was established to undertake "A Joint Study of **EURO**pean Cooperative **M**easures for **A**eronautical **R**esearch and **T**echnology", abbreviated as EUROMART.

Activities were directed towards European collaborative programmes on aerothermodynamics, materials, acoustics, computation, airborne equipment and systems, all-electric aircraft, CAD-CAM/CIM and laminar-flow control, addressing the aircraft, engine and equipment industries. All these programmes are seen in the light of enhancing and expanding the technological capabilities of the European Aeronautics Industry, whilst, at the same time, concentrating on key technology areas.

As a result of the study, the Commission initiated a collaborative research programme on aeronautics, addressing the needs of aircraft, engine and equipment industries. Following the decision of the Council of the European Communities, the Pilot Phase of the collaborative aeronautics programme was launched in 1989 as Area 5 of the generic research program on industrial and materials technologies BRITE/EURAM in order to help to redress an imbalance in research activities at community level. Solely addressing civil applications, the specific activities relating to aeronautics covered four work areas, namely aerodynamics, acoustics, airborne systems and equipment, and propulsion systems.

An evaluation process was carried out in all four areas, and this led to quite a number of proposals submitted to the Commission of the European Communities. One of the aerodynamics proposals selected in the Pilot Phase was EUROVAL, a "**EURO**pean Initiative on **VAL**idation of CFD Codes", officially launched in February 1990 and incorporating 16 partner organisations from industries (including small and medium enterprises), universities and research establishments in 11 European countries.

Generally, the aim of this project was to work towards an improvement and enhancement of the quality of existing computational fluid dynamics codes based on a careful validation against measurements. It was planned as, and in fact turned out to be, a strong collaborative project, standing well apart from what is sometimes called an "Olympics of CFD Codes", aimed at generating high-quality numerical results. Special emphasis was placed on the assessment of turbulence-models on the basis of two- and three-dimensional flow investigations.

To achieve this goal, selected test cases had to be solved with well defined, mandatory constraints. Work was divided into seven tasks, ranging from 2D boundary layer predictions through 3D Navier-Stokes investigations to direct numerical simulations, and covering transonic airfoil and wing flow as well as maximum- and high-lift problems, wind-tunnel interference investigations and vortex breakdown studies. The aim of all working tasks was to accelerate the development and improvement of computational fluid dynamics codes and, moreover, to increase the expertise on fast and robust numerical methods, which are increasingly becoming powerful tools in a design context by lowering costs and providing information in off-design conditions where measurements are difficult to conduct.

Thanks are due to all partners who have contributed to the success of the EUROVAL initiative with extraordinary team spirit, thus making it a pleasure for the editors to present the technical outcome in this book.

Thanks are also due to the Siemens-Nixdorf company which has sponsored the work in EUROVAL by providing partners with free-of-charge CPU time on a VP-600 supercomputer and supporting partners during the time of three one-week working periods in Karlsruhe, Germany.

Last but not least, the editors are grateful to E.H. Hirschel as the general editor of the "Notes on Numerical Fluid Mechanics", as well as to the Vieweg Verlag, for the opportunity to publish the results of EUROVAL in this scientific book series.

April, 1992

F. Brandsma,	Amsterdam
E. Elsholz,	Bremen
W. Haase,	Friedrichshafen
M. Leschziner,	Manchester
D. Schwamborn,	Göttingen

Table of Contents

		Page
1	**Introduction**	1
	1.1 Main Validation Topics	1
	1.2 Technical Results	2
	1.3 Cooperative Strategy	3
	1.4 Conclusions	5
2	**Structure of the EUROVAL Project**	7
	2.1 List of Tasks and Brief Descriptions	7
	2.2 Task Matrix	9
	2.3 Application List	9
	2.4 List of Partners	10
3	**Description of Turbulence Models**	13
	3.1 Algebraic Models	13
	3.1.1 Introduction	13
	3.1.2 The Baldwin-Lomax [BL] Model (1978)	14
	3.1.3 The Granville [GR] Model (1987)	15
	3.1.4 The Cebeci-Smith [CS] Model (1974)	16
	3.1.5 The Goldberg [GB] Backflow Model (1986)	18
	3.2 One-half Equation Turbulence Models	20
	3.2.1 Introduction	20
	3.2.2 The Johnson-King [JK] Model (1985)	20
	3.2.3 The Johnson-Coakley [JC] Model (1990)	22
	3.2.4 The Horton [HO] Model (1990)	24
	3.3 One-Equation Models	26
	3.3.1 Introduction	26
	3.3.2 The Wolfshtein [W] Model (1969)	26
	3.3.3 The Hassid-Poreh [HP] Model (1975)	27
	3.4 Two-Equation Models	28
	3.4.1 Introduction	28
	3.4.2 The Jones-Launder and Launder-Sharma [JL] Models (1972, 1974)	28

Table of Contents (continued)

                                                                                                  Page

      3.4.3 The Chien [C] Model (1982) 29
      3.4.4 The Lam-Bremhorst [LB] Model (1981) 30
      3.4.5 The "Standard" High-Reynolds-number
           Jones-Launder [SJL] Model (1972) 31

  3.5 Reynolds-Stress Models 32
      3.5.1 Introduction 32
      3.5.2 The Gibson-Launder-Rodi
           Reynolds-Stress [RS] Model (1978) 32

  3.6 Calculation of Boundary Layer Length Scales 34

4 Numerical Methods Used by Partners 37

  4.1 Description of Method Used by
      Analysis Systems Research Ltd. (S. Perrakis, M. Tzitzidis) 37
      4.1.1 Introduction 37
      4.1.2 Method 37
      4.1.3 Boundary Conditions 38
      4.1.4 Grid Adaption 38
      4.1.5 Modifications to the 'Standard' Turbulence Models 38
      4.1.6 Computers and CPU-Times 38

  4.2 Description of Methods Used by British Aerospace (J. J. Benton) .. 39
      4.2.1 Introduction 39
      4.2.2 2D Multiblock Navier-Stokes 39
           4.2.2.1 Summary 39
           4.2.2.2 Transport Equations 39
           4.2.2.3 Discretisation 40
           4.2.2.4 Artificial Dissipation 42
           4.2.2.5 Time Integration 43
           4.2.2.6 Multigrid Acceleration 44
           4.2.2.7 Turbulence Models 45
           4.2.2.8 Boundary Conditions 47
           4.2.2.9 Integrated Forces 48
           4.2.2.10 Computational Performance 48
      4.2.3 3D Multiblock Viscous-Coupled Euler 48

  4.3 Description of Method Used by CAPTEC (E. A. Cox) 49
      4.3.1 Introduction 49

Table of Contents (continued)

	Page
4.3.2 Method	49
4.3.3 Grid Adaptation	49
4.3.4 Modifications of the Cebeci-Smith Model	50
4.4 Description of Methods Used by Construcciones Aeronáuticas S.A. - CASA (A. Abbas, J.J. Guerra)	51
4.4.1 Introduction	51
4.4.2 Navier-Stokes Approach	51
4.4.2.1 Governing Equations	51
4.4.2.2 Numerical Formulation	52
4.4.2.3 Artificial Dissipation	53
4.4.2.4 Turbulence Model Application	54
4.4.2.5 Numerical Details	54
4.4.3 Euler Approach	55
4.4.3.1 Governing Equations	55
4.4.3.2 Dissipation Model	56
4.4.3.3 Numerical Details	57
4.5 Description of Method Used by CERFACS (L. Davidson, A. Kourta, L.Tourette)	58
4.5.1 Introduction	58
4.5.2 Method 1 (Single Element and Bumps)	58
4.5.3 Method 2 (Multi-Element)	59
4.5.4 Method 3 (DLR-F5 Wing)	60
4.5.4.1 Mesh	60
4.5.4.2 The Mean-Flow Equations and their Approximation	62
4.5.4.3 The k-ε Turbulence Model and its Approximation	65
4.6 Description of Method Used by CFD norway (I. Øye, E. Ørbekk, H. Nørstrud)	69
4.6.1 Introduction	69
4.6.2 Governing Equations	69
4.6.3 Numerical Method	70
4.6.4 Modifications to 'Standard' Turbulence Models	71
4.6.5 Mesh Generation	71
4.7 Description of Methods Used by Deutsche Airbus (E. Elsholz)	72
4.7.1 Introduction	72
4.7.2 Navier-Stokes Approach	72

Table of Contents (continued)

	Page
4.7.3 Boundary Layer Approach	75
4.7.3.1 'Locally Infinite Swept Wing' (LISW) Boundary Layer Method	75
4.7.3.2 The 3D Boundary Layer Method	76
4.8 Description of Method Used by DLR (D. Schwamborn)	78
4.8.1 Introduction	78
4.8.2 The Block Structured Navier-Stokes Method	78
4.8.3 Boundary Conditions	79
4.8.4 Performance of the Code	80
4.8.5 Turbulence Models	80
4.8.5.1 The Baldwin-Lomax Model (BL)	80
4.8.5.2 The Granville Modification of the Baldwin-Lomax Model (GR)	81
4.8.5.3 The Johnson-King Model (JK)	81
4.8.5.4 The Johnson-Coakley Model (JC)	81
4.9 Description of Method Used by Dornier (W. Haase, W. Fritz)	82
4.9.1 Introduction	82
4.9.2 The Boundary Layer Method	82
4.9.2.1 Numerical Approach	82
4.9.2.2 Turbulence Model Implementation	83
4.9.3 The Navier-Stokes Method	83
4.9.3.1 Governing Equations	83
4.9.3.2 Finite Volume Method	84
4.9.3.3 Filtering Technique	84
4.9.3.4 Convergence Acceleration	85
4.9.3.5 Steady State	85
4.9.3.6 Boundary Conditions	85
4.9.3.7 Mesh Generation	86
4.9.3.8 Modifications to Original Versions of Turbulence Models	87
4.10 Description of Method Used by NLR (F.J. Brandsma)	89
4.10.1 Introduction	89
4.10.2 Governing Equations	89
4.10.3 Discretization in Space	90
4.10.4 Time Integration	92
4.10.5 Convergence Acceleration	93

Table of Contents (continued)

	Page
4.10.6 Boundary Conditions	93
4.10.7 Turbulence Models	94
4.10.8 Parameter Settings and Code Performance	95
4.11 Description of Method Used by RAE (P.D. Smith)	97
4.11.1 Governing Equations	97
4.11.2 Numerical Method	98
4.11.3 Initial and Boundary Conditions	99
4.11.4 Turbulence Models	99
4.12 Description of Method Used by SAAB (B. Arlinger, T. Larsson)	100
4.12.1 Introduction	100
4.12.2 Numerical Method (SAAB)	100
4.12.3 Turbulence Modelling	100
4.12.3.1 The Baldwin-Lomax (BL) Model	101
4.12.3.2 The Two-Layer k-ε Model	101
4.12.4 Grid Generation	101
4.12.5 Numerical Method (HUT)	102
4.13 Description of Method Used by TU Berlin (L. Xue, T. Rung, F. Thiele)	103
4.13.1 Transformation and Discretisation of the Governing Equations	103
4.13.2 Space Marching Integration Scheme	105
4.13.3 Turbulence Models	106
4.14 Description of Method Used by TU Denmark (J. Michelsen, J.N. Sørensen)	107
4.14.1 2D Navier-Stokes Method for Incompressible Flow	107
4.14.2 3D Navier-Stokes Method for Direct Simulation of Vortex Breakdown	109
4.15 Description of Method Used by UMIST (G. Page, M.A. Leschziner)	111
4.15.1 Governing Equations	111
4.15.2 Discretisation	111
4.15.3 Time Marching	113
4.15.4 Time-Stepping Control	114
4.15.5 Artificial Dissipation	115

Table of Contents (continued)

	Page
4.15.6 Multigrid Acceleration	116
4.15.7 Turbulence Models	117
4.16 Description of Method Used by VUB (M. Mulas, Ch. Hirsch)	118
4.16.1 Introduction	118
4.16.2 Cell-Centered Finite Volume Formulation	118
4.16.3 High Resolution TVD Schemes	119
4.16.4 Time Integration	121
4.16.5 Observations	121

5 Application-Oriented Discussion of Results ... 123

- 5.1 Airfoil Flow RAE-2822 Cases 9 and 10 ... 125
  - 5.1.1 Introduction ... 127
  - 5.1.2 Overview on Investigations ... 127
  - 5.1.3 Mandatory Mesh and Parameter Set Establishment, Effect of Camber Correction ... 129
  - 5.1.4 Effect of Turbulence Modelling: Mandatory Test Cases ... 131
    - 5.1.4.1 Algebraic Turbulence Models ... 134
    - 5.1.4.2 Half-Equation Turbulence Models ... 136
    - 5.1.4.3 Transport and Reynolds Stress Turbulence Models (C), (SJL-W), (RS) ... 137
  - 5.1.5 Effect of Mesh: Density, Extension and Far-Field Condition ... 138
    - 5.1.5.1 Mesh Density Effects ... 138
    - 5.1.5.2 Mesh Extension Effects ... 141
    - 5.1.5.3 Effects of Far-Field Boundary Condition ... 141
  - 5.1.6 Effect of Mach Number ... 142
  - 5.1.7 Effect of Numerical Incidence ... 144
  - 5.1.8 Summary / Conclusions ... 146
- 5.2 ONERA Bumps A and C ... 185
  - 5.2.1 Introduction ... 187
  - 5.2.2 Overview of Contributors and Contributions ... 188
  - 5.2.3 Turbulence Models ... 190
  - 5.2.4 Numerical Algorithms ... 192
  - 5.2.5 Grids ... 192
  - 5.2.6 Boundary Conditions ... 193
  - 5.2.7 Resource Requirements ... 194

Table of Contents (continued)

	Page
5.2.8 Results	195
5.2.8.1 Introductory Comments	195
5.2.8.2 Delery Case A	196
5.2.8.3 Delery Case C	201
5.2.9 Conclusions	204
5.3 Maximum Lift Investigations for A-Airfoil	**267**
5.3.1 Partners Versus Methods and Models	269
5.3.2 Discussion of Results	269
5.3.3 Limits of Applicability	270
5.3.4 Turbulence Model Validation	270
5.3.5 Matrix of Distinct Values	270
5.4 High-Lift Investigations for NLR 7301 Flapped Airfoil	**305**
5.4.1 Introduction	307
5.4.2 Partners and Methods	307
5.4.3 Test Cases	309
5.4.4 Computational Grids	310
5.4.5 Discussion of Results	311
5.5 Two-Dimensional Boundary Layers	**349**
5.5.1 Introduction	351
5.5.2 Test Cases	351
5.5.3 Turbulence Models	352
5.5.4 Discussion of Results	352
5.5.4.1 Stanford IDENT-1300	352
5.5.4.2 Stanford IDENT-1400	353
5.5.4.3 Stanford IDENT-2200	354
5.5.4.4 Stanford IDENT-2400	355
5.5.4.5 Stanford IDENT-2600	356
5.5.5 Conclusions	356
5.6 Three-Dimensional Boundary-Layers	**389**
5.6.1 Introduction	391
5.6.2 Partners and Methods	391
5.6.3 Turbulence Models	391
5.6.4 Discussion of Results	392
5.6.4.1 Infinite Swept Wing Test Case BEEL72	392

Table of Contents (continued)

	Page
5.6.4.2 Cylinder on Flat Plate Test Case DEFE77	393
5.6.5 Conclusions	393
5.7 DLR-F5 Wing Test Case	**405**
5.7.1 Introduction	407
5.7.2 Results from Euler and Navier-Stokes Calculations	408
5.7.3 Results from Boundary-Layer Calculations	417
5.7.4 Comparison Between Boundary-Layer and Navier-Stokes Results	419
5.7.5 Concluding Remarks	420
5.8 2D and 3D Wind-Tunnel Simulations	**489**
5.8.1 Introduction	491
5.8.2 2D Tunnel Simulation: RAE 5225 Aerofoil	492
5.8.2.1 Aerofoil Characteristics and Wind-Tunnel Measurements	492
5.8.2.2 Wind Tunnel Corrections	493
5.8.2.3 Mesh Generation	494
5.8.2.4 Flow Calculation and Evaluation of Corrections	494
5.8.2.5 Results	496
5.8.2.6 Conclusions	497
5.8.3 3D Tunnel Interference Simulation: DLR-F5 Wing	497
5.8.3.1 The DLR-F5 Wing and Wind-Tunnel Measurements	497
5.8.3.2 Mesh Generation	498
5.8.3.3 Flow Calculation	498
5.8.3.4 Results	499
5.8.3.5 Conclusions	500
5.8.4 Overall Conclusion	500
5.9 Vortex Breakdown	**513**
5.9.1 Introduction	515
5.9.2 Test Cases	515
5.9.3 Discussion and Results	512
5.9.3.1 Vortex Embedded in a Constant Axial Flow Field	516
5.9.3.2 Vortex Embedded in Plane Shear Flow	516
5.9.4 Conclusion	517
6 Reference List	**523**

1 Introduction

The design process of future aircraft requires extensive research and development work in the area of aerodynamics. To accomplish this task in a cost-effective manner, high-quality ground-test facilities (wind tunnels) and very high performance numerical algorithms must be available and must operate in tandem. The success of this combination will, to a large extend, depend on the availability of fast and robust numerical methods supporting sound mathematical models of the physical processes at work. Together with a rapid increase in computer power and speed, today's Computational Fluid Dynamics (CFD) methods are becoming increasingly powerful tools in the design context, lowering costs and providing information on processes in critical flow regions and in off-design conditions, where measurements are difficult to conduct.

Theoretical methods often suffer from a lack of realism, a problem rooted principally in weakness of the model of turbulence being used. Therefore, numerical methods need to be carefully validated against measurements in order to quantify and improve their applicability and accuracy. Further efforts are subsequently needed to improve code efficiency, robustness and user-friendliness. To achieve a better understanding of turbulence models and their influence on the numerical solutions, a range of models, based both on algebraic and on differential formulations, are being investigated. For the design process of single airfoils and even more so in the case of high-lift multi-element airfoils, it is of utmost industrial importance to compute accurately maximum lift situations. In addition to the aforementioned investigations, and for the purpose of method evaluation, calculations with both free-air and tunnel boundary conditions are being compared in order to derive corrections allowing free-flight conditions to be simulated on the basis of test calculations for wind-tunnel configurations. With respect to three-dimensional wing flows, a first step is undertaken in the EUROVAL project to validate Navier-Stokes and boundary-layer codes against each other, compensating for incomplete experimental data.

1.1 Main Validation Topics

The main topics of the validation project cover two- and three-dimensional aerodynamic flows with particular emphasis being placed on:

- Comparison of different 2D Navier-Stokes methods against measurements, with special attention paid to the exploration of applicability under realistic flight conditions, i.e. with respect to different meshes, boundary conditions and turbulence models.
- Maximum lift for single airfoils and high-lift characteristics for two-element airfoils using algebraic and two-equation turbulence models.
- Validation of 2D CFD methods with respect to physical modelling, i.e. validation with respect to commonly used turbulence models based on boundary-layer and Navier-Stokes approaches.
- Comparison of different 3D Euler codes with measurements for weak-interaction situations.

- Validation of 3D CFD methods with respect to turbulence modelling based on boundary-layer approaches and Navier-Stokes methods.
- Wind-tunnel-wall interference problems and extrapolation of related results to free flight in 2D and 3D conditions.
- Vortical flows including vortex breakdown computed by solving the full Navier-Stokes equations in cylindrical coordinates.

1.2 Technical Results

After the two-year BRITE/EURAM Pilot Phase, a great deal of high-quality results may be claimed to have been generated in the EUROVAL project for several task interdependent applications. A discussion of the results obtained is presented in detail in Chapter 5.

Single Element Airfoil Flow

The main investigations of airfoil flows centre around two different geometries, the RAE-2822 and the ONERA A-airfoil. While the RAE-2822 geometry serves as a test case for validating different boundary conditions, meshes and, above all, turbulence models, the A-airfoil is used for maximum-lift investigations, i.e. for predicting flows near or at stall, this being of prime industrial importance. These results, in several cases performed on mandatory meshes, form the basis for further investigations, leading to improvements of codes, boundary conditions, mesh structures and turbulence models.

Two-Element Airfoil Flow

In order to improve multi-block Navier-Stokes methods, as a contribution towards efficient computational design tools for high-lift configurations, these methods are applied to the flow around a flapped NLR-7301 airfoil at incidences of 6° and 13.1°.

ONERA Channel Flow

An important task in the EUROVAL project is turbulence model validation. Shock/boundary-layer interference problems serve as challenging test cases for validation, as the shock position in internal choked flow is very sensitive to the resolution of the boundary layer interacting with the shock. Thus, comparisons between different codes and turbulence models rely on the specification of a target which has to be reached by all methods. In the present collaborative effort, a specific Mach number was defined as a particular target, in order to enable a comparison of all results relative to the same shock location. To achieve a high degree of comparability, mandatory meshes were distributed and had to be used by all partners involved.

3D Flow over DLR-F5 Wing

The validation of 3D Navier-Stokes codes with respect to turbulence modelling is an undertaking which is viewed as a high-priority area in the airframe industry because of the need to gain a better insight into complex flow domains and to improve the numerical codes with respect to emerging novel computer architectures. Moreover, for a not too complex geoemetry, like a wing, it proves effective to use and test boundary-layer methods, which are not afflicted by artificial (numerical) viscosity or explicitly added dissipation. Additional to the complex Navier-Stokes computations, inviscid (Euler) computations have been performed as well as 3D boundary-layer investigations. Altogether, this forms an excellent basis for an improved view of performance and accuracy of the different codes employed.

Boundary Layer Flows

As mentioned above, boundary-layer methods are free from numerical dissipation and allow, therefore, a fair assessment of turbulence models. Both 2D and 3D cases have been investigated. The former group consisted of the "Stanford '68" flows, while the latter group included a flow past an infinitely-swept wing (BEEL72) and a flow around the base of a cylinder resting on a flat plate (DEFE72). The results demonstrate the ability of boundary-layer methods to reflect the performance of different turbulence models in the absence of numerical dissipation. The methods allow fast, extensive and cost-effective tests of adjusted, improved or new turbulence models, although it is restricted to attached or mildly detached flows. Nevertheless, it can be assumed that models which fail in the context of boundary layer calculations cannot be satisfactorily used in Navier-Stokes solvers.

Wind-Tunnel Interference

For the purpose of method evaluation, the solutions from CFD codes with free air outer boundary conditions are commonly compared with experimental data from conventional wind tunnels. The tunnel working section, however, is restricted in size so that an aerodynamic blockage effect will result and the tunnel operates at Reynolds numbers some way short of flight. The physical tunnel restrictions imply a need to accurately asses the interference effects in order to relate tunnel results to free air.

Based on transonic conditions, two- and three-dimensional simulations - for the RAE 5225 airfoil and the DLR-F5 wing, respectively - were performed using a viscous coupled Euler approach.

Vortex Breakdown

It is well known that in highly rotating flows, a sudden appearance of a stagnation point followed by reversed axial flow may occur. This phenomenon, termed "vortex breakdown", was first observed in vortices shed from delta wings. Later investigations demonstrated that vortex breakdown also appears in other flows, e.g. in combustion chambers, cyclone separators or in swirling flows in divergent tubes. For this study, an isolated vortex embedded in a cylindrical flow domain was chosen and computed by solving the Navier-Stokes equations by direct numerical simulation, i.e. without the use of a turbulence model.

1.3 Cooperative Strategy

The structure of EUROVAL with its four working groups has proven itself to be very successful as regards cooperation and understanding between the partners. Moreover, a strong spirit of active collaboration and interdependence had been established which led to activities, ranging from an information-exchange strategy to specific joint effects involving the exchange of complete Navier-Stokes codes and mesh generators, the provision of related training, and the prescription of meshes to be used mandatorily by all partners for agreed computations. Consequently, this cooperation may be said to have contributed towards creating a more effective European CFD community, in which each partner now possesses enhanced capabilities.

Exchange of Data and CFD Codes

In order to enable a more precise comparison of results, the use of mandatory test cases led to a close cooperation between partners involved in the same working group(s). Additionally to mandatory meshes, provided by potent partners, an exchange of so-called code kernels took place, i.e. routines for the calculation of turbulence properties and the computation of boundary length scales have been exchanged.

In two cases not only code kernels were provided but complete Navier-Stokes codes have been released together with a mesh generation system and a two-days "training on the job". This initiative has to be seen in the light of an increasing European spirit, to support partners in a way which will allow for a more close European cooperation, reaching out for an increase in European competetiveness. As a consequence of this software release, close communication between the partners took place, which also resulted in an information "backflow", i.e. the partner releasing the code achieved an extended knowledge about his own code.

In addition to software exchanges, computed flow field data - from airfoil applications - were transmitted to the EUROMESH group for initial testing of mesh adaption strategies.

Workshops

Communication between EUROVAL and other BRITE/EURAM aeronautic themes took place, particularly with EUROMESH (mesh generation for CFD purposes) and with the project focussing on "Investigation of Supersonic Flow Phenomena". Concerning EUROMESH, a direct exchange of data was arranged in order to start initial work on mesh adaption, and a symposium was organized which combined all researchers from EUROMESH and EUROVAL, resulting in a fruitful information exchange. The contact with partners from "Supersonic Flow Phenomena" resulted in a workshop on turbulence modelling for the above-mentioned ONERA bump test cases.

Synergy

The EUROVAL initiative has generated a "synergetic" environment which involved Siemens-Nixdorf offering computer resources at no costs on one of its most advanced Giga-Flop machines at the University of Karlsruhe, Germany, thus helping the partners to overcome difficulties arising from limited local computer allocations.

Moreover, the use of the Siemens-Nixdorf machine, potentiated by an overwhelming support of Siemes-Nixdorf personnel during the time of three working periods (4 to 5 days, normally), led to first re-engineering attempts on the partners' CFD codes. In some cases, this re-engineering effort led to run-time reductions of nearly 50 %, which proves to be a lot. The advantage of code re-design has also to be seen in the light of improved code knowledge and is not limited to the Siemens-Nixdorf computer environment but to a class of computers possessing vector capabilities; a type of computer which can be accessed nowadays by most partner organizations.

In any follow-on phase, it is intended to incorporate Siemens-Nixdorf as an associate partner, in an effort directed towards vectorization and parallelization of CFD codes, as well as re-engineering of codes designed for improved run-time performance. The company has also undertaken to contribute additional (free-of-charge) CPU resources.

1.4 Conclusions

It is pleasing to state that the EUROVAL project, focusing on the validation of CFD codes for two- and three-dimensional flows, has proven to be very successful in terms of its technical outcome, the level of cooperation established, and its enhancement of European collaboration and understanding. The work during the two years of EUROVAL ranged from transonic airfoil flow investigations, maximum- and high-lift computations, through turbulence model validation of Navier-Stokes and boundary-layer flows, to 2D and 3D wind-tunnel interference problems, and to the investigation of 3D vortex breakdown. Although it is impossible to cover the large variety of still unsolved CFD problems completely, the work in the, so-called, BRITE/EURAM Pilot Phase is seen as an important step towards a substantial strengthening of the capabilities of Europe's Fluid Dynamics Community with consequent benefits to the technological prowess of Europe's aircraft industries.

2 Structure of the EUROVAL Project
2.1 List of Tasks and Brief Descriptions

In the following, a list of tasks performed in the context of the EUROVAL Project is presented. This list corresponds directly to task headings which were defined in the Technical Project Document. In order to provide the reader with more insight and a better understanding of the programme, brief comments are added under the headings of the different tasks.

1. Comparison of different 2D methods against measurements.
   1.1 Suitability of 2D NS methods for transonic airfoil-flow prediction and investigation of numerical parameters.

   Assuming typical transport wing section geometries at design and off-design conditions, Navier-Stokes codes are tested in order to determine practical utility and capability, numerical sensitivity and cost efficiency. This is seen in the context of comparing different codes as regards performance, leading to possible future improvements.

   1.2 Suitability of 2D NS methods for transonic-flow prediction with respect to explore the applicability under design and off-design conditions.

   With the large number of Navier-Stokes codes in existence today, there is a lively debate on which method represents the best approach for each aerodynamic problem examined. A direct comparison with test results from wind tunnels and flight experiments in collaboration with other partners working on related topics exhibits ways of improving numerical methods.

2. High-lift characteristics for airfoil flow.
   2.1 Maximum lift for single-element airfoils.

   One issue of industrial importance in relation to Navier-Stokes codes is their ability to predict maximum lift. For one airfoil, several points on a polar are calculated and compared with experimental data. Furthermore, Reynolds-number effects are investigated in order to validate the capability of existing codes for high lift and to demonstrate the ability of those methods to simulate both free-air and wind-tunnel flow conditions.

   2.2 Multi-element airfoil flow.

   One of the main objectives of this task is to investigate the application (and the limits) of zero- and two-equation turbulence models in Reynolds-averaged Navier-Stokes solvers for predicting (high-lift) flows about two-element airfoils. The ultimate goal is to enhance existing methods in order to arrive at efficient design tools for high-lift configurations.

3. Validation of 2D CFD methods with respect to physical modelling.
   3.1 Comparison of different turbulence models using 2D boundary-layer methods. An evaluation of turbulence models using Navier-Stokes methods is time-consuming and difficult because of the presence of artificial viscosity and the associated need for fine meshes. With coarse grids, artificial diffusion can easily swamp the real viscous process and render meaningless an exercise on testing turbulence models. Therefore, a turbulence model validation proc-

ess starts with 2D boundary-layer algorithms, which are free from numerical viscosity and allow an efficient and fast model testing.

3.2 Validation of 2D Navier-Stokes codes with respect to turbulence modelling.

The use of Navier-Stokes methods for flow prediction plays an important role in industrial practice and is becoming a standard approach. Often, however, the predictive accuracy is poor and depends strongly on the turbulence model in use. Detailed investigations are performed, comparing different turbulence models against one another (and against boundary-layer solutions) with mandatory meshes used by all partner.

4. Comparison of different 2D (and 3D) Euler codes and methods against measurements.

Navier-Stokes solutions and Euler solutions, suffer from artificial viscosity. An Euler approach offers the opportunity, however, to investigate the influence of artificial viscosity on the solution because its performance is not obscured by physical diffusion. Thus, total-pressure-loss areas (with losses not caused by shocks) can be correlated to an improper use of artificial dissipation and/or insufficient flow resolution.

5. Validation of 3D CFD methods with respect to turbulence modelling.

5.1 Validation of 3D boundary-layer codes.

A comparison of 3D Navier-Stokes methods with respect to different turbulence models is extremely costly; alternatively, the 3D boundary layer-equations can be solved. These techniques are free from numerical viscosity and explicitly added numerical dissipation, and they allow a reliable assessment of turbulence models. Results thus obtained, guide the use of the models in Navier-Stokes approaches.

5.2 Validation of 3D Navier-Stokes codes for flows around wings.

Although a validation process of 3D Navier-Stokes methods is costly, as pointed out above, it is now widely used in industry in order to improve and shorten the design process for complete aircraft configurations. Due to the extreme flow complexity, the validation effort requires high-quality experiments to judge the accuracy of the numerical solutions in terms of global and local quantities.

6. Tunnel interference and extrapolation to free flight.

6.1 Tunnel simulation on the basis of a 2D approach.

Calculations for both free-air and tunnel-boundary conditions are compared in order to achieve corrections to free-stream (free-flight) conditions. These calculations cover the use of different boundary conditions (tunnel wall - free flight) and a study at different Reynolds numbers to demonstrate the sensitivity of predictions on the Reynolds number.

6.2 Tunnel simulation on the basis of a 3D approach.

For the purpose of method evaluation, the solutions from CFD codes with free-air boundary conditions are commonly compared with experimental data obtained from wind tunnel tests. The physical tunnel restrictions (e.g. aerodynamic blockage) raise the need for an accurate assessment of interference effects in order to relate tunnel results to free-air conditions.

7. Vortex breakdown in free vortex systems.
   In highly rotating flows, a stagnation point followed by reversed axial flow can occur. This phenomenon, known as vortex breakdown, is observed in vortices shed from delta wings and can also appear in combustion chambers, cyclone separators and swirl flow in diverging ducts. In this task, the three-dimensional structures of vortex breakdown are resolved by a Navier-Stokes approach (direct numerical simulation). In particular, an isolated vortex embedded in a cylindrical flow domain is considered.

2.2 Task Matrix

The task matrix provides an overview of partners allocated to the work packages mentioned in the previous section.

In the course of the work it turned out, however, that some partners were willing to upgrade their contribution relative to the original proposal. Such partners were given the opportunity to redeploy or, indeed, add man-power. Instances in which the latter occurred are identified by (X) in the matrix below.

Partner	1.1	1.2	2.1	2.2	3.1	3.2	4	5.1	5.2	6.1	6.2	7
Analysis		X										
British Aerospace			X	X		X				X	X	
CAPTEC	X											
CASA		(X)					X		X			
Cerfacs			X	X		X			X			
CFD norway				X					X			
Deutsche Airbus	X				(X)	(X)		X				
DLR						X			X			
Dornier	(X)				X	X	X					
NLR		(X)	X	X		(X)						
RAE					(X)			X				
Saab			X	X								
TU Berlin								X				
TU Denmark						X						X
UMIST						X						
VU Brussel	X					X			X			

2.3 Application List

It must be stressed that this is not simply an inversion of the matrix presented in the previous section. It turned out at an early stage of the EUROVAL project that certain tasks could, rationally, be rearranged to form so-called application areas. This approach offers the advantage of increasing transparency, from a technical point of view. It also allows a logical and structured documentation.

Thus, numbers appearing in front of application-areas in the table below identify application related sections in Chapter 5. The table also identifies partners who have authored the application-oriented section and partners who were responsible for cross-plotting results.

Chapter-no. and Application	Task	Partner responsible for text	plots
5.1 RAE-2822 Airfoil Test Cases	1.1, 1.2, 3.2	Deutsche Airbus	Analysis
5.2 ONERA Bumps 'A' and 'C'	3.2	UMIST	VUB
5.3 ONERA A-Airfoil	2.1	Cerfacs	NLR
5.4 NLR-7301 Flapped Airfoil	2.2	NLR	Saab
5.5 2D Boundary Layers	3.1	Dornier	Dornier
5.6 3D Boundary Layers	5.1	RAE	TUB
5.7 DLR-F5 Wing Test Case	4, 5.1, 5.2	DLR	TUB
5.8 Tunnel Simulations	6.1, 6.2	BAe	BAe
5.9 Vortex Breakdown	7	TU Denmark	TU Denmark

2.4 List of Partners

In the following, a complete set of all partners involved in EUROVAL is presented, together with the assignment to applications and tasks, in order to identify responsibility for work packages and to allow direct contact in case of questions and/or requests.

Organisation Responsible Person(s)	Application (in Chapter)	Task
Analysis System Research Ltd. 2, Papadiamantopoulou St. GR-11528 Athens A. Karidis, St. Perrakis, M. Tzitzidis	5.1	1.2
British Aerospace Regional Aircraft Ltd Research / Aerodynamics Department Woodford - Stockport Cheshire SK7 1QR England J. Benton, D. King	5.1, 5.2, 5.3 5.4, 5.8	2.1, 2.2, 3.2, 6.1, 6.2
CAPTEC 3, St. James' Tce. Malahide CO. Dublin Ireland T. Cox, P. Duffy, F. Kennedy	5.1	1.1

Organisation Responsible Person(s)	Application (in Chapter)	Task
CASA Dept. Aerodinamica P. John Lennon s/n Getafe - Madrid Spain A. Abbas	5.1, 5.7	1.2, 4
Cerfacs 42, Avenue Gustave Coriolis F-31057 Toulouse L. Davidson, A. Kourta, L. Tourette	5.2, 5.3 5.4 5.7	2.1, 3.2 2.2 5.2
CFD norway as Teknostallen Professor Brochsgt. 6 N-7030 Trondheim H. Nørstrud, I. Øye	5.4, 5.7	2.2, 5.2
Deutsche Airbus Dept. EF11 Hünefeldstr. 1-5 W-2800 Bremen E. Elsholz D. John	5.1, 5.5, 5.7 5.6, 5.7	1.1, 3.1, 3.2, 5.1 5.1
Deutsche Forschungsanstalt für Luft- und Raumfahrt Inst. f. Theoretische Strömungsmechanik Bunsenstr. 10 W-3400 Göttingen D. Schwamborn	5.7	3.2, 5.2
Dornier Luftfahrt GmbH Postfach 1303 An der B31 W-7990 Friedrichshafen 1 W. Fritz, W. Haase	5.4 5.1, 5.2, 5.5	2.2 1.1, 3.1, 3.2
Nationaal Lucht- en Ruimtevaartlaboratorium Aerodynamics Department Anthony Fokkerweg 2 1059 CM Amsterdam The Netherlands F. Brandsma	5.1, 5.3	1.2, 2.1, 3.2

Organisation Responsible Person(s)	Application (in Chapter)	Task
DRA/RAE Aerodynamics Department Farnborough Hants GU14 6TD England P. D. Smith	5.5, 5.6, 5.7	3.1, 5.1
Saab-Scania Aerospace Division Linköping Sweden B. Arlinger, T. Larsson	5.3, 5.4	2.1, 2.2
TU Berlin Sekr. HFI Straße des 17. Juni 135 W-1000 Berlin 12 L. Xue, T. Rung, F. Thiele	5.6, 5.7	5.1
TU Denmark Dept. of Fluid Mechanics Lundtoftevej 100 DK-2800 Lyngby P. Larsen, J. Michelsen, J. N. Sørensen, M.L. Hansen	5.1 5.9	3.2 7
University of Manchester - UMIST Dept. of Mech. Engineering P.O. Box 88 Sackville Street Manchester M60 1QD England M. Leschziner, G. Page	5.1, 5.2	3.2
Vrije Universiteit Brussel Faculteit Toegepaste Watenschappen Dienst Stromingsmechanica Pleiniaan 2 B-1050 Brussel Ch. Hirsch, M. Mulas, Ch. Lacor	5.1, 3.2	1.1, 3.2, 5.2

3 Description of Turbulence Models

In this section, the reader will find detailed descriptions of the turbulence models investigated by the EUROVAL partners in different tasks for a variety of applications. The turbulence models presented hereafter are given in their original form, hence enabling a potential user to get a complete set of all relevant informations from the current chapter.

For modifications of the 'standard' turbulence model descriptions listed below, the reader is referred to chapter 4 where all partners present their numerical methods together with a discussion of special features of their specific turbulence model implementation(s).

It should be noted at this point that the equation numbers, related to the description of a distinct turbulence model, are combined with a letter code, i.e. an abbreviation of the model, in order to allow for the highest possible transparency and an easy referencing.

3.1 Algebraic Models

The following sections are split into the main categories of commonly used turbulence models, namely algebraic, 1/2-equation, 1-equation and 2-equation eddy viscosity models, as well as Reynolds stress models.

3.1.1 Introduction

The algebraic turbulence models described herein are two-layer eddy-viscosity models. This means that according to the eddy-viscosity concept, in the stress terms of the Navier-Stokes equations for laminar flow, the molecular viscosity μ is replaced by

$$\mu = \mu + \rho v_t \tag{3.1}$$

while in the heat-flux terms, $k/c_p = \mu/Pr$ is replaced by

$$\frac{\mu}{Pr} = \frac{\mu}{Pr} + \frac{\rho v_t}{Pr_t} \tag{3.2}$$

with $v_t = \mu_t/\rho$. The Prandtl numbers are chosen to be $Pr=0.72$ for laminar and $Pr_t=0.90$ for turbulent flows.

The second coefficient of viscosity,

$$\lambda = -\frac{2}{3}\mu \tag{3.3}$$

which is related to a zero value for the bulk viscosity, is treated differently. Some partners have applied the eddy-viscosity concept also to λ following a replacement of λ by the molecular viscosity μ; others leave the second coefficient of viscosity unchanged, i.e. use the molecular viscosity in equ. (3.3).

All models described in the following are two-layer models, i.e. They employ different approaches for the near wall and the outer region of the boundary layer.

3.1.2 The Baldwin-Lomax [BL] Model (1978)

The Baldwin-Lomax turbulence model (Baldwin-Lomax, 1978) is a two-layer algebraic eddy-viscosity model and patterned after that of Cebeci-Smith. However, it avoids the need of finding the edge of the boundary layer and the displacement thickness.

The eddy-viscosity v_t is given by

$$v_t = \begin{cases} v_{ti} & \text{for } y \leq y^* \\ v_{to} & \text{for } y > y^* \end{cases} \tag{BL1}$$

where y^* is the smallest, normally wall-nearest, value of y at which the eddy-viscosity values from the inner and outer region are identical.

In the inner region, the Prandtl-van Driest formulation is used by replacing, for Navier-Stokes applications, the normal derivative of the velocity profile, du/dy, by the absolute value of the vorticity

$$v_{ti} = l^2 |\omega| \tag{BL2}$$

with

$$l = Ky[1 - e^{-y^+/A^+}] . \tag{BL3}$$

$|\omega|$ is the magnitude of the vorticity which, for three-dimensional flows, reads

$$|\omega| = \sqrt{\left[\frac{\partial u}{\partial y} - \frac{\partial v}{\partial x}\right]^2 + \left[\frac{\partial v}{\partial z} - \frac{\partial w}{\partial y}\right]^2 + \left[\frac{\partial w}{\partial x} - \frac{\partial u}{\partial z}\right]^2} \tag{BL4}$$

and

$$y^+ = \frac{\sqrt{\rho_w |\tau_w|}\, y}{\mu_w} \tag{BL5}$$

with K=0.4 and A^+=26. The subscript "w" denotes values to be taken at walls. For the outer layer, alternate expressions for the eddy-viscosity are proposed:

$$v_{to} = k\, C_{CP}\, F_{Kleb}\, F_{WAKE} \tag{BL6}$$

with

$$F_{WAKE} = \begin{cases} y_{max} F_{max} \\ \overline{or} \\ C_{WK}\, y_{max}\, u_{Diff}^2 / F_{max} \end{cases} \text{the smaller} \tag{BL7}$$

with C_{CP}=1.6, C_{WK}=0.25.

At this point it should be noted that the original value for C_{WK} (=0.25), defined in the 1978-Baldwin-Lomax paper, has been changed for some computations to C_{WK}=1.0. The higher values yielded considerably stronger interaction, but also gave rise to unrealistic features in the velocity field; see section 5.2 for further discussion according to the ONERA bump test cases.

The Clauser parameter, k, is generally assigned to be a constant, k=0.0168, although it varies slightly in the low-momentum Reynolds number range. A modification to the Clauser constant arises in the following section, 3.1.2, where the description of the Granville turbulence model is presented.

The smallest value for F_{WAKE} in equ. (BL7) has to be taken. The quantities F_{max} and y_{max} are determined from the function

$$F = y\,|\omega|\,[1 - e^{-y^+/A^+}] \,. \tag{BL8}$$

In wakes, the exponential term is set to zero. The quantity F_{max} is the maximum value of F that occurs in the velocity profile and, consequently, y_{max} defines the y-location where F equals F_{max}.

In equ. (BL7), U_{Diff} is the difference between maximum and minimum velocity in the profile,

$$U_{Diff} = U_{max} - U_{min} \tag{BL9}$$

where U_{min} is taken to be zero except in wakes. Furthermore, the intermittency factor F_{Kleb} is given by

$$F_{Kleb} = \frac{1}{1 + 5.5\,(C_{KLEB}\,y/y_{max})^6} \tag{BL10}$$

with C_{KLEB}=0.3.

To simulate transition from laminar to turbulent flows, Baldwin & Lomax proposes to set ν_t equal to zero everywhere in the 'profile', i.e. along the corresponding grid line emanating from the solid surface, for which the (tentatively) computed value of ν_t, using equation (BL1) given above, is less than a specified value:

$$\rho\nu_t = 0 \quad \text{if} \quad (\rho\nu_t)_{max,\,in\,profile} < \mu_\infty\,C_{MUTM} \tag{BL11}$$

with C_{MUTM}=14.

3.1.3 The Granville [GR] Model (1987)

The Granville model (Granville, 1987) is a modification to the Baldwin-Lomax model presented in the section given above.

Baldwin-Lomax introduced in the outer layer formulation two constants, C_{KLEB} and C_{CP} in equations BL6, BL7 and BL10. As shown by Granville, however, these two factors should actually be functions with C_{KLEB} depending on the Coles' wake factor Π and C_{CP} depending on C_{KLEB} in the following way:

$$C_{KLEB} = \frac{4}{9}\frac{1 + 6\Pi}{1 + 4\Pi} \tag{GR1}$$

$$C_{CP} = \frac{3-4C_{KLEB}}{2C_{KLEB}(2-3C_{KLBE}+C_{KLEB}^3)} \quad . \tag{GR2}$$

These equations are obtained from a comparison between the outer layer formulations of the Baldwin-Lomax and the Cebeci-Smith models, assuming an outer similarity law for the turbulent boundary layer, (Granville, 1976), and its validity up to the wall.

For equilibrium pressure gradients, Π is a constant in the streamwise direction and can be empirically correlated to the Clauser pressure-gradient parameter. Granville fits the latter to the Baldwin-Lomax model and derives an explicit formula for C_{KLEB} as a function of a modified Clauser pressure-gradient parameter β where

$$c_{KLEB} = \frac{2}{3} - \frac{0.01312}{0.1724 + \beta} \tag{GR3}$$

with

$$\beta = -\frac{y_{max}}{u_\tau}\frac{dU}{dx} \quad . \tag{GR4}$$

Thus the equations presented above open up the possibility of introducing flow-dependent variables into the Baldwin-Lomax model, which are indirectly based on experimentally confirmed similarity laws.

For low Reynolds numbers, Granville proposes an additional modification in the equation for the outer-layer eddy-viscosity by changing the Clauser constant, k=0.0168,

$$k = 0.0168\left[1 + \left(1100\frac{\nu}{C_{CP}}F_{WAKE}\right)^2\right] \quad . \tag{GR5}$$

Thus, where ν is the kinematic viscosity and F_{WAKE} is identical to equ. (BL7) in the Baldwin-Lomax model, the Clauser factor k varies with the displacement-thickness Reynolds number, $Re_\theta = u\delta^*/\nu$.

3.1.4 The Cebeci-Smith [CS] Model (1974)

The Cebeci-Smith (1974) turbulence model is a two-layer algebraic eddy-viscosity model. It uses a Prandtl-van Driest formulation for the inner and a Clauser formulation together with Klebanoff's intermittency function for the outer region.

The eddy-viscosity ν_t is given by

$$\nu_t = \begin{cases} \nu_{ti} & \text{for } y \leq y^* \\ \nu_{to} & \text{for } y > y^* \end{cases} \tag{CS1}$$

where y* is the smallest value of y at which the eddy-viscosity values from the inner and outer region are identical.

In the inner region, the Prandtl-van Driest formulation is used by replacing, for

Navier-Stokes applications, the normal derivative of the velocity profile by the vorticity:

$$v_{ti} = l^2 |\omega| \qquad (CS2)$$

with

$$l = Ky[1 - e^{-y^+/A^+}]. \qquad (CS3)$$

$|\omega|$ is the magnitude of the vorticity which, for three-dimensional flows, reads

$$|\omega| = \sqrt{\left[\frac{\partial u}{\partial y} - \frac{\partial v}{\partial x}\right]^2 + \left[\frac{\partial v}{\partial z} - \frac{\partial w}{\partial y}\right]^2 + \left[\frac{\partial w}{\partial x} - \frac{\partial u}{\partial z}\right]^2} \qquad (CS4)$$

and

$$y^+ = \frac{\sqrt{\rho_w |\tau_w|}\, y}{\mu_w} \qquad (CS5)$$

with K=0.4 and A$^+$=26. The subscript "w" denotes values to be taken at walls. As mentioned above, in boundary layer applications the normal derivative of the resulting velocity profile should be used.

For the outer layer, Clauser's formulation together with Klebanoff's intermittency function is used:

$$v_{to} = k U_e \delta_i^* F_{Kleb} \qquad (CS6)$$

with the incompressible boundary layer displacement thickness

$$\delta_i^* = \int_0^\delta [1 - \frac{u}{U_e}]\, dy \qquad (CS7)$$

and δ being the boundary layer thickness. Although varying slightly in the low-momentum-Reynolds-number range, the Clauser parameter K (CS6) is generally taken to be a constant, k=0.0168, and the Klebanoff intermittency function is

$$F_{Kleb} = \frac{1}{1 + 5.5\,(y/\delta)^6}. \qquad (CS8)$$

3.1.5 The Goldberg (GB) Backflow Model (1986)

The Goldberg backflow model (Goldberg, 1986) has been developed especially for (and is restricted to) regions of separated flow. It therefore should always be combined with another turbulence model. The model has been successfully used in combination with the Baldwin&Lomax turbulence model as well as with a k-ε model (Goldberg and Chakarvarthy, 19787).

Based on experimental observations of separated turbulent flows, turbulent kinetic energy, k, and its dissipation, ε, are prescribed analytically inside separation bubbles. For k a Gaussian distribution in the direction normal to the wall, n, is assumed between the wall and the "backflow edge" (the edge of the separation bubble), n_b, with

$$\frac{\rho k}{\rho_b k_b} = \frac{e^\varphi \left[1 - e^{-\varphi (n/n_b)^2} \right]}{e^\varphi - 1} = G(s,n), \qquad 0 \leq n \leq n_b \qquad \text{(GB1)}$$

where φ is a constant (see below), s is the distance along the surface, ρ is the density, and the subscript b denotes values evaluated at $n = n_b$. For the dissipation, the following analytical expression has been derived,

$$\left(\frac{\rho}{\rho_b}\right)^{3/2} \varepsilon = \frac{[k_v G(s,n)]^{3/2}}{n_v}, \qquad 0 \leq n \leq n_b, \qquad \text{(GB2)}$$

where the main considerations have been that
- the turbulence length scale inside the separation bubble should be proportional to the height of the bubble
- the dissipation should be constant through the viscous sublayer outside the bubble (i.e. for $n_b \leq n \leq n_v$)
- continuity of ε across $n = n_b(s)$
- agreement with the boundary conditions of the high turbulence Reynolds number k-ε equations set at the edge of the viscous sublayer (i.e. $n = n_v$)

In equation (GB2), the value of k at the viscous sublayer edge follows from

$$\beta \equiv \frac{k_v}{k_b} = 1 + \frac{[(n_v/n_b)^2 - 1]\varphi}{e^\varphi - 1}, \qquad \text{(GB3)}$$

and the edge of the viscous sublayer is determined as

$$\frac{n_v}{n_b} = 1 + 20 \left(\frac{v_w}{u_s n_b}\right) C_\mu^{1/4}, \qquad \text{(GB4)}$$

where C_μ is a constant, v_w is the kinematic viscosity at the wall, and u_s is the square-root of the maximum Reynolds shear stress in the profile (usually located outside the separation bubble). This maximum is assumed to coincide with the location of the maximum of the normal-to-wall mean-velocity gradient, and when an algebraic turbulence model is used outside the separation bubble, u_s is obtained from

$$u_s = \sqrt{(-\overline{u'v'})_{max}} = \sqrt{v_{t,m}(\partial u_t/\partial n)_{max}} \qquad \text{(GB5)}$$

with u_t the streamwise mean velocity, and $v_{t,m}$ the eddy viscosity at the location where $\partial u_t/\partial n$ reaches its maximum value.

From the equations above, it may be observed that the turbulent length scale inside the bubble is concequently given as $L = n_v/\beta^{3/2}$.

Finally, the eddy viscosity within the backflow region is assumed to be of the form $v_t = f(n/n_b) k^2/\varepsilon$, which leads to

$$v_t = \frac{u_s n_v \left(\dfrac{\rho_w G(s,n)}{\rho}\right)^{1/2} (A\dfrac{n}{n_b} + B)}{2\sqrt{2}\beta^2}, \qquad \text{(GB6)}$$

where A and B are given by

$$A = -(\tfrac{1}{2}C_\mu^*)^{9/5}, \quad B = (\tfrac{1}{2}C_\mu^*)^{3/5} - A. \qquad \text{(GB7)}$$

The constants used in the model are

$$C_\mu = 0.09, \quad C_\mu^* = 0.7, \quad \varphi = 0.5. \qquad \text{(GB8)}$$

3.2 One-half Equation Turbulence Models

3.2.1 Introduction

The models described so far, assume that the turbulent shear stress depends only on local properties of the mean flow and thus they are referred to as equilibrium algebraic or zero-equation models. In contrast to those, models which try to account for history effects by use of a simplified transport equation are called 1/2-equation models. The latter designation is taken for Johnson and King (Johnson-King, 1985) who describe their model as a 1/2-equation Reynolds-stress eddy-viscosity model.

As the above algebraic models the 1/2-equation models use also an two-layer approach for the eddy viscosity and the overall approach to the viscosity and heat-conductivity coefficients is as described in chapter 3.1.1.

3.2.2 Johnson-King [JK] Model (1985)

The Johnson-King turbulence model (Johnson-King, 1985, Johnson, 1987, King, 1987) tries to account for effects of convection and diffusion of the Reynolds shear stress, although the model employs an eddy-viscosity distribution across the boundary layer. This distribution depends, however, on the local maximum shear stress, which is obtained from an ordinary differential equation (ODE) derived from the turbulence kinetic-energy equation.

The eddy-viscosity distribution v_t for the JK model is similar to that of the above models in that there is an inner and an outer layer formulation. But, instead of switching from the inner to the outer where the two distributions become equal the first time, an exponential blending

$$v_t = v_{to}[1 - \exp(-v_{ti}/v_{to})] \tag{JK1}$$

is introduced which makes v_t dependent on the outer distribution v_{to} across almost the entire boundary layer. This outer distribution is identical to that of the CS model (CS6) except for an additional parameter $\sigma(s)$ which is a measure for how far the flow is away from equilibrium ($\sigma = 1$). Thus the equation for v_{to} is

$$v_{to} = (k\, U_e\, \delta_i^*\, F_{Kleb}) \cdot \sigma(s) \tag{JK2}$$

with F_{Kleb} the Klebanoff intermittency function (CS8) and the Clauser parameter k=0.0168.

The inner turbulent eddy-viscosity distribution takes the form

$$v_{ti} = D^2 K y (-\overline{u'v'}_m)^{1/2} \tag{JK3}$$

which differs from the CS model (CS2, CS3) in employing the maximum turbulent shear stress $(-\overline{u'v'})_m^{1/2}$ as the velocity scale instead of $Ky|\omega|$. In the first publication of Johnson and King (Johnson-King, 1985) the van Driest damping term is the same as in the original CS and BL model except that they use a coefficient $A^+ = 15$ instead of $A^+ = 26$ to account for the different dependency of v_{ti} on y in their model. In later publications (Johnson, 1987, King, 1987), however, they replace the wall friction velocity $(\tau_w/\rho_w)^{1/2}$ by $(-\overline{u'v'})_m^{1/2} = (\tau/\rho)_m^{1/2}$ which is advantageous if flows with separation are considered:

$$D = 1 - \exp(-y(-\overline{u'v'}_m)^{1/2}/vA^+) \,. \tag{JK4}$$

Thus the inner layer formulation (JK3) is strongly dependent on the maximum Reynolds shear stress. Johnson and King developed an equation for $(-\overline{u'v'}_m)$ from the equation of turbulence kinetic energy by restriction to the path of maximum kinetic energy assuming that this path is nearly aligned with the flow direction (denoted with x in the following) and nearly perpendicular to the wall-normal direction (denoted with y). We will not repeat their derivation of this equation, but refer the reader to the original papers and present only the final result:

$$(-\overline{u'v'}_m)^{1/2} = L_m \left(\left.\frac{\partial u}{\partial y}\right|_m - \frac{C_{dif}(-\overline{u'v'}_m)^{1/2}}{a_1(0.7\delta - y_m)} |1 - \sigma(x)^{1/2}| - \frac{u_m}{a_1(-\overline{u'v'}_m)} \frac{d(-\overline{u'v'}_m)}{dx} \right). \quad (JK5)$$

The terms in this ODE represent from left to right: dissipation, production, diffusion and convection. The index m denotes that the corresponding quantity is to be evaluated at the position of the maximum Reynolds shear stress, and L_m is the dissipation length scale given by

$$\begin{aligned} L_m &= 0.4\, y_m & y_m \leq 0.225\, \delta, \\ L_m &= 0.09\, \delta & y_m > 0.225\, \delta, \end{aligned} \quad (JK6)$$

and a_1 is the ratio of maximum Reynolds shear stress to maximum turbulent kinetic energy which has been observed to be between 0.2 and 0.3, experimentally. In (Johnson-King, 1985, Johnson, 1987) a value of $a_1 = 0.25$ is proposed and the modelling constant C_{dif} is set to .5.

Assuming equilibrium the last two terms in (JK5) would cancel such that $L_m(\partial u/\partial y)|_m$ can be interpreted as being the equilibrium value of the maximum shear stress. The latter quantity is thus obtained from equations (JK1 - JK3) by setting σ to one and replacing $(-\overline{u'v'}_m)^{1/2}$ by $(-\overline{u'v'}_{m,eq})^{1/2}$, i.e.

$$\nu_{t,eq} = \nu_{to,eq}[1 - \exp(\nu_{ti,eq}/\nu_{to,eq})]$$

$$\nu_{to,eq} = 0.0168\, U_e\, \delta_i^* F_{Kleb} \quad (JK7)$$

$$\nu_{ti,eq} = D^2\, \kappa\, y\, (-\overline{u'v'}_{m,eq})^{1/2}.$$

Comparing the definition of ν_{to} and $\nu_{to,eq}$ from equations (JK2) and (JK7) we find the definition of $\sigma(x)$:

$$\sigma(x) = \nu_{to}/\nu_{to,eq} \quad (JK8)$$

which has already been used in (JK5).

For the ease of integration of equation (JK5) Johnson and King introduce a change of variables by substituting

$$g = (-\overline{u'v'}_m)^{-1/2} \quad \text{and} \quad g_{eq} = (-\overline{u'v'}_{m,eq})^{-1/2} \quad (JK9)$$

which together with a reordering of terms results in

$$\frac{dg}{dx} = \frac{a_1}{2 u_m L_m}\left[(1 - \frac{g}{g_{eq}}) + \frac{C_{dif} L_m}{a_1(0.7\delta - y_m)} |1 - \sigma^{1/2}|\right]. \quad (JK10)$$

Prior to the solution of this equation an equilibrium value g_{eq} has to be obtained. This results in the following procedure to compute the final eddy-viscosity distribution (and the corresponding flow result):

1. One obtains a maximum shear stress $(-\overline{u'v'}_m)$ for each x, e.g. from a calculation with a Cebeci Smith model.
2. The equilibrium form of the JK model (JK7) is applied iteratively until the flow solution is converged. The Reynolds stress is then computed from

$$(-\overline{u'v'}) = v_t \frac{\partial u}{\partial y} \qquad (JK11)$$

and from this the maximum value $(-\overline{u'v'}_{m,eq})$.

3. Now equations (JK1) through (JK3) are employed instead of (JK7) to compute the non-equilibrium eddy-viscosity and the maximum shear stress starting with $\sigma = 1$.
4. Integration of equation (JK10) results also in a maximum shear stress which will be different in general.

This difference can be made to vanish by the use of σ. From equations (JK1) to (JK3) one has $v_t|_m$, $v_{to}|_m$ and $v_{ti}|_m$ in step (3) while equation (JK11) allows to compute an eddy viscosity $\tilde{v}_t$ from the results of the integration in step (4) if one employs $(\partial u/\partial y)_m$, i.e. the values of the velocity gradient at y_m.

Johnson (Johnson, 1987) suggests to do two iterations of the following algorithm to obtain a new σ (the index m is left out for sake of simplicity):

$$v_{to}^{(n+1)} = v_{to}^{(n)} \tilde{v}_t / v_t^{(n)}$$

$$v_t^{(n+1)} = v_{to}^{(n+1)}\left[1 - \exp(-v_{ti}/v_{to}^{n+1})\right] \qquad (JK12)$$

whereupon

$$\sigma(x) = \sigma(x) \cdot v_{to}^{(3)} / v_{to}^{(1)} . \qquad (JC13)$$

With the new σ a new eddy-viscosity distribution is computed and step (3) and (4) are repeated in the course of the flow calculation until convergence is obtained.

3.2.3 Johnson-Coakley [JC] Model (1990)

Johnson and Coakley (1990) tried to implement some improvements to the JK model after a few deficiencies of the latter model had become obvious, as e.g. in flows with shocks but no separation. Most of their modifications affect only the formulation for the inner layer eddy-viscosity but there are also two other changes. The first of those is the exchange of the blending function between the inner and outer layer formulation where the exponential in equ. (JK1) is replaced by a hyperbolic tangent:

$$v_t = v_{to} \tanh(v_{ti}/v_{to}) \qquad (JC1)$$

which results in a larger value of v_t at and near the position where $v_{ti} = v_{to}$ and should improve the skin friction results in favourable pressure-gradient situations. The second change is to the diffential equation for the maximum shear stress where the diffusion term is now set to zero if σ is less than one i.e. $|1 - \sigma^{1/2}|$ in (JK5, JK10) is replaced by

$$\max(0, \sigma^{1/2} - 1).$$

The reason for this is, that the diffusion term was originally added to enhance the solution in regions of flow recovery (i.e. $\sigma > 1$) while it was believed that its presence would be negligible for retarded flow. Johnson and Coakley claim, however, that the latter is not always true and that the diffusion term does not improve the solution there and should be omitted there since its validity is far from established.

The remaining changes are all with respect to the inner layer eddy-viscosity viscosity formulation (JK3). The first of these is the introduction of a scaling factor $\sqrt{\rho_m/\rho}$ in (JK3) and in the van Driest damping (JK4) to account for density effects, although this should mainly be of importance in high supersonic flows. Another change to the damping term is to use the larger of u_τ and $\sqrt{\rho_m/\rho}\,(-\overline{u'v'}_m)^{1/2}$ instead of the maximum shear stress alone which as the exchange of the blending function should result in better prediction of skin friction under favourable or zero pressure-gradient conditions.

The last change to the JK model introduced by Johnson and Coakley is the formulation of a new velocity scale in the inner layer formulation which is based on the assumption that the law-of-the-wall is more valid in the immediate neighbourhood of the wall for attached flow than the original JK formulation or the mixing length theory. For separated flow the new formulation reduces essentially to the JK version which has seen proved to be superior there.

The inner layer eddy-viscosity is now

$$v_{ti} = D^2 k\, y\, u_s \tag{JC2}$$

with the new velocity scale

$$u_s = \sqrt{\rho_w/\rho}\, u_\tau\, [1 - \tanh(y/L_c)] + \sqrt{\rho_m/\rho}\, u_m \tanh(y/L_c) \tag{JC3}$$

where L_c is based on the dissipation length scale L_m (equation (JK6)), namely

$$L_c = \frac{\sqrt{\rho_w}\, u_\tau}{\sqrt{\rho_w}\, u_\tau + \sqrt{\rho_m}\, u_m}\, L_m\,. \tag{JC4}$$

The first term of u_s in (JC3) leads to a Clauser formulation of the eddy viscosity which is smoothly blended into the compressible JK formulation represented by the second term. Instead of a Clauser formulation a mixing length approach seems equally reasonable and can be obtained by replacing $\sqrt{\rho_w/\rho}\, u_\tau$ in (JC3) by $ky|du/dy|$.

Johnson and Coakley claim that both approaches yield similar results if one uses an $A^+ = 17$ in the van Driest damping for the Clauser and JK formulation and a $A^+ = 26$ in the mixing length approach.

For computational purposes a lower limit of e.g. .005 has to be imposed on L_c in order to prevent an overflow error in the tanh calculation if u_τ goes to zero or becomes negative in separated regions.

3.2.4 The Horton [HO] Model (1990)

The Horton model (1990) resembles somewhat that of Johnson-King (1985). It employs a non-equilibrium variant of the Cebeci-Smith model (1970) such that the outer eddy viscosity relation is modified by the incorporation of a multiplying factor σ(x), x being the wall-tangent boundary layer coordinate, whose value is determined by solving a first-order ordinary differential equation (ODE). This has the form of a modified rate equation, and is derived from Horton's (1969) entrainments history equation.

A hyperbolic-tangent smoothing function is used to blend the inner and outer eddy viscosity formulations together. Specifically, the eddy viscosity reads:

$$v_t = v_{to} \tanh(v_{ti}/v_{to}) \tag{HO1}$$

where

$$v_{ti} = k^2 y^2 D^2 \left|\frac{\partial u}{\partial y}\right| \tag{HO2}$$

and

$$v_{to} = \sigma(x) \, \alpha_{eq} \, U_e \, \delta_i^* . \tag{HO3}$$

The parameters in equations (HO2) and (HO3) are defined as follows:

$$D = 1 - \exp(-y^+/A^+) \tag{HO4}$$

$$\delta_i^* = \int_0^\infty [1 - \frac{u}{U_e}] \, dy \tag{HO5}$$

$$y^+ = \frac{y \, u_\tau}{v_w}, \quad u_\tau = \sqrt{\frac{\tau_w}{\rho_w}} \tag{HO6}$$

with the constants to be taken as A^+=26, k=0.40 and α_{eq}=0.0168.

The evolution of the parameter σ(x) is determined from the solution of the following ordinary differential equation

$$\delta_i^* \frac{d}{dx}(as) = C_1 s (1-a) \tag{HO7}$$

where

$$s = \left[\left\{\frac{\partial(u/U_e)}{\partial(y/\delta_i^*)}\right\}_{0.5}\right]^{0.69} \tag{HO8}$$

and $a = \sigma^{0.69}$. C_1 is an empirical constant and set provisionally to C_1=0.03, hence, it should be noted that a calibration against strongly non-equilibrium boundary layer data has to be established. The notation $\{\ \}_{0.5}$ indicates that the bracketed quantity is evaluated at $y/\delta=0.5$, where the boundary layer thickness, δ, is defined as the value of y where $u/U_e=0.995$.

For equilibrium flows, s is a constant when equation (HO7) has the solution a=1, so that the outer eddy viscosity has its equilibrium value for all x, as required. Perturbations to this stable solution decay locally as $\exp(-C_1 x/\delta_1^*)$.

In the absence of detailed information, the parameter 'a' would normally be assumed to have the equilibrium value of unity at the start of a boundary layer calculation.

The quantity 's', i.e. the non-dimensional rate-of-strain at the mid-height of the boundary layer to the power 0.69, is determined from the mean velocity field at each x-station.

A Klebanoff intermittency function is not applied to to v_t in the above presented description, but could be applied - if desirable - to the composite eddy viscosity, v_t.

3.3 One-Equation Models

3.3.1 Introduction

Dimensional reasoning dictates that the eddy viscosity be proportional to a turbulent velocity scale and a length scale. The former is almost invariably characterized by the rms value of the turbulence energy 'k'. With the length scale denoted by 'ℓ', the eddy viscosity thus arises as:

$$\mu_t = \rho C_\mu k^{1/2} \ell \qquad (3.4)$$

where C_μ is either a numerical 'constant' or some function of a turbulence Reynolds number, as will be detailed below in relation to specific models.

The length scale is frequently represented indirectly by a related variable of the general form $k^m \ell^n$, where 'n' and 'm' depend on the particular modelling approach adopted. The most frequently used length-scale variable is the turbulence dissipation rate 'ε', related to 'k' and 'ℓ' via:

$$\varepsilon = C_D k^{3/2} / \ell \qquad (3.5)$$

or a similar form. A fact favouring the choice of 'ε' over any other combination of 'k' and 'ℓ' is that 'ε' appears naturally as the sink term in the transport equation for the turbulence energy 'k'. Such a transport equation is a key ingredient in all one-equation models. Closure requires a further equation for 'ε' or 'ℓ'. In the present class of models, the latter is described by an empirically-based algebraic expression, the precise form of which varies from model to model.

3.3.2 The Wolfshtein [W] Model (1969)

This model describes the variation of turbulence energy by the transport equation:

$$\frac{\partial \rho U_j k}{\partial x_j} = \frac{\partial}{\partial x_j}\left[(\mu + \frac{\mu_t}{\sigma_k})\frac{\partial k}{\partial x_j}\right] + P_k - \rho\varepsilon \qquad (W1)$$

where P_k is the rate of turbulence-energy production,

$$P_k = \frac{\partial U_i}{\partial x_j}\left[\mu_t\left(\frac{\partial U_i}{\partial x_j} + \frac{\partial U_j}{\partial x_i} - \frac{2}{3}\delta_{ij}\frac{\partial U_k}{\partial x_k}\right) - \frac{2}{3}\delta_{ij}\rho k\right] \qquad (W2)$$

in which the subscripts i,j,k denote Cartesian tensors and δ_{ij} is the unit tensor.

The length scale is described by an algebraic relation. However, a distinction is made between the length scale pertaining to the eddy viscosity ($\ell=\ell_\mu$) and that used to characterize dissipation ($\ell=\ell_D$). The respective scales are given by:

$$\ell_\mu = C_\ell y \, [1\text{-}exp \, (\text{-}A_\mu R_t)] \qquad \text{(W3)}$$

$$\ell_D = C_\ell y \, [1\text{-}exp \, (\text{-}A_D R_t)] \qquad \text{(W4)}$$

where $R_t = \rho k^{1/2} y/\mu$ is the turbulent Reynolds number. The length scales ℓ_μ and ℓ_D are then used in relations (3.4) and (3.5), respectively, to determine the eddy viscosity and the rate of dissipation.

The constants appearing in this model are:

$$\sigma_k=1., \quad \kappa=0.42, \quad C_\mu=0.09, \quad C_\ell=\kappa C_\mu^{-3/4}, \quad C_D=1., \quad A_\mu=0.014, \quad A_D=0.20.$$

3.3.3 The Hassid–Poreh [HP] Model (1975)

This model is, essentially, a corrected version of Wolfshtein's proposal. The correction reflects the recognition that, at the wall, the rate of dissipation must be balanced by viscous diffusion of turbulence energy - a constraint readily derived from equation (W1).

As in Wolfshtein's model, the turbulence energy is described, here too, by the transport equation:

$$\frac{\partial \rho U_j k}{\partial x_j} = \frac{\partial}{\partial x_j}\left[(\mu + \frac{\mu_t}{\sigma_k})\frac{\partial k}{\partial x_j}\right] + P_k - \rho\varepsilon \qquad \text{(HP1)}$$

where P_k is again given by (W2). To represent correctly the near-wall behaviour, the rate of dissipation is expressed as a sum of two contributions:

$$\varepsilon = \frac{C_D f_D k^{3/2}}{\ell} + \frac{2\mu k}{\rho \ell^2} \qquad \text{(HP2)}$$

where the latter contribution is arranged to be consistent with the fact that k decays quadratically in the viscous sublayer as the wall is approached.

Here too, the length scale obeys an algebraic prescription, namely:

$$\ell = C_\ell f_D \, \delta \, \tanh\left(\frac{\kappa y}{C_\ell \delta}\right) \qquad \text{(HP3)}$$

where $f_D=[1\text{-}exp\,(\text{-}A_\mu R_t)]$,

$R_t = \rho k^{1/2} \ell y/\mu$,

and δ is the boundary-layer thickness.

The constants appearing in the above model are:

$$\sigma_k=1., \quad C_\mu=0.22, \quad C_\ell=0.085, \quad C_D=0.164, \quad A_\mu=0.0119, \quad \kappa=0.42.$$

3.4 Two-Equation Models

3.4.1 Introduction

In this class of models, the length scale - or rather a related variable - is governed by a transport equation additional to that describing the turbulence energy. In all models featuring below, the length-scale variable is 'ε'. Hence, the eddy viscosity arises from:

$$\mu_t = \rho C_\mu k^2 / \varepsilon \tag{3.6}$$

or a very similar form.

Models are described as being "low-Reynolds-number" or "high-Reynolds-number" variants. The former provide a description of turbulence processes across the entire flow domain, including the semi-viscous (low-Re) near-wall region. The latter, in contrast, do not account for the interaction between turbulence and fluid viscosity, and do not, therefore, apply to the near-wall region. Rather, they must be used in conjunction with an "auxiliary" near-wall model, which may either be a simplified low-Re variant or which is based on so-called "wall laws", the latter being semi-empirical algebraic formulae derived on the assumption of a logarithmic velocity variation close to the wall.

3.4.2 The Jones–Launder and Launder–Sharma [JL] Model (1972, 1974)

These two models are, essentially, identical in form, the only point of difference lying in one attenuation function, as identified below. Both models are based on two transport equations, one for turbulence energy and the other for dissipation. A peculiar feature of the present formulation, in contrast to other two-equation variants, is its use of a modified rate of dissipation '$\tilde{\varepsilon}$', defined by:

$$\tilde{\varepsilon} = \varepsilon - 2\nu \left(\frac{\partial k^{1/2}}{\partial x_j} \right). \tag{JL1}$$

The rationale of this modification is rooted in analytical considerations which suggest that '$\tilde{\varepsilon}$' vanishes at the wall; that is, the dissipation ε is finite at the wall, approaching the value of the second term on the r.h.s. of equation (JL1). Analogous considerations have motivated Hassid & Poreh (1972) to formulate a corrected variant of Wolfshtein's one-equation model (1969) (see above). Hence, with definition (JL1) adopted, homogeneous boundary conditions may be applied to both 'k' and '$\tilde{\varepsilon}$', yielding superior numerical characteristics. It is also '$\tilde{\varepsilon}$', rather than 'ε', which is used in determining the eddy viscosity.

With the above modification introduced into the turbulence-energy equation, there results:

$$\frac{\partial \rho U_j k}{\partial x_j} = \frac{\partial}{\partial x_j}\left[(\mu + \frac{\mu_t}{\sigma_k})\frac{\partial k}{\partial x_j}\right] + P_k - \rho\tilde{\varepsilon} - 2\rho\nu\left(\frac{\partial k^{1/2}}{\partial x_j}\right)^2 \quad \text{(JL2)}$$

$$\frac{\partial \rho U_j \tilde{\varepsilon}}{\partial x_j} = \frac{\partial}{\partial x_j}\left[(\mu + \frac{\mu_t}{\sigma_\varepsilon})\frac{\partial \tilde{\varepsilon}}{\partial x_j}\right] +$$
$$+ C_1 \frac{\tilde{\varepsilon}}{k} P_k - C_2 \frac{\rho\tilde{\varepsilon}^2}{k}\left[1 - 0.3\exp(-R_t^2)\right] - 2\mu\mu_t\left(\frac{\partial^2 U_i}{\partial x \partial x_j}\right)^2 \quad \text{(JL3)}$$

where $R_t \equiv k^2/\nu\tilde{\varepsilon}$ is the turbulent Reynolds number and P_k is the production rate of turbulence energy, as given by relation (W2). The eddy viscosity is evaluated from a modified version of relation (3.6), namely:

$$\mu_t = \rho f_\mu C_\mu k^2/\tilde{\varepsilon} \quad \text{(JL4)}$$

where $f_\mu = \exp[-3.4/(1+0.02R_t)^2]$ for the Launder/Sharma model, and

$f_\mu = \exp[-2.5/(1+0.02R_t)]$ for the Jones/Launder variant.

The constants appearing in the above models take the following values:

$\sigma_k = 1., \ \sigma_\varepsilon = 1.3, \ C_\mu = 0.09, \ C_1 = 1.44, \ C_2 = 1.92.$

3.4.3 The Chien [C] Model (1982)

As the previous model, that here consists of two transport equations, one for turbulence energy and the other for the rate of dissipation:

$$\frac{\partial \rho U_j k}{\partial x_j} = \frac{\partial}{\partial x_j}\left[(\mu + \frac{\mu_t}{\sigma_k})\frac{\partial k}{\partial x_j}\right] + P_k - \rho\tilde{\varepsilon} - \frac{2\mu k}{y^2} \quad \text{(C1)}$$

$$\frac{\partial \rho U_j \tilde{\varepsilon}}{\partial x_j} = \frac{\partial}{\partial x_j}\left[(\mu + \frac{\mu_t}{\sigma_\varepsilon})\frac{\partial \tilde{\varepsilon}}{\partial x_j}\right] +$$
$$C_1 \frac{\tilde{\varepsilon}}{k} P_k - C_2 \frac{\rho\tilde{\varepsilon}^2}{k}\left[1 - 0.22\exp[-(R_t/6)^2]\right] - \frac{2\mu\tilde{\varepsilon}}{y^2}\exp(-C_4 y^+) \quad \text{(C2)}$$

where $R_t = \rho k^2/\mu\tilde{\varepsilon}$ and

$y^+ = \rho k^{1/2} y/\mu$

are forms of the turbulent Reynolds number and P_k is given by relation (W2). Here again, the eddy viscosity is evaluated from:

$$\mu_t = \rho f_\mu C_\mu k^2 / \tilde{\varepsilon} \qquad (C3)$$

where $f_\mu = 1 - exp(-C_3 y^+)$.

A noteworthy feature here is the last term in equation (C1). Yet again, this term is intended to account for the finite level of dissipation very close to the wall, an issue already considered above in relation to the models of Hassid/ Poreh, Launder/Sharma and Jones/Launder.

The constants appearing in the above equations take the following values:

$$\sigma_k = 1., \quad \sigma_\varepsilon = 1.3, \quad C_\mu = 0.09, \quad C_1 = 1.35, \quad C_2 = 1.8, \quad C_3 = 0.0115, \quad C_4 = 0.5.$$

3.4.4 The Lam–Bremhorst [LB] Model (1981)

In this variant, the effects of viscosity (except diffusion) are introduced only via corrective terms to the ε-equation and through the constituitive eddy-viscosity relations. Thus, the k and ε equations read here:

$$\frac{\partial \rho U_j k}{\partial x_j} = \frac{\partial}{\partial x_j}\left[(\mu + \frac{\mu_t}{\sigma_k})\frac{\partial k}{\partial x_j}\right] + P_k - \rho\varepsilon \qquad (LB1)$$

$$\frac{\partial \rho U_j \varepsilon}{\partial x_j} = \frac{\partial}{\partial x_j}\left[(\mu + \frac{\mu_t}{\sigma_\varepsilon})\frac{\partial \varepsilon}{\partial x_j}\right] +$$

$$C_1\left[1 + (\frac{0.05}{f_\mu})^3\right]\frac{\varepsilon}{k} P_k - C_2 \frac{\rho\varepsilon^2}{k}\left[1 - exp(-R_t^2)\right] \qquad (LB2)$$

where $R_t = \rho k^2 / \mu\varepsilon$,

$f_\mu = exp[1 - exp(-0.0165 y^+)]^2 (1 + 20.5/R_t)$,

$y^+ = \rho k^{1/2} y / \mu$,

and P_k is given, as usual, by relation (W2). As in previous variants, the eddy viscosity is evaluated from:

$$\mu_t = \rho f_\mu C_\mu k^2 / \varepsilon \, . \qquad (LB3)$$

The constants appearing in the above equations take the following values:

$$\sigma_k = 1., \quad \sigma_\varepsilon = 1.3, \quad C_\mu = 0.09, \quad C_1 = 1.44, \quad C_2 = 1.92.$$

3.4.5 The "Standard" High–Reynolds–number Jones–Launder [SJL] Model (1972)

The previously reported Jones/Launder model is applicable regardless of the Reynolds number. The model may, therefore, be applied across the semi-viscous near-wall sublayer. There are circumstances in which it is advantageous, mainly in terms of computational economy, to restrict the application of the model to the fully turbulent regime away from the immediate vicinity of walls. The task of resolving near-wall processes is then delegated to an auxiliary model which must be 'coupled' to the former. This approach allows a formal simplification of the Jones/Launder variant reported above. In particular, the relations involving exponential functions and fluid viscosity may be omitted. Moreover, the distinction between 'ε' and '$\tilde{\varepsilon}$' becomes irrelevant. The version arising therefrom is often referred to as the "standard k-ε model". The equations forming this model are:

$$\frac{\partial \rho U_j k}{\partial x_j} = \frac{\partial}{\partial x_j}\left(\frac{\mu_t}{\sigma_k}\frac{\partial k}{\partial x_j}\right) + P_k - \rho\varepsilon \qquad \text{(SJL1)}$$

$$\frac{\partial \rho U_j \varepsilon}{\partial x_j} = \frac{\partial}{\partial x_j}\left(\frac{\mu_t}{\sigma_\varepsilon}\frac{\partial \varepsilon}{\partial x_j}\right) + C_1 \frac{\varepsilon}{k} P_k - C_2 \frac{\rho\varepsilon^2}{k} \qquad \text{(SJL2)}$$

$$\mu_t = \rho C_\mu k^2/\varepsilon \qquad \text{(SJL3)}$$

with the constants being:

$\sigma_k=1.$, $\sigma_\varepsilon=1.3$, $C_\mu=0.09$, $C_1=1.44$, $C_2=1.92$.

3.5 Reynolds-Stress Models

3.5.1 Introduction

This class of models does not involve the eddy-viscosity hypothesis. Any model variant consists, rather, of a set of equations, either differential or algebraic, which determine directly the individual components of the Reynolds stress tensor. These equations are modelled variants of exact parent forms which may be derived by suitably manipulating the Navier-Stokes and Reynolds equations. The key advantage of Reynolds-stress models is their ability to capture anisotropy and the disproportionately sensitive response of the stresses to curvature-related secondary strains.

3.5.2 The Gibson–Launder–Rodi [RS] Model (1978)

The Reynolds-stress model used in the present study arises from a simplification of the transport closure of Gibson and Launder (1978), applicable to fully turbulent conditions only (i.e. high Re number). This consists of a set of equations for all Reynolds-stress components $\overline{u_i u_j}$:

$$\underbrace{\frac{\partial \rho U_k \overline{u_i u_j}}{\partial x_k}}_{C_{ij}} = D_{ij} + P_{ij} - \underbrace{\frac{2}{3} \rho \delta_{ij} \varepsilon}_{\varepsilon_{ij}} + \phi_{ij} \qquad \text{(RS1)}$$

in which

$$D_{ij} = \frac{\partial}{\partial x_k} \left[C_k \rho \, \overline{u_k u_\ell} \, \frac{k}{\varepsilon} \, \frac{\partial \overline{u_i u_j}}{\partial x_\ell} \right], \qquad \text{(RS2)}$$

$$P_{ij} = -\rho \, \overline{u_i u_k} \, \frac{\partial U_j}{\partial x_k} - \rho \, \overline{u_j u_k} \, \frac{\partial U_i}{\partial x_k}, \qquad \text{(RS3)}$$

$$\phi_{ij} = \phi_{ij1} + \phi_{ij2} + \phi_{ijw1} + \phi_{ijw2} \qquad \text{(RS4)}$$

$$\phi_{ij1} = -C_1 \rho \, \frac{\varepsilon}{k} \left(\overline{u_i u_j} - \frac{\delta_{ij}}{3} \overline{u_k u_k} \right) \qquad \text{(RS5)}$$

$$\phi_{ij2} = -C_2 \rho \left[P_{ij} - \frac{\delta_{ij}}{3} P_{kk} \right] \qquad \text{(RS6)}$$

$$\phi_{ijw1} = C_1' \rho \frac{\varepsilon}{k} \left[\overline{u_k u_m} \, n_k n_m \, \delta_{ij} - \frac{3}{2} \overline{u_k u_i} \, n_k n_j - \frac{3}{2} \overline{u_k u_j} \, n_k n_i \right] f \qquad \text{(RS7)}$$

$$\phi_{ijw2} = C_2' \rho \left[\phi_{km2} \, n_k n_m \, \delta_{ij} - \frac{3}{2} \phi_{ik2} \, n_k n_j - \frac{3}{2} \phi_{jk2} \, n_k n_i \right] f . \qquad \text{(RS8)}$$

In the last two contributions, n_i is the wall-normal unit vector in the direction i and $f = C_\mu^{0.75} k^{1.5} / \varepsilon \kappa \Delta n$ with Δn being the wall-normal distance.

The above set of partial differential equations may be simplified to become 'algebraic' by approximating stress convection and diffusion in terms of convection and diffusion of turbulence energy (which is governed by its own partial differential equation). Here, the form proposed by Rodi (1972) has been adopted:

$$C_{ij} - D_{ij} \leftarrow 0.5 \frac{\overline{u_i u_j}}{k} (C_{kk} - D_{kk}) = \frac{\overline{u_i u_j}}{k} (0.5 P_{kk} - \rho \varepsilon) \qquad \text{(RS9)}$$

in which the double subscript 'kk' identifies a contraction of the associated tensor with superscripts 'ij'.

Finally, the dissipation rate ε is governed by:

$$\frac{\partial \rho U_j \varepsilon}{\partial x_j} = \frac{\partial}{\partial x_k} \left((\mu + C_\varepsilon \frac{\rho \overline{u_i u_j} \, k}{\varepsilon}) \frac{\partial k}{\partial x_k} \right) + C_1 \frac{\varepsilon}{k} \, 0.5 P_{kk} - \rho C_2 \frac{\varepsilon^2}{k} . \qquad \text{(RS10)}$$

The model constants are:

$C_\mu = 0.09$, $\kappa = 0.42$, $C_k = 0.22$, $C_\varepsilon = 0.12$, $C_1 = 1.8$, $C_2 = 0.6$, $C_1' = 0.5$, $C_2' = 0.3$.

3.6 Calculation of Boundary Layer Length Scales

Approach Used for 2D Applications

For Navier-Stokes computations, it is not straightforward to calculate the edge of the boundary layer, δ, and from this U_e and the incompressible displacement thickness, δ_i^*, directly from the data. Different approaches are used by the EUROVAL partners in this respect, and descriptions on how the viscous-layer thickness is determined may be found in chapter 4, in which the partner-related method descriptions are presented.

Most of the partners confronted with the necessity to derive the boundary-layer length scales as input for a turbulence model implementation or merely for postprocessing the results for a proper comparison with the experimental findings, are taking the approach by Stock and Haase (1989).

This method relies on the assumption that computed Navier-Stokes profiles are comparable to Coles' boundary layer profiles. From these profiles it can be shown that $(y|du/dy|)_{max}$ occurs at a relative wall distance inside the viscous layer for all possible flow situations, accelerated or decelerated flows and attached or detached layers.

Using that feature when analysing the Navier-Stokes mean flow data, it is now possible to compute the desired length scale for the outer layer. Evaluating the wall distance y_{max} for which $y|du/dy|$ - or the F-function in the Baldwin-Lomax model, equ. (BL8) - has its maximum, delivers the boundary layer thickness to $\delta = 1.548 \cdot y_{max}$.

However, a careful compilation of various Navier-Stokes calculations has shown (Stock and Haase, 1989) a systematic underprediction of the viscous layer thickness when using $\delta = 1.548 \cdot y_{max}$. By analysing numerous computed velocity profiles, it was seen that by increasing the constant 1.548 by 25% yielded a well predicted viscous layer thickness in all flow cases. Consequently, the boundary layer thickness equals

$$\delta = 1.936 \cdot y_{max} \ . \tag{3.7}$$

With the knowledge of the boundary layer thickness, δ, all other boundary layer length scales can be easily computed by numerical integration of the corresponding boundary layer profiles.

Approach Used for 3D Wing Calculations

For computations concerning the DLR-F5 wing in chapter 5.7, all partners who employed Navier-Stokes codes used the following procedure - as suggested by DLR - to derive the boundary layer thickness, δ:

1. The position y_{umax} of $|u_{max}|$, with u_{max} being the resultant velocity, is calculated for each local velocity profile normal to the wall.

2. In the following, the vorticity distribution and its maximum value, $|\omega_{max}|$, are calculated for the same position.

3. Scanning the vorticity distribution, starting at the wall, it has to be checked whether

$$|\omega| > 0.001 \, |\omega_{max}| \quad for \quad y < y_{umax} \, . \tag{3.8}$$

If this becomes true for all y, y_{umax} is taken as the boundary layer thickness, δ. Otherwise, i.e. if $|\omega| \leq 0.001 \cdot |\omega_{max}|$ is found in the profile, δ is set to the last position of y where equation (3.8) is still fulfilled.

Although this procedure seems to be somewhat arbitrary, it provided very reasonable answers for the DLR-F5-wing computations, as it can be seen in chapter 5.7, comparing the Navier-Stokes results with corresponding boundary layer calculations. Of course, it can not be expected that the presented procedure, at least with the same fraction of $|\omega_{max}|$, will hold in any other application, too.

4 Numerical Methods Used by Partners

4.1 Description of Method Used by Analysis Systems Research Ltd.
(S. Perrakis, M. Tzitzidis)

4.1.1 Introduction

In the EUROVAL project Analysis Systems Research took part in the work on the validation of CFD codes with respect to finding their possible limits of applicability. ASR used a 2D finite volume Navier-Stokes code named SIFLO, kindly provided by Dornier, on the RAE 2822 aerofoil under several and varied flow conditions.

4.1.2 Method

The 2D unsteady compressible Navier-Stokes equations in integral form

$$\frac{\partial}{\partial t} \iint_{Vol} U \, dVol + \int_{S} H \cdot n \, dS = 0$$

with $U = (\rho, \rho u, \rho v, E)^T$ being the dependent variables vector
n being the cell normal unit vector
S being the cell surface area
H being the tensor of viscous fluxes

are used by SIFLO, together with the perfect gas equation of state, in order to solve the flow around a body. Sutherland's law is used for viscous laminar flows and the Prandtl number is 0.72. SIFLO uses the finite volume method which assumes constant values for all variables in a small cell where the governing equations are applied. The resulting set of ordinary equations is solved with a 3 stage Runge-Kutta time stepping method,

$$u(0) = u(n)$$
$$u(n) = u(0) - a_v \, P \, u(v-1), \quad v = 1,3$$
$$u(n+1) = u(3)$$

with coefficients $a_1=0.6$, $a_2=0.6$ and $a_3=1.0$.

This scheme is stable for a Courant number of 1.8 maximum. Blended second and fourth order artificial dissipation, Jameson et al (1981) is used to avoid odd-even decoupling with second order dissipation being active in areas with large pressure gradients and fourth order dissipation being active everywhere else in the flow. This artificial dissipation is applied once to give the best damping possible. Residual averaging, the implicit collection of residuals information enhances the convergence speed dramatically by allowing the use of a higher Courant number, reaching up to a Courant number of 3.5.

4.1.3 Boundary Conditions

The boundary conditions used by the code at the far field boundaries are Riemann invariants. For the solid wall boundaries no-slip conditions are used and the flow is assumed to be adiabatic with zero pressure gradient in the wall normal direction. At the outflow boundary, linear extrapolation of density and mass fluxes, with fixed pressure are used. Flow density is also linearly extrapolated. Lower and upper wake cells overlap creating a wake boundary. These conditions were used throughout the work performed for project EUROVAL.

4.1.4 Grid Adaption

Grids used were provided by BAe to all partners for commonality purposes. This grid was modified by Analysis Systems Research because there appeared some severe convergence problems. The modifications consisted of 'thinning out' the grid in the direction vertical to the surface by removing every other grid point. This had as a result a 256X33 grid that finally gave a correct convergence.
In house generated grids of size 193X65 were also used in several runs beyond the runs that were to be performed under the same flow conditions and using the BAe grids by all partners. This was due to the fact that SIFLO seemed inclined to reach convergence more easily when using the grid generator which was specifically designed for it, and time constraints became very important.

4.1.5 Modifications To The 'Standard' Turbulence Models

Turbulence modelling used the Cebeci-Smith algebraic model described in detail in chapter 3. The version of the model used includes the modifications in the turbulent length scale, particularly the displacement thickness, determination proposed by Stock and Haase (1989) described in section 3.6

4.1.6 Computers And CPU Times

For the calculations performed for EUROVAL a Schneider 386/25 PC was initially used. This machine gave CPU times of about 27 hours which was reduced to about 6 hours when the converged solution of one run was used as an initial guess for the next run. Later a SUN SparcStation 2 was used and the CPU times were reduced considerably. A typical CPU time on the SUN was of the order of 5 hours and if the converged solution of one run was used as an initial guess for the next run, the CPU time decreased to about 1.5 hours.

4.2 Description of Methods Used by British Aerospace (J.J.Benton)

4.2.1 Introduction

The flow solvers used by British Aerospace to produce results for EUROVAL fall into two categories. A 2D multiblock Reynolds-averaged Navier-Stokes code is used for the aerofoil and bump flow cases described in Sections 5.1 to 5.4, while a 3D multiblock Euler method with viscous coupling is used for the 2D and 3D wind tunnel simulation investigation in Section 5.8. These two codes are described below.

4.2.2 2D Multiblock Navier-Stokes

4.2.2.1 Summary

This is a structured grid method in which complex geometry is accommodated by use of a multiblock approach. Euler or Navier-Stokes solutions can be selected in each block. Finite volume central differencing is employed with added dissipation. Convergence to the steady state is found by explicit Runge-Kutta local time stepping with implicit residual smoothing. A multigrid acceleration scheme is incorporated. The discretisation of viscous terms aims to promote damping of odd/even spurious modes. Several turbulence models of zero, one, or two equation eddy-viscosity type are incorporated, with low Reynolds number formulations near wall boundaries. Constraints on the block structure are that a block edge may have only a single boundary condition, and neighbouring block edges must have a one-to-one correspondence in grid dimension and spacing.

4.2.2.2 Transport Equations

A zonal facility allows either Euler or Reynolds-averaged Navier-Stokes solutions to be selected in each grid block. For cases with a far-field free air boundary the multiblock grid would typically be arranged to provide an outer region of Euler blocks. The transport equations are in the form of conservation laws integrated over a finite volume ie. a grid cell area. The conserved variables for the mean flow are mass, momentum and total energy, all per unit volume. These four equations express the time dependence of the volume integral over the cell of the conserved quantity in terms of flux integrals along the cell boundaries:

$$\frac{\partial}{\partial t} \int_{Vol} U dVol + \int_{S} \mathbf{H} \cdot \mathbf{n} dS + D_2 + D_4 = 0. \qquad (1)$$

$U = [\rho, \rho U, \rho V, \rho E]^T$ being the four conserved variables, $\mathbf{H}$ contains the corresponding cell face fluxes, S is the cell face area and $\mathbf{n}$ its outward unit normal vector. D_2 and D_4 are the added second and fourth difference artificial dissipation terms described in Section 4.2.2.4. The equation of state for a perfect gas is used to evaluate pressure from energy and so close the equations set:

$$p = (\gamma - 1) [\rho E - \tfrac{1}{2}\rho (U^2 + V^2)]. \qquad (2)$$

Non-dimensional forms of the above equations are used in the actual code.

$\mathbf{H}$ involves convective and, for viscous flows, diffusive components. The latter derive from the action of stress and heat conduction on the cell faces from both molecular and turbulent action. The viscous terms are represented in full; thin-layer approximations are not used. The molecular viscosity (μ) of the external flow is fixed by a Reynolds number and its variation with temperature is determined from Sutherland's law. The molecular heat conduction coefficient is $k = c_p\mu/Pr$ with the Prandtl number (Pr) = .72 and γ = 1.4. Section 4.2.2.7 describes the turbulent contributions μ_t and k_t which are derived from a turbulence model.

One and two equation turbulent transport models require the addition of similar conservation equations for turbulence energy (ρk), and in the case of two equation models, the energy dissipation rate ($\rho \varepsilon$), both expressed per unit volume. The detailed form of these depends on the turbulence model in use and the relevant equations may be found in differential form in Sections 3.3 and 3.4. Volume integrals of source terms now appear in addition to convective and diffusive flux integrals.

4.2.2.3 Discretisation

The following describes the discretisation of the full equations as used to define the final solution on the grid input into the code as data. This distinction is necessary as this is simply the finest grid of a multigrid sequence; variations in discretisation are employed on the coarser grid levels which are used purely for acceleration of convergence and do not affect the fine grid solution. Such multigrid issues are described in Section 4.2.2.6.

Discretisation of the conservation equations (1) defined in Section 4.2.2.2 follows the semi-discrete methods used by Jameson, Schmidt & Turkel (1981) for Euler equations in which space and time discretisation can be considered separately as follows. The conserved variables are stored as cell quantities ie. effectively at cell centres, and these and any source terms in the equations are assumed constant over the cell so that volume integrals reduce to the product of a cell centre value and cell volume. This results in equations for the cell centre values of the conserved variables U:

$$\frac{\partial}{\partial t} U Vol = - \int_S \mathbf{H}.\mathbf{n} dS - D_2 - D_4 \qquad (3)$$

which are integrated to the steady state by the explicit 4-stage Runge-Kutta method described in Section 4.2.2.5.

This procedure requires evaluation of the right hand side **H** flux integral term on each time step and here central differencing methods in physical space are used which are of second order accuracy for uniform grids and smooth regions of flow. The explicit nature of the time integration is reflected in the use of the solution at the current time to evaluate these terms. The convective components of **H** are evaluated at the cell face centres from values of the conserved variables there which, following Jameson, are approximated as simple averages of the two cell centre values that share that face.

For viscous flows the diffusive fluxes in **H** are also required and these involve the calculation of flow gradients from which the fluxes due to stress and conduction are evaluated, again at the cell face centres. Gauss' method is used to evaluate gradients $\partial/\partial x$ and $\partial/\partial y$ of velocity and temperature at the cell face centres by integration around the boundaries of auxiliary cells, as indicated in Fig.1. This integral is evaluated as the sum, over the four faces, of the face centre value of the velocity or temperature multiplied by the x or y projection of the face area.

As may be seen two orientations of auxiliary cell are required, one for gradients on

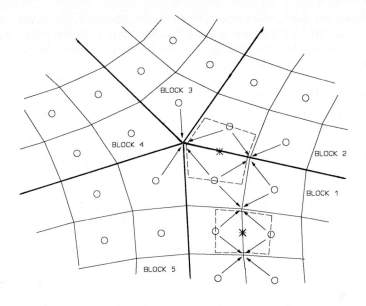

Fig.1 Calculation of gradients at cell face centres.
o cell centres --- auxiliary cells
* face centres → 4-point averaging

constant i faces, and one for constant j faces. Two faces of each auxiliary cell use unaveraged cell centre values and in this way contribute to a coupling between odd/even points through the fluxes generated by real viscosity, thus helping to damp out spurious +/- modes. Such a feature should be beneficial when it is required to minimise or switch off the added artificial dissipation, and should be particularly effective if the diffusion coefficient is large such as when it includes turbulent viscosity. The two faces concerned are the ones that largely determine the gradient in the direction normal to the real cell face, thus maximising the above coupling effect when the flux boundary integral is formed. A cell centre method such as the present one facilitates this optimal selection of the auxiliary cell.

The auxiliary cells require special treatment at singular points in the multiblock grid structure ie. block corners at which three or five blocks meet. This situation is illustrated in Fig 1. Boundary conditions for the block are handled for inviscid solutions by including two layers of halo cells surrounding the block which take values from the immediate neighbouring block or via physical boundary conditions, no values being required in halo corners. However the gradient calculation requires this to be extended to the storage of values in the halo corners for use in the integration round the auxiliary cells. In fact two values must be stored there, corresponding to values in the halos of the neighbouring blocks on each of the two edges forming the corner. One corner value is used for evaluation of gradients on the adjacent i constant cell face and the other is used for the j constant face. The two values will be identical for a non-singular block corner.

At cell faces on the zonal boundary between viscous and inviscid grid blocks, diffusive fluxes for both mean flow and turbulent transport are made zero. After evaluation of all the cell face fluxes the boundary integral on the right hand side of equation (3) is evaluated by summing, over the four faces of the main cell, the scalar product of face centre flux and face area.

4.2.2.4 Artificial Dissipation

The method follows Jameson, Schmidt & Turkel (1981) in adding artificial dissipation of first order to avoid overshoots at shocks and third order to damp spurious modes in the solution arising from odd/even decoupling. The first order term is in the form of a second difference of the conserved variables and the third order term is a fourth difference in a form used by Swanson and Turkel (1987). Pressure is added into the differencing for the energy equation resulting in ρH rather than ρE being used. Jameson's second difference of pressure sensor is used to detect shocks and switch on the second differencing there while also switching off the fourth differencing. Additional bounds on this pressure sensor and on Mach number are added in this code to ensure that occurrence of second differencing away from shock waves or within the boundary layer is minimised. Various scalings of the dissipation are available, the form shown below using an anisotropic power law ratio of eigenvalues is typical and is used in all results.

The second difference term $D_{2,i}$ in cell i is constructed as follows for the i direction, with $\lambda_{j,i}$ = convective flux eigenvalue in j direction in cell i, and $W_i = \rho, \rho U, \rho V$, or ρH:

For cells i: $\quad v_i = \left| \dfrac{p_{i+1} - p_i + p_{i-1}}{p_{i+1} + p_i + p_{i-1}} \right| \qquad \phi_i = \left[\max\left(1, \dfrac{\lambda_{j,i}}{\lambda_{i,i}}\right) \right]^{\frac{1}{4}}.$

An average is taken for cell faces i+½: $\quad F_{i+1/2} = \tfrac{1}{2}(\phi_i \lambda_{i,i} + \phi_{i+1} \lambda_{i,i+1}).$

The dissipation coefficient at the cell face is: $\varepsilon_{2,i+1/2} = d_2 \, v_{i+1/2}$ where $d_2 = 1.25$. $v_{i+1/2}$ is taken as a maximum of v_i over three cells. In addition ε_2 is scaled to zero unless $v > .005$ in nearby cells and also as Mach number passes below approximately 0.5. Then dissipation in cells i is:

$$D_{2,i} = -\nabla_i (\varepsilon_{2,i+1/2} \, F_{i+1/2} \, \Delta_{i+1/2} W_i). \qquad (4)$$

The fourth difference term $D_{4,i}$ in cell i has the following form for the i direction:

For cells i: $\quad \phi_i = \left[\max\left(1, \dfrac{\lambda_{j,i}}{\lambda_{i,i}}\right) \right]^{\frac{1}{2}}$

$$\varepsilon_{4,i} = \max\left[0, d_4 - 3 \max(\varepsilon_{2,i-1/2}, \varepsilon_{2,i+1/2}) \right] \qquad d_4 = 1/128$$

$$F_i = \phi_i \lambda_{i,i}$$

and the dissipation in cells i is:

$$D_{4,i} = \nabla_i \Delta_{i+1/2} (\varepsilon_{4,i} \, F_i \, \nabla_i \Delta_{i+1/2} W_i) \qquad (5)$$

which is the non-dispersive form suggested by Swanson and Turkel (1987).

The fourth difference dissipation in the normal to wall direction may be switched off over a specified depth of cells adjacent to the wall, and the process may be continued into the associated wake blocks. The depth of cells used would typically place this boundary outside all but the thickest parts of the boundary layer. At the boundary of this zero region the dissipation is scaled up to its normal value over several cells. The resultant large number of cells over which dissipation is not applied can lead to stability problems in the time integration when starting from free-stream conditions but these are very effectively overcome by the residual smoothing used in this code and which is described below. No problems of odd/even wiggles have been encountered when dissipation is removed in this way.

Two tests for adverse effects arising from the added dissipation were performed. Firstly a flat-plate laminar boundary layer test case at M=.15 with and without wall-normal dissipation shows this dissipation to have a significant effect on the velocity profile for a grid with 12 cells across the boundary layer. Using 22 cells this degradation is negligible, illustrating the third order nature of the term. Without normal dissipation the results show good agreement with the Blasius profiles of u and du/dy for both grid densities.

Secondly a high Reynolds number transonic aerofoil calculation (RAE2822 case 9) using Chien's low-Re_t k-ε turbulence model and between 20 (leading edge) and 40 (trailing edge) cells in the boundary layer was used to compare results with the normal-to-wall dissipation zeroised for 30 cells and with it taken to the wall. Very small differences in Cf values near the shock are seen, with only slightly larger differences in velocity and turbulence profiles. This suggests that normal-to-wall dissipation is not a critical issue but it should be recalled that the levels of dissipation used in the code are relatively low anyway.

When turbulent transport equations are included, dissipation is added in the same form as for the mean flow. As the turbulence variables can undergo sudden changes where fixed transition is implemented, second differencing is used in the stream direction for several cells upstream and downstream of the transition point, with the corresponding fourth differencing suppressed.

4.2.2.5 Time Integration

The steady state is found by explicit 4-stage Runge-Kutta time integration following Jameson, Schmidt and Turkel (1981) and Jameson and Baker (1984). The coefficients of the scheme are 1/4, 1/2, .55, and 1.0. Methods of this type yield approximately second order time accuracy in time. However, time accuracy is not sought so the maximum time step is determined locally from the stability limit for each cell. Thus $\Delta t = \text{CFL} \, \Delta t_p$ where the limiting CFL number is determined by the integration scheme and Δt_p is a propagation time for the cell combining acoustic, convection and viscous effects. Δt_p is taken as the minimum of separate evaluations for the i and j directions. Though a separate, larger, time step could be evaluated for the turbulent transport if present by omitting or reducing the acoustic term, this has generally led to instability in the present multigrid code. Thus a common step size is used for mean flow and turbulent transport.

The Runge-Kutta scheme requires the addition of dissipation as described previously in order to remain stable, and allow Δt to exceed Δt_p by the CFL factor. Stability analysis using linear equations (Jameson and Baker, 1984) gives the CFL limit as 2.0 for the above 4 stage scheme if dissipation is calculated at the first stage only, which is typically done in the code due to the expense of computing the dissipation. This CFL limit is further reduced by the non-linear nature of the actual transport equations, particularly if the turbulent dissipation rate equation is present.

The CFL limit is increased by use of implicit residual smoothing (Jameson & Baker 1983). It is found best to include also the time step/cell volume term in the smoothed term. The residuals are smoothed at the second and fourth stages of the 4 stage scheme. A zero smoothing flux boundary condition is imposed at all block boundaries to maintain conservation, by making the first difference across the boundary zero. Directionally scaled coefficients are used by means of an adaptation of the scheme suggested by Radespiel et al (1989). The use of scaled coefficients reduces unnecessary smoothing in the coarser grid direction and so should improve convergence rate. A CFL number of 3 to 4 is typically used with this scheme.

Time integration is started from free-stream conditions with sudden imposition of boundary conditions, or from a previous flow field output. A tendency to lightly damped large scale oscillatory behaviour in the flow field following such a free-stream start can be very rapidly damped out by adding enthalpy damping (Jameson at al, 1981). This is removed after 50 multigrid cycles in viscous blocks so that it does not affect the final steady state, as the assumption $H = H_\infty$ is not valid in viscous regions. However, it can be retained in the outer Euler grid blocks, if any.

Block halo cells are updated from neighbouring blocks at the beginning of the time step, while halo cells at physical boundaries are updated by the relevant boundary condition procedure at each stage of the Runge-Kutta scheme.

4.2.2.6 Multigrid Acceleration

The multigrid scheme follows the method of Jameson & Baker (1984). In this, a coarse grid correction to the fine grid solution commences with the solution restricted onto a coarser grid, volume weighted for conservation, together with the summed residuals. This restricted solution is then immediately used to calculate residuals on the coarse grid which are subtracted from the restricted residuals to leave a difference value referred to as a forcing function or truncation error for the coarse grid. This now remains constant for this coarse grid until a restriction from the finer grid onto it again takes place.

A time step update is now performed on the coarse grid taking the restricted solution as the initial value. The time integration differs from the fine grid procedure in that the truncation error is added to the residuals before use in the Runge-Kutta update stage, thus forcing the coarse grid solution to differ from that which would otherwise be obtained. For the first stage only, the result of this addition is to recover the restricted residuals. A single complete time step (4 stages) is performed on each block before moving to the next block.

Having obtained the updated coarse grid solution, the original restricted solution from the finer grid is subtracted from it and the resultant differences, or corrections, are interpolated back onto the finer grid where they can be added to the fine grid solution to complete the coarse grid correction for that grid level. It can be seen that if the fine grid residuals are zero then no correction is generated.

The above represents a 2-level V-cycle. This is extended to a 3-level W-cycle in typical applications of the present code, with 2 to 4 time steps on the coarsest grid level. A time step, or update, on any grid level always spans all blocks.

Correction boundary values in the block halo cells are needed to complete the interpolation of corrections onto the finer grid block edge cells. For block to block continuity boundaries the corrections are simply computed in the halo cells as for the in-block cells, with both coarse and fine solutions extracted from the neighbouring block. At physical boundaries use of the coarse grid solution halo values resulting from application of boundary conditions would perturb the fine grid solution and so the halo cell corrections are found by first order extrapolation from the interior and factored, or are given zero values, thus severely under-relaxing the corrections in the adjacent cells. The interpolation of corrections is bi-linear, with cell volume ratios used as the weighting in this code. The resultant fine grid corrections are not smoothed. They are factored by 0.7 (less for k and ε equations) before adding to the fine grid solution.

The code allows grid coarsening to proceed in one block dimension after the other has reached a lower limit. This is useful in allowing a deeper level of multigridding for blocks that have many more cells in one direction than the other. If both dimensions have limited, no further updates are performed for the block at this or coarser levels.

The coarse grid transport equations do not have to be an accurate model of the flow in the manner of the finest grid level. For single aerofoil cases the use of convection only, with dissipation described below, has proved to be robust. Residuals for restriction make use of the dissipation component and, if necessary, the turbulent viscosity value evaluated in the preceding Runge-Kutta update. For complex or stepped geometries it may be necessary for all residual components to be recomputed at the end of the time step and also to update the block halo values. Diffusion on the coarse grids also may be beneficial in such cases, with turbulent viscosity interpolated from the fine grid. The coarse grid added dissipation follows the second differencing used on the fine grid but without the shock sensor and with $d_2=0.1$. It can easily destabilise the code if too large. No enthalpy damping is used on coarse grid levels as this would perturb the fine grid steady state solution.

Boundary conditions for all coarse grid equations are implemented as they would be for the fine grid. However the normal momentum equation used to evaluate surface pressure for the inviscid mean flow (Euler) equations is not used on the coarse grids. The Runge-Kutta time updates on the coarse grids are exactly as for the fine grid.

4.2.2.7 Turbulence Models

Algebraic (zero equation) and one or two equation transport models of turbulence are included in the code, all using the concept of eddy viscosity (μ_t) to obtain the Reynolds stresses. These models are:

(i) Baldwin-Lomax (1978) algebraic model.
This model is described in Section 3.1.2. In the implementation in this code no interpolation is used in finding Fmax, while a simple linear in-cell interpolation is used to refine ymax. The maximum viscous stress in the profile, rather than the wall value, is used in calculating y^+. No blending is used between inner and outer layers or between boundary layer and wake. The outer layer constant C_{wk} is set to the value 1.0 unless otherwise stated in the discussion of results in Chapter 5. The 1978 published value of 0.25 leads to unsteady solutions in transonic cases on typical aerofoil grids which are fairly coarse in the streamwise direction through the shock. Tests on the ONERA bump case (Section 5.2) show that setting $C_{wk}=0.25$ dramatically increases the viscous interaction leading to good prediction of the separation velocity profile, but also a severe failure of the boundary layer to recover downstream. It is also notable that

steady results were obtained on this grid which is very fine in the streamwise direction.

(ii) k-ε 2-equation high-Re$_t$ standard model with 1-equation sublayer.
The k-ε model (Jones and Launder, 1972) follows the description in Section 3.4.5. The sublayer is the k-ℓ formulation by Wolfshtein (1969) as described in Section 3.3.2. It is required to handle the low turbulent Reynolds number (Re$_t$) wall region and is applied over a fixed depth of near wall cells such that the interface with the k-ε model occurs at a y^+ value in the range 60 to 200 (except at shock waves or separation points). The high-Re$_t$ k-ε model is used alone in the wake with no blending. The Wolfshtein model is also used over the full field in order to provide starting conditions for the k-ε model and has proved to be very robust in this role.

(iii) k-ε 2-equation low-Re$_t$ Chien model (1982).
This model follows the description in Section 3.4.4. The dissipation equation is reformulated in terms of $\tilde{\varepsilon}$ which goes to zero at the wall and an extra term is then added to the k equation to recover the full dissipation in the wall (low-Re$_t$) region. Thus both k and $\tilde{\varepsilon}$ are integrated to the wall. The extra low-Re$_t$ terms are dropped in the wake with no blending, resulting in a reversion to the standard high-Re$_t$ k-ε model. This process is facilitated by the multiblock structure in which block boundaries at the boundary-layer/wake interface have their halo cells adjusted to provide the correct upstream or downstream boundary condition for ε or $\tilde{\varepsilon}$. The use of Chien's model makes implementation of the multigrid scheme more difficult because of this change of variable: geometry-dependent blending functions could help but should be avoided in a general geometry multiblock method such as the present one.

For the transport models, the turbulent length scale (ℓ) is limited to a user specified maximum value to determine a minimum level of dissipation related to the local turbulent energy:

$$\varepsilon_{min} = \frac{k^{3/2}}{\ell_{max}} \to 0 \quad as \quad k \to 0. \tag{6}$$

This serves to help stabilise the code and prevent build up of turbulence energy outside the viscous regions. A typical value for ℓ_{max} is 0.05 x aerofoil chord.

Fixed transition is implemented by factoring to zero the turbulent viscosity in the case of Baldwin-Lomax, or the source terms (production and dissipation) in the k and ε transport equations in the case of one or two equation models, in the regions required to have laminar boundary layers. This switch is smoothed over four cells in the surface direction, centred on the specified transition location. For the case of k-ε transport models sudden disturbances often occur around the transition point and can disrupt convergence. This situation is improved by local use of the one equation model in the transition zone. Also for the case of transport models it is necessary to seed the turbulence to allow for very low or zero values of free-stream turbulence. This is done by imposing a low minimum value of turbulent viscosity in the transition region, typically .01 x molecular viscosity, but not in the very near wall cells as turbulent viscosity has to be allowed to go to zero there.

In these eddy viscosity models turbulent heat transfer is modelled by use of a turbulent Prandtl number Pr$_t$=0.9 with the turbulent conduction coefficient $k_t = c_p \mu_t / Pr_t$, following the molecular expression.

4.2.2.8 Boundary Conditions

Field boundaries include the far-field typical of free-air calculations, and inlet/outlet surfaces for channel flow. Various options are provided for determining the values required in the block halo cells. Inflow or outflow and subsonic or supersonic conditions are determined for each cell face from the flow normal to the boundary. For subsonic conditions, options include 1D Riemann invariants and various combinations of prescribed free-stream values and extrapolation from the interior.

The Riemann invariant boundary condition is intended to avoid wave reflection and is similar to that used by Jameson and Baker (1983,1984) and Thomas and Salas (1985). A far-field circulation velocity may be added to the exterior (free-stream) using the linearised compressible model following Thomas and Salas. The circulation is determined from lift due to pressure calculated as part of the convergence measures typically every 5 or 10 cycles. The invariants give the velocity normal to the boundary and the local sound speed. For an inflow velocity the tangential velocity and entropy at the boundary take exterior values. For outflow they take the interior cell value. Density at the boundary is found from the sound speed and entropy and the pressure can then be determined. A free-stream turbulence level may be prescribed and is applied at inflow boundaries, a typical value of zero being used.

At walls, viscous boundary conditions consist of a zero normal pressure gradient with the wall pressure simply assigned the same value as the corresponding cell centre in view of the very thin cells there, and the adiabatic condition of zero normal temperature gradient. Turbulent kinetic energy (k) is zero, and the adjusted turbulent dissipation ($\tilde{\epsilon}$) is zero for the Chien turbulence model.

At inviscid walls the normal momentum equation is used to determine wall pressure, density is set to give the same entropy as at the first cell centre (implying a non-adiabatic wall), the normal velocity is zero, and the tangential velocity at the wall is chosen to give the same total enthalpy as the first cell centre. Only wall pressure affects the Euler equations but other quantities are fixed in this way to allow flexible implementation of dissipation differencing at the wall and to provide sensible values when post-processing the flow field.

Boundary conditions for the added artificial dissipation are critical in minimising the generation of spurious entropy. Of those described by Swanson and Turkel (1987) for the fourth difference dissipation their option B1 is used in the present code and shows a good performance. Thus second difference = 0 in the first cell and third difference = 0 at the boundary cell face. The effect of this scheme on long wavelength disturbances is similar to the dissipation on interior cells, which is important for minimum distortion of the flowfield. However it offers a significantly reduced dissipation on short wavelengths at the boundary. The procedure is applied at the edge of the dissipative region when zeroising the dissipation in near-wall cells for viscous flows, though typically the dissipation is scaled up from zero over several cells from this point, making the boundary treatment insignificant. It is of interest to note that though Swanson and Turkel do not recommend their option A2 due to the risk of an undamped mode, it is has given slightly less spurious entropy than B1 in this code, and has the advantage of twice the dissipation on the highest frequency +/- mode. It has the form: second difference in first halo cell = second difference in first interior cell.

For the second difference dissipation the boundary treatment consists of setting the second difference = 0 in the first interior cell, a procedure that avoids a strong long wavelength effect and is essential in avoiding spurious entropy. This is not done on the coarse grids for multigrid which have only second differencing for stability. No distinction is made between walls and field type boundaries when applying the dissipation boundary conditions.

4.2.2.9 Integrated Forces

Lift and drag are found by double precision summation over surface cell faces. Force is evaluated from the surface boundary values of the conserved variables as determined by the boundary condition procedures, and area is the cell face area.

4.2.2.10 Computational Performance

For Navier-Stokes multiblock cases with 3-level W-cycle multigrid around 60 to 80 Mflops are achieved on a CRAY YMP/1, corresponding to 4.0E-5 to 9.0E-5 sec /cell /multigrid cycle depending on grid and turbulence model. For non-multigrid runs 70 to 90 Mflops are achieved corresponding to 2.0E-5 sec/cell/cycle. For a transonic aerofoil with the Baldwin-Lomax turbulence model 200 multigrid cycles give good convergence. Low speed cases take longer to obtain good convergence of boundary layer profiles.

4.2.3 3D Multiblock Viscous-Coupled Euler

This flow solver is used in all the wind tunnel simulations presented in Section 5.8. Like the Navier-Stokes code described in Section 4.2.2 it is a structured grid method in which complex geometry is accommodated by use of a 3D implementation of the same multiblock approach. The Euler equations are solved in each block and viscous effects are accounted for by means of coupled boundary layer solutions operating on a separate surface grid.

The Euler solver again follows Jameson et al (1981) in using finite volume central differencing in physical space with added third order dissipation to smooth spurious modes and first order at shocks to avoid wiggles. Convergence to the steady state is found by explicit 4-stage Runge-Kutta local time stepping with implicit residual smoothing. Multigrid is not used in this code. Riemann invariant boundary conditions are applied at the far field, and wall pressure is found from the normal momentum equation. The viscous effect at the wall is incorporated by means of extra transpiration terms in the wall boundary condition.

Various boundary layer solvers are included but in the EUROVAL results only a 2D integral method is used (Lock and Williams, 1987). This employs the lag-entrainment model of Green et al (1973), extending it to handle separated flow. In the multiblock Euler solver various forms of coupling of the boundary layer to the inviscid flow are available but for EUROVAL only a quasi-simultanious technique is used (Veldman, 1980; King and Williams, 1988). In this the coupling scheme makes use of a simplified form of the inviscid equations which are solved simultaneously with the equations for the viscous flow. A fixed transition point is used in the results presented, with the laminar boundary layer ahead of transition solved by Thwaites' (1949) method.

The boundary layer solver operates on strip-like surface grids covering the wing and extending into the wake. These are taken from the surface grid on which the multiblock Euler 3D field grid is based. When this field grid is not of this regular form over the whole surface, such as where a C-grid interfaces with an O-grid at a wing tip resulting in a singular point in the surface grid near the tip, the viscous coupling is switched off over the non-regular part of the resultant surface grid.

4.3 Description of Method used by CAPTEC (E. A. Cox)

4.3.1 Introduction

The Navier-Stokes equations are solved using a finite volume formulation with the time integration carried out using a three stage multistep Runge-Kutta procedure. The two layer algebraic turbulence model developed by Cebeci and Smith (1974) is implemented. A detailed description of this model is given in section (3.1.4).

4.3.2 Method

The finite volume method is used to solve the Navier-Stokes in integral conservation form. The spatial domain is divided into quadrilateral cells and a system of ordinary differential equations is constructed by applying the integral equations to each cell. A three stage Runge-Kutta time stepping method is used to then integrate this system of time dependent ordinary differential equations. Implicit residual averaging (see Jameson & Schmidt (1985)), introduced during the stages of the Runge-Kutta marching scheme increases the maximum allowable Courant number. For all computations the Courant number was chosen to be 3.5. A blend of second and fourth order artificial viscosity terms is used to both enhance dissipation where there are strong pressure gradients, and to remove induced numerical oscillations that would prevent convergence to a steady state. The fourth order artificial dissipation term is present throughout the entire spatial domain but is turned off in the neighbourhood of large pressure gradients to prevent overshooting near shock fronts.

For stationary flow, accuracy in time is not a requirement, and local time steps are used to accelerate convergence. This leads to significant improvements in convergence rates. The steady state is defined by variations in the force coefficients, and pressure on the solid surface and wake of less than 0.05%. In addition we require that the number of supersonic points remains unchanged over 10 iterations.

No slip boundary conditions are imposed on the solid wall; subsonic flow is assumed and Riemann invariants are imposed as farfield boundary conditions. The wake boundary is given by overlapping upper and lower wake volumes, and at the outflow boundary the density and mass fluxes are extrapolated linearly.

4.3.3 Grid Adaptation

A C-type mesh is used for the spatial discretisation, with geometric stretching in the surface normal direction and a first mesh size normal to the wall giving y^+ values (see CS5) of order unity.

Three different grids were generated for use with both cases 9 and 10 separately. The first, grid A, contains 192 cells in total in the wrap around direction and 64 cells in the wall normal direction. There were 24 cells in the boundary layer on the upper surface of the airfoil. In grid B the number of cells in the wrap around direction is increased to 256, with the cells in the wall normal direction left unchanged. The finest grid, grid C, leaves the wrap around cells at 256 but increases the wall normal cells to 128, locating 40 of the cells in the boundary layer. These grids were used to investigate the sensitivity of the computed flow to mesh refinement inside the boundary layer and grid refinement in the wrap around direction.

4.3.4 Modifications of the Cebeci-Smith model

A modification of the two layer turbulence model developed by Cebeci and Smith (1974) (and described in 3.1.4) is used. This involves using the boundary layer profiles of Coles to determine the boundary layer thickness δ. The details of this approach are found in Stock and Haase (1989) and leads to a boundary layer thickness given by equation (3.7), namely

$$\delta = 1.936 y_{max}.$$

In attached and slightly separated wall layers the eddy viscosity is given by v_t in (CS1) with v_t in the inner boundary layer given by (CS2) and v_t in the outer layer given by (CS6). The term U_e in (CS7) denotes the velocity U at the edge of the boundary layer δ where δ is given by equation (3.7). The value y_{max} in (3.7) is the value of y where F in eqn (BL8) is a maximum. F_{max} is determined from (BL8) using second order interpolation. The displacement thickness δ_i^* in (CS7) is evaluated by numerically integrating the velocity profiles.

In the wake we assume that the eddy viscosity is given by the outer eddy viscosity (CS6) for wall layers. The thickness of the wake upper and lower layers are determined by equation (3.7) and the term δ_i^* in (CS7) is evaluated by numerical integration of the velocity profiles.

Acknowledgements

The assistance of the consortium members in the BRITE EURAM Aeronautics programme EUROVAL is gratefully acknowledged. In particular we wish to thank the Dornier company for our use of their proprietary codes.

4.4 Description of Methods Used by Construcciones Aeronáuticas S.A. (CASA)
(A. Abbas and J.J. Guerra)

4.4.1 Introduction

In the past few years, substantial progress has been achieved in the development of efficient numerical schemes for solving the Euler and Navier-Stokes equations. Accuracy and robustness of the numerical schemes have also continued to improve. Strong emphasis that has been placed on sharp representation of shock waves, which is reflected in the Euler solutions, now is focused on the accuracy of viscous flow calculations where additional attention is required.

In this section, solution methods are presented for the two-dimensional compressible Reynolds averaged Navier-Stokes equations and for the three-dimensional Euler equations. These methods have been validated, within the frame work of the EUROVAL project, against the experimental data of the RAE-2822 airfoil and the DLR-F5 wing (tunnel case) respectively.

4.4.2 Navier-Stokes Approach

A method for the calculation of compressible turbulent flows using general unstructured grids is described. The Reynolds averaged Navier-Stokes equations are solved by means of a cell-vertex finite volume spatial discretisation and an explicit time stepping scheme. Artificial dissipation is introduced to damp oscillations and to ensure convergence to steady state solution. An empirical remedy for the reduction of the numerical dissipation influence in the viscous wall region is employed. The Baldwin-Lomax turbulence model is used for the calculation of turbulence quantities. The method is applied to predict the turbulent flow around the RAE-2822 airfoil.

4.4.2.1 Governing Equations

Viscous compressible fluid flows are governed by the Navier-Stokes equations, which express the conservation principle for mass, momentum and energy. The time dependent Reynolds averaged equations in mass averaged form is used to enable the computation of turbulent flows. The additional Reynolds stress terms are modelled by introducing a turbulent eddy viscosity μ_t. For a two-dimensional flow domain with volume Ω and surface boundary $\delta\Omega$, the mean flow equations can be given in the following integral form :

$$\frac{\partial}{\partial t}\int_\Omega W \, d\Omega + \int_{\delta\Omega}(F \, dx - G \, dy) = 0 \qquad (1)$$

where $W = (\rho, \rho u, \rho v, \rho e)$ is the time averaged conserved variables vector. The flux vectors F and G are split into the inviscid and viscous contributions, denoted by the superscripts I and V, respectively:

$$F = F^I + F^V$$
$$G = G^I + F^V$$

where

$$F^I = (\rho u, \rho u^2 + p, \rho uv, \rho uh)$$
$$F^V = (0, \sigma_{xx}, \sigma_{xy}, (u\sigma_{xx} + v\sigma_{xy} + q_x))$$
$$G^I = (\rho v, \rho uv, \rho v^2 + p, \rho vh)$$
$$G^V = (0, \sigma_{xy}, \sigma_{yy}, (u\sigma_{xy} + v\sigma_{yy} + q_y)).$$

The stress and heat flux elements are denoted by σ and q respectively. h is the total enthalpy per unit mass. The effective viscosity is equal to the laminar viscosity μ, given by Sutherland's law, plus the turbulent eddy viscosity μ_t. The standard Baldwin-Lomax turbulence model (see section (3.1)) is employed to calculate the viscosity μ_t. The boundary conditions applied at the solid surface are zero normal flow, no slip and adiabatic wall. Non-reflecting boundary conditions are applied at the far field, based on the introduction of Riemann invariants for a one-dimensional flow normal to the outer boundary of the flow domain.

4.4.2.2 Numerical Formulation

The computational domain is subdivided into a set of polygonal cells. The numerical formulation of equation (1) involves performing flux balances for each polygonal domain which encloses each node within the triangulation (Jameson, Baker and Weatherill (1986)). Applying the finite volume spatial discretisation, the inviscid and viscous contributions to the contour integral become:

$$\frac{dW}{dt} + \frac{1}{\Omega}(Q_k^I + Q_k^V) = 0 \qquad (2)$$

where W is the conserved variables for the K-th node which defines the polygonal domain. The fluxes Q_k^I and Q_k^V are given by:

$$Q_k^I = \sum_{i=1}^{n} (F_i^I \Delta y_i - G_i^I \Delta x_i) \qquad (3)$$

$$Q_k^V = \sum_{i=1}^{n} [F_i^V \Delta y_i - G_i^V \Delta x_i]. \qquad (4)$$

The summation process, in equation (3) and equation (4) over all polygonal domains is implemented by computing fluxes across every edge in the triangulation and sending the contributions to the nodes whose polygonal boundary contains the edge. The convective fluxes at the i-th edge are calculated, see Fig. 1, as follows:

$$F_i^I = \frac{1}{2}(F_l^I + F_r^I)$$
$$G_i^I = \frac{1}{2}(G_l^I + G_r^I).$$

The viscous components of the summation in equation (4) involve the evaluation of derivatives of the conserved variables on each polygonal contour. These derivatives are computed using the auxiliary control volume of Fig. 1. The derivative at i is the sum of four terms as:

$$\frac{\partial u}{\partial x}\bigg|_i = \frac{1}{2A}((u_2 + u_l)(y_l - y_2) + (u_1 + u_l)(y_1 - y_l)$$
$$+ (u_r + u_1)(y_r - y_1) + (u_2 + u_r)(y_2 - y_r))$$

where A is the sum of areas of the two adjacent triangles which define the auxiliary control volume.

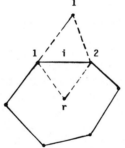

Fig. 1 Polygonal Domain and Auxiliary Control Volume (---)

4.4.2.3 Artificial Dissipation

Although equation (1) contains dissipative terms, it is found that for stability of the numerical scheme this dissipation must be augmented by artificial dissipation terms. Here we employ an artificial dissipation in the form used by Mavriplis (1988), consisting of a blending of second and fourth difference operators. The basic idea of accumulated edge differences is used. This involves a term generated from a weighted difference between variables at a given node and its nearest neighbours. The weighting is adapted to gradients in the flow variables. The artificial dissipation term, D_K, is added explicitly to equation (2) which becomes:

$$\frac{dW_k}{dt} + \frac{1}{\Omega_k}(Q_k^I + Q_k^V - \phi D_k) = 0. \tag{5}$$

In the viscous regions adjacent to the solid surface it is important to ensure that the artificial dissipation does not dominate the real viscous effects. Therefore, it has been found necessary to introduce modifications to the dissipation terms in these regions.

The explicitly added artificial dissipation model used in the present method has the possibility of monitoring its magnitude for each individual cell by introducing an adequate scaling of the dissipation term. The present method employs a scaling parameter ϕ, as a function of the ratio of the laminar viscosity to the eddy viscosity to reduce the level of artificial dissipation in the real viscous regions. This scaling parameter is given by :

$$\phi = 1 - e^{-(\beta\mu/\mu_t)}. \tag{6}$$

In the present calculations β is a constant which takes the value of 25.

The discretisation procedure outlined leads to a set of coupled differential equation, equation (5). The steady state solution of this equation is obtained by marching in time usig an explicit 3-stage scheme. Local time stepping and residual smoothing are utilised to accelerate convergence to a steady state.

4.4.2.4 Turbulence Model Application

The efficient calculations of turbulent flows on unstructured grids using algebraic models is not straight-forward because the direction normal to the aerofoil surface is no longer a coordinate axis, while the value of μ_t at a point in a streamwise station is a function of the flow variables at all points in the same station.

Therefore, to calculate μ_t at a grid node one needs to draw the normal from this node to the surface and interpolate all flow variables along this normal. This may result very expensive, both in storage and CPU time, because it is only valid for the considered node and has to be repeated for each grid node.

In the present application we consider a discrete set of normals, to the aerofoil surface and the vortex sheet, to compute and store all the elements they get through. These geometrical information are calculated and stored only once and then used at each time step to calculate the eddy viscosity on each normal. The eddy viscosity at each node is then obtained interpolating the values for the nearest two points on the two neighbouring normals.

4.4.2.5 Numerical Details

For the present method a general unstructured triangular mesh of 16923 nodal points with 337 points on the airfoil surface and 50383 elements is used. For the accurate simulation of the viscous flow in the wall region, and to be able to compare with other partner calculations, a regular structured grid is built around the airfoil in near wall region. This regular sub-grid has 25 layers with the same distribution as in the mandatory grid generated by BAe for the airfoil group.

For the present calculations, the wake is initially positioned on a straight line in the direction of the unperturbed flow. This approach results in a less accurate turbulent viscosity profiles between the upper and lower parts of the wake. To correct this effect, two approaches have been used. In the first approach the wake is located on the stream line coming out of the trailing edge of the airfoil. This stream line is calculated using Euler solution and does not changed for the rest of the time steps. In the second approach, the wake is first located on a straight line coming out of the trailing edge. This line is updated every few iterations to be located on the points of maximum viscosity on the profiles across the wake.

The first approach is proved to be effective and cheap.

4.4.3 Euler Approach

A method for the solution of the three-dimensional steady compressible Euler equations on structured grids is described. The solution is achieved by advancing the unsteady form of the equations to steady state by means of an explicit time stepping scheme (Jameson (1985)). A cell-centred finite volume spatial discretisation technique is utilised. The convergence is accelerated by the use of local time stepping and the incorporation of implicit residual averaging. The resulting scheme is stabilised by the application of an artificial dissipation operator which is formed as a blended combination of second and fourth differences. Modifications of the basic dissipation model is also discussed.

4.4.3.1 Governing Equations

The three-dimensional compressible Euler equations for a flow domain of volume Ω and associated boundary $\delta\Omega$ can be expressed in the form :

$$\frac{\partial}{\partial t} \int_\Omega W \, d\Omega + \int_{\delta\Omega} F \, ds = 0 \qquad (7)$$

where $W = (\rho, \rho u, \rho v, \rho w, \rho e)$ is the density, rectangular components of momentum referred to a Cartesian System x, y, z fixed in space and total energy. The quantity F represents the net flux of W transported across, plus the pressure acting on it, the closed surface $\delta\Omega$. For a perfect gas, the total energy can be expressed as:

$$e = \frac{P}{(\gamma-1)\rho} + \frac{1}{2}(u^2 + v^2 + w^2).$$

In order to apply the finite volume technique a structured grid of cells packed so that they discretise the flow field is constructed. Since equation (7) is valid for any arbitrary volume it also holds locally for each individual cell (i,j,k) in the grid. For one such cell a semi-discretisation of equation (7) leads to :

$$\frac{\partial}{\partial t}(\Omega_{i,j,k} \, W_{i,j,k}) + Q(W)_{i,j,k} = 0. \qquad (8)$$

$Q_{i,j,k}$ represents the net flux out the cell which is balanced by the rate of change of W in the cell whose volume is Ω. This flux is given by :

$$Q_{i,j,k} = \sum_{\text{all faces}} F \cdot S$$

where F is the flux at the center of a cell face and S denotes the cell face area. The value of F at the cell face is taken as the average of F at the cell centers on either sides of the cell face.

4.4.3.2 Dissipation Model

In order to suppress the tendency for odd and even point decoupling and to capture shock waves without any overshoots, it is necessary to add a dissipative term to equation (8). The basic model used was first introduced by Jameson, Schmidt and Turkel (1981) in conjunction with Runge-Kutta explicit schemes. Adding this dissipative term equation (8) becomes :

$$\frac{\partial}{\partial t} (\Omega_{i,j,k} \ W_{i,j,k}) + Q(W)_{i,j,k} - D(W)_{i,j,k} = 0 \qquad (9)$$

where D(w) is the artificial dissipation term which is constructed so that it is third order in smooth regions of flow. That is :

$$D(W) = (D_x^2 + D_y^2 + D_z^2 - D_x^4 - D_y^4 - D_z^4) (W) \qquad (10)$$

where:

$$D_x^2 (W) = \nabla_x (\lambda_{i+½,j,k} \cdot \epsilon^{(2)}_{i+½,j,k}) \Delta_x W_{i,j,k} \qquad (11)$$

$$D_x^4 (W) = \nabla_x (\lambda_{i+½,j,k} \cdot \epsilon^{(4)}_{i+½,j,k}) \Delta_x \nabla_x \Delta_x W_{i,j,k} \ . \qquad (12)$$

Δ_x and ∇_x are forward and backward difference operators associated with the x direction. λ is the cell variable scaling factor. The coefficients $\epsilon^{(2)}$ and $\epsilon^{(4)}$ are made proportional to the normalized second difference of the pressure as follows:

$$\epsilon^{(2)}_{i+½,j \ k} = K^{(2)} \ max(\vartheta_{i,j,k}, \vartheta_{i+1,j,k})$$

$$\vartheta_{i,j,k} = | \frac{P_{i+1,j,k} - 2P_{i,j,k} + P_{i-1,j,k}}{P_{i+1,j,k} + 2P_{i,j,k} + P_{i-1,j,k}} |$$

$$\epsilon^{(4)}_{i+½,j \ k} = max(0, (K^{(4)} - \epsilon^{(2)}_{i+½,j \ k})) \ .$$

$K^{(2)}$ and $K^{(4)}$ are constants with typical values of 1/4 ad 1/256 respectively. The operators in equation (10) for y and z directions are constructed in a similar manner.

The present method employs the modification proposed by Swanson and Turkel (1987) to the fourth difference term, equation (12). This term is replaced by the following:

$$D_x^4(W) = (\nabla_x \Delta_x)(\lambda_{i,j,k} \cdot \epsilon^{(4)}_{i,j,k} \nabla_x \Delta_x) W_{i,j,k} . \qquad (13)$$

This modified fourth difference term is only dissipative and does not produce any dispersive term. For this modification λ and $\epsilon^{(4)}$ are calculated at nodes rather than at cell faces.

The other modification considered in the present method is for the cell scaling factor λ. Large cell distortions, cells with high aspect ratio, can produce slow convergence and low accuracy of the numerical scheme. For such case, an anisotropic model for the evaluation of the cell scaling factor together with the introduction of functions of the cells aspect ratio have been suggested. That is:

$$\lambda_{i+\frac{1}{2},j,k} = \frac{1}{2}((\bar{\lambda}_x)_{i,j,k} + (\bar{\lambda}_x)_{i+1,j,k})$$

where

$$(\bar{\lambda}_x)_{i,j,k} = \beta_{i,j,k}(r) \cdot (\lambda_x)_{i,j,k}$$

and

$$\beta_{i,j,k}(r) = 1 + r^\alpha_{i,j,k}$$

$$r = \left(\frac{\lambda_x^2}{\lambda_x^2 + \lambda_y^2 + \lambda_z^2}\right)^{\frac{1}{2}}.$$

In the present calculation a value of 0.8 is used for the constant α.

The discretised governing equations, equation (9) are integrated in time to steady state using 4-stage explicit time stepping scheme. Contribution from dissipation terms is evaluated at the first stage and then held constant for the remaining stages of the time step. Convergence to steady state is accelerated by using local time step and by the application of a residual averaging procedure.

4.4.3.3 Numerical Details

For the present calculations a mandatory grid generated by DLR for the F5 wing is used. This grid was produced for the solution of Navier-Stokes equations. For the Euler solution this mandatory grid is modified eliminating the first seven planes in the viscous region around the wing surface and wake. In the wake region, from the wing trailing edge to the tunnel exit section, every second plane normal to the wake is eliminated. The surface mesh is maintained to 161x41. The total number of grid nodes became 185x26x41.

4.5 Description of Method Used by CERFACS
(L. Davidson, A. Kourta, L. Tourrette)

4.5.1 Introduction

In the Euroval project CERFACS used three different numerical methods. Two methods are based on the Runge-Kutta scheme and the last one on the MacCormack scheme. For the single element Airfoil flow and The ONERA channel flow the two dimensional Runge-Kutta method is used. It will be called method 1. For the two element Airfoil flow, the Mac-Cormack multi-block method is used and specified here as method 2. For the 3D flow over DLR-F5 wing the three dimensional Runge-Kutta scheme is used and will be described like method 3.

4.5.2 Method 1 (Single Element and Bumps)

Numerical Method

The code solves the continuity equation ρ, the momentum equations $\rho u, \rho v$, and the equation for total energy ρe_o; the pressure is calculated using the gas law. The main features of the code are:
- explicit
- compressible
- time-marching
- cell-centered
- finite volume
- central differencing
- local time stepping
- four stage Runge-Kutta scheme for the mean flow equations
- the k and ε equations are solved using a semi-implicit solver (hybrid central/upwind scheme, ADI)
- fourth-order numerical non-homogenous dissipation term in all mean flow equations proposed by Swanson and Turkel (1987); the cell-aspect ratio is taken into account through the directional convective local time step t^d, and the numerical dissipation $\mathbf{F}_N$ is decreased in the boundary layers by scaling $\mathbf{F}_N$ with the local Mach-number.

Modifications of the 'Standard' Turbulence Models

Wall-correction in RS used in the single element airfoil flow computation:

The simplified form of the near-wall correction term

$$\Phi'_{ij,1} = c'_1 \frac{\varepsilon}{k}(\overline{u_n^2}\delta_{ij} - \frac{3}{2}\overline{u_n u_i}\delta_{nj} - \frac{3}{2}\overline{u_n u_j}\delta_{ni})f(\frac{\ell_t}{x_n})$$

$$\Phi'_{ij,2} = c'_2(\Phi_{nn,2}\delta_{ij} - \frac{3}{2}\Phi_{ni,2}\delta_{nj} - \frac{3}{2}\Phi_{nj,2}\delta_{ni})f(\frac{\ell_t}{x_n})$$

$$P_{ij} = -\overline{u_i u_k}\frac{\partial U_j}{\partial x_k} - \overline{u_j u_k}\frac{\partial U_i}{\partial x_k}$$

was used in the single element airfoil flow computation, which is based on the assumption that the walls are parallel to the Cartesian velocity components. This assumption seems to be reasonable. However, tests were also carried out using the general formulation, and it was found that the contribution from $\overline{u^2}$ in the expression for $\Phi'_{12,1}$ gave a large amplifying effect, which almost cancelled the damping effect due to the $\overline{uv}$-term. The wall correction term $\Phi'_{12,1}$ has the form

$$\Phi'_{12,1} = -c'_1 \frac{\varepsilon}{k}\frac{3}{2}[\overline{uv} + (\overline{u^2} + \overline{v^2})n_x n_y]f(\frac{l_t}{x_n})$$

and it becomes almost zero on the suction side near the trailing edge (in the separation region). This is due to that the product $n_x n_y$ is not negligible (up to 0.2), and that the normal stress $\overline{u^2}$ is much larger than the shear stress. For these reasons the simplified formulation, which does damp the shear stress in the separation region, was used.

Baldwin-Lomax model used in the ONERA channel flow calculation:

The constant C_{WK} in the wake function in Eq. (BL7) was set to zero, i.e. $C_{WK} = 0$.

4.5.3 Method 2 (Multi-Element)

The numerical method is a version of the explicit MacCormack scheme. This method has a predictor-corrector structure. It is a second order accurate method. The scheme is a finite volume, cell centre method. The multiblock technique is used. Briefly, the algorithm is as follows for advancing the numerical solution in time.

For the predictor step:

$$\Delta W^{(n)} = -\Delta t \left(\frac{\partial F^{(n)}}{\partial x} + \frac{\partial G^{(n)}}{\partial y}\right)$$

$$W^{\overline{(n+1)}} = W^{(n)} + \Delta W^{(n)}$$

and for the corrector step:

$$\Delta W^{(n+1)} = -\Delta t \left(\frac{\partial F^{\overline{(n+1)}}}{\partial x} + \frac{\partial G^{\overline{(n+1)}}}{\partial y}\right)$$

$$W^{(n+1)} = \frac{1}{2}(W^{(n)} + W^{\overline{(n+1)}} + \Delta W^{(n+1)}).$$

In the above, for each time step, forward approximations are used in the predictor step and backward approximations are used in the corrector step. This choice could have been reversed or one forward or one backward approximation could have been used in each. There are four different choices that should be cycled through during the course of calculation. The forward or backward approximations are used for the inviscid part. For the viscous terms the central difference is used.

The method is explicit, hence the CFL number must be less than one to satisfy the stability criteria. For convergence acceleration, the local time step is used.

In order to improve numerical efficiency, the second order accurate flux splitting has been included in the method. The flux splitting was motivated by a need for a better description of discontinuities and a more rigorous treatment of the boundary conditions. The flux splitting used here is close to the one developed by Steger and Warming. These procedures take into consideration the direction of information travel to compute fluxes and therefore can approximate the physics of the governing equations more realistically.

Let A' the Jacobian of the inviscid part of the flux vector with respect to W.

$$A' = \frac{\partial F'}{\partial W}$$

Because F' is homogeneous of degree one in the element of W: F'=A'W.
The Jacobian matrix A' can be diagonalized by similarity transformation S as follows

$$A' = S^{-1}\Delta S.$$

In general some of the elements of diagonalized matrix above are positive and others are negative. Their signs detremine the direction of information travel. If we let Δ_+ and Δ_- be two diagonal matrices containing respectively the positive and negative elements of the diagonalized matrix above, we can define the following matrices.

$$A'_+ = S^{-1}\Delta_+ S$$
$$A'_- = S^{-1}\Delta_- S$$
$$A' = A'_+ + A'_-$$
$$F' = A'W = A'_+ W + A'_- W.$$

The first or second order flux splitting approximation can be used.

4.5.4 Method 3 (DLR-F5 Wing)

4.5.4.1 Mesh

An O-O type mapping relates a single block in the computational domain (ξ,η,ζ) to the region of the physical space (x,y,z) bounded by the wing surface, the wind tunnel walls and the inflow and outflow boundaries. As pointed out in Eriksson (1982), such a topology enables a natural grid point concentration in the vicinity of the wing but a major drawback is the poor quality of the mesh near the leading edge, the trailing edge and the tip where highly distored cells are encountered.

The mesh contains 193, 49 and 41 nodes in the chordwise I, near normal J and spanwise K directions respectively, i.e. a total of 387737 points.

Wing

The wing surface mesh is generated using a geometry generator supplied by H. Sobieczky (1985), slightly modified in order to enable control over the spanwise grid point distribution. Grid points are clustered at the leading edge, the trailing edge and the wing tip. As the wing tip and trailing edge sections were blunt and open, the surface had to be closed and the resulting mesh had to be smoothed in order to avoid undesired vortices due to unnatural sharp edges.

Outer Boundary

The outer boundary is defined by the five surfaces of the wind tunnel box which correspond to $\eta = 1$ in the computational space (the splitter-plate corresponds to $\zeta = 0$). Two parabolic singularities are located at the wind tunnel wall opposite to the splitter plate, the so-called "singular surface". The mesh is generated by two-boundary transfinite interpolation between the wing and outer boundary meshes.

Grid Smoother

One of the drawbacks of transfinite interpolation is that irregularities at the boundaries propagate into the interior region of the mesh. As the finite volume method used here is sensitive to such defects, the trailing edge, the wing tip and the singular surface have been smoothed using a neighbourhood averaging method based on a Laplace-like operator.

The mesh has been designed by Tony Lindeberg (1987) for the International Workshop on the F5 wing that was held on September 30 to October 2, 1987 in Göttingen (Kordulla, 1987).

4.5.4.2 The Mean-Flow Equations and their Approximation

Governing Equations

The fluid flow is governed by the conservation laws for mass, momentum and total energy. We assume that the fluid is an ideal gas obeying to Newton's and Fourier's laws. External forces and heat sources are neglected.

Initial Value and Boundary Conditions

At $t = 0$, the conservative variables are set equal to their freestream values.

On the wing surface, which is assumed to be adiabatic, the no-slip condition holds. The pressure is obtained by neglecting the viscous terms in the wall normal momentum equation:

$$\mathbf{V} = \mathbf{0}, \quad \frac{\partial T}{\partial n} = 0, \quad \frac{\partial p}{\partial n} = 0. \tag{1}$$

At the inflow surface, we apply absorbing boundary conditions based on the theory of Riemann invariants for locally one-dimensional inviscid flow. At the outflow surface, we have used two types of boundary conditions:
- absorbing boundary conditions based on the Riemann invariants as for the inflow plane and
- static pressure prescribed the value 0.99 p_{inflow} (mandatory outflow boundary condition) together with zeroth-order extrapolation of the outgoing Riemann invariants.

At the wind tunnel walls and at the splitter plate, there is no boundary layer resolution in the mesh and the no-slip boundary condition is replaced by

$$V.n = 0. \tag{2}$$

The O-O mesh topology introduces periodic boundaries. They extend from the trailing edge and tip of the wing in the positive x and y directions, respectively. At periodic boundaries, grid points on upper and lower surfaces are mapped onto each other. The conditions on the conservative variables are:

$$q_{IMAX+1/2,J+1/2,K+1/2} = q_{1+1/2,J+1/2,K+1/2}, \tag{3}$$

$$q_{I+1/2,J+1/2,KMAX+1/2} = q_{IMAX-I+1/2,J+1/2,KMAX-1/2}. \tag{4}$$

Space Discretization

The compressible Navier-Stokes equations are discretized in hexahedrons using the finite volume technique. Integrated over an arbitrary stationary cell V_P with boundary ∂V_P and outer normal unit vector **n**, they read:

$$\frac{d}{dt}\int_{V_P} q \, dV + \int_{\partial V_P} H(q).n \, dA = 0. \tag{5}$$

The surface integral in (5) over the boundary of cell V_P is approximated by assuming the mean-value of the flux tensor on each side to be equal to the arithmetic average of the flux tensor in the adjacent cells:

$$\int_{\partial V_P} H.n \, dA \approx \sum_{k=1}^{6} H_{Pk} . \int_{\partial V_{Pk}} n \, dA, \tag{6}$$

where

$$H_{Pk} = \frac{1}{2}(H_P + H_k). \tag{7}$$

∂V_{Pk} denotes the common part of the boundaries of P and its neighbouring cell k.

With the conservative variables given, all terms of the flux tensor are readily available in cell P, except for the gradients of the velocity components and temperature. Following the definition of the conservative variables as cell averages, the gradients in cell P are defined by:

$$(\nabla \phi)_P = \frac{\int_{V_P} \nabla \phi \, dV}{\int_{V_P} dV}, \tag{8}$$

where ϕ = u, v, w or T. Using the gradient theorem, the volume integral in (8) can be expressed by a surface integral, which is approximated similarly to (6):

$$(\nabla\phi)_P = \frac{\int_{\partial V_P} \phi n dA}{\int_{V_P} dV} \approx \frac{\sum_{k=1}^{6} \phi_{Pk} \int_{\partial V_{Pk}} n dA}{\int_{V_P} dV}, \quad (9)$$

where

$$\phi_{Pk} = \frac{1}{2}(\phi_P + \phi_k). \quad (10)$$

$(\nabla . \mathbf{V})_P$ is evaluated similarly to $(\nabla\phi)_P$.

Numerical Damping

The spatial discretization results in the physical difference operator F_{PH} defined by the negative right hand side of (6) divided by the cell volume. In order to damp unphysical oscillations due to flow discontinuities and high frequency waves, numerical damping terms $F_N(q)$ are added to F_{PH}. They contain nonlinear second-order differences sensed by the discretized second derivative of the pressure and linear fourth-order differences of the conservative variables:

$$F_N(q) = \frac{CFL}{\Delta t}\{\chi(\delta_I[s_I(p)\delta_I] + \delta_J[s_J(p)\delta_J] + \delta_K[s_K(p)\delta_K])$$

$$- \Lambda(\delta_I^4 + \delta_J^4 + \delta_K^4)\}q, \quad (11)$$

with CFL the maximum CFL number used (see below) and Δt the time step. The constants χ and Λ have respective values 0.05 and 0.01. The sensors s_I, s_J and s_K are of similar form, e. g.

$$[s_I(p)]_{I+1,J+1/2,K+1/2} = \frac{(\mu_I|\delta_I^2 p|)_{I+1,J+1/2,K+1/2}}{\max_{I',J',K'} |\delta_I^2 p|}. \quad (12)$$

Operators δ_I and μ_I are defined by:

$$(\delta_I\phi)_{I+1,J+1/2,K+1/2} = \phi_{I+3/2,J+1/2,K+1/2} - \phi_{I+1/2,J+1/2,K+1/2}, \quad (13)$$

$$(\mu_I\phi)_{I+1,J+1/2,K+1/2} = \frac{\phi_{I+1/2,J+1/2,K+1/2} + \phi_{I+3/2,J+1/2,K+1/2}}{2}, \quad (14)$$

and similarly for J and K.

Boundary conditions at the wing surface and farfield are applied to the numerical damping operator $F_N(q)$ in order to ensure its dissipative property also there.

Time Integration

The problem to solve reads:

$$\frac{d}{dt}q = F(q), \tag{15}$$

where

$$F = F_{PH} + F_N. \tag{16}$$

This large system of first order differential equations is solved for steady state by the following three stage Runge-Kutta type scheme:

$$\begin{cases} q^{(0)} = q^n, \\ q^{(1)} = q^{(0)} + \Delta t\, F(q^{(0)}), \\ q^{(2)} = q^{(0)} + \frac{\Delta t}{2}[F(q^{(0)}) + F(q^{(1)})], \\ q^{n+1} = q^{(0)} + \frac{\Delta t}{2}[F(q^{(0)}) + F(q^{(2)})]. \end{cases} \tag{17}$$

Stability

The linear L^2-stability of Runge-Kutta schemes applied to the spatial semi-discrete approximation (15) of the compressible Navier-Stokes equations has been studied for a scalar linear model equation (Müller & Rizzi., 1987). The stability condition reads:

$$\Delta t \leq \min(\Delta t_I, \Delta t_V), \tag{18}$$

with

$$\Delta t_I = CFL\, vol\, [|\mathbf{V}.\mathbf{S}_I| + |\mathbf{V}.\mathbf{S}_J| + |\mathbf{V}.\mathbf{S}_K| + c(|\mathbf{S}_I| + |\mathbf{S}_J| + |\mathbf{S}_K|)]^{-1} \tag{19}$$

and

$$\Delta t_V = \frac{1}{2}|RK|vol^2 [\nu(|\mathbf{S}_I|^2 + |\mathbf{S}_J|^2 + |\mathbf{S}_K|^2) + 2\nu(|\mathbf{S}_I.\mathbf{S}_J| + |\mathbf{S}_I.\mathbf{S}_K| + |\mathbf{S}_J.\mathbf{S}_K|)$$

$$+ ((\lambda+\mu)/\rho)(|\mathbf{S}_I||\mathbf{S}_J| + |\mathbf{S}_I||\mathbf{S}_K| + |\mathbf{S}_J||\mathbf{S}_K|)]^{-1} \tag{20}$$

where c is the speed of sound, $v = \max(\mu, \lambda+2\mu, \gamma\mu/Pr)/\rho$, vol the cell volume, S_I the area vector in I-direction, etc. The stability bounds RK and CFL are chosen so that all complex number z with $RK \leq \text{Re}(z) \leq 0$ and $|\text{Im}(z)| \leq \text{CFL}$ lie inside the stability region of time integration scheme (17). The factor 1/2 in (20) leaves space on the negative real axis of the stability region to accomodate the numerical damping contribution (Müller & Rizzi, 1987).

Turbulence is accounted for through the concept of turbulent viscosity. In the Navier-Stokes equations, the molecular viscosity coefficient μ_l is replaced by $\mu_l + \mu_t$ and in the heat flux terms, the ratio μ_l/Pr is replaced by $\mu_l/Pr + \mu_t/Pr_t$ with the turbulent Prandtl number $Pr_t = 0.9$. We have used two turbulence models, namely the Baldwin-Lomax model (3.1.2) and a two-layer k-ε model described below.

4.5.4.3 The k-ε Turbulence Model and its Approximation

Governing Equations

We use a two-layer model which combines the standard k-ε model (3.4.4) with a simpler but more reliable one-equation model (Chen & Patel, 1987). In this approach, the flow domain is divided into two regions:
• the near-wall region (region 1) where Wolfshtein's one-equation model (3.3.2) is employed to account for the wall proximity effects,
• the off-wall region (region 2) where the standard k-ε model is used.
This two-layer model has been introduced by Chen and Patel in order to accurately resolve the near-wall region in flows involving three-dimensionality, unsteadiness and separation and is intermediate in complexity to the wall-function method, on the one hand, and the low-Reynolds-number models, on the other.

In two-layer modeling, the two models have to be matched at some location and this should be placed near the edge of the viscous sublayer, i.e. in a region where viscous effects have become negligible. Iacovides and Launder (1990) use ten nodes in region 1 and report that the matching takes place in a y^+ region of 80 to 120. Chen and Patel (1987) match along a grid line where the minimum value of $R_t = \rho k^{1/2} y/\mu_l$ (turbulent Reynolds-number based on the distance to the wall) is of the order 250 but they found that their results were not sensitive to the matching criterion as long as R_t remains greater than 200. According to Rodi (1991), $R_t = 250$ corresponds roughly to $y^+ = 135$ in normal boundary-layer flow. Moreover, Rodi does not match the two models at a preselected grid line but at a location where a certain criterion is satisfied. He has tested the two criteria $v_t/v \geq 30$ and $l_\mu/(C_l y) \geq 0.95$ and reports that they lead to a switching between the models at $y^+ = 80$-90.

Initial Value

At the beginning of a k-ε computation, k and ε are prescribed their freestream values k_∞ and ε_∞. We have chosen $k_\infty = 10^{-8}$ and $\varepsilon_\infty = 10^{-10}$.

Boundary Conditions
- wing surface: k = 0, ε = 0,
- inflow: k = k$_\infty$, ε = ε$_\infty$,
- outflow, splitter-plate and wind-tunnel walls: $\frac{\partial k}{\partial n} = 0, \frac{\partial \varepsilon}{\partial n} = 0$.

Methodology
A k-ε solution is computed in the following way:
1) obtain a converged Baldwin-Lomax solution,
2) freeze ρ, $\rho\vec{V}$ and ρE and solve the k-ε model starting from (k,ε) = (k$_\infty$,ε$_\infty$) until convergence is reached,
3) start the coupling process, solving the mean flow equations and the turbulence model one after the other (uncoupled approach).

Space Discretization
Following the lines drawn by Davidson (1990), we use the implicit hybrid scheme described by Patankar (1980) for solving a three-dimensional advection-diffusion equation with forcing terms. In this scheme, the convective part is approximated by centered finite differences when diffusion is large enough (i.e. the Peclet number Pe is smaller than two) and by upwind biased finite differences when convection dominates.

The general form of the k and ε equations reads:

$$\frac{\partial}{\partial t}(\rho\phi) + \nabla.(\rho\phi V) - \nabla.(\Gamma\nabla\phi) = S, \tag{21}$$

where ϕ stands for k or ε and Γ for $\mu_l + \frac{\mu_t}{\sigma_\phi}$. S denotes the source term.

The source term S is linearized with respect to ϕ as follows:

$$S(\phi) = S_1\phi + S_0, \tag{22}$$

with $S_1 \leq 0$ and $S_0 \geq 0$. More precisely, we have:

$$S_0 = \mu_t G - \frac{2}{3}\rho k(\nabla.V)^-, \tag{23}$$

$$S_1 = \begin{cases} \frac{-C_\mu\rho^2 k}{\mu_t} - C_\mu\frac{\rho k}{\varepsilon}\frac{2}{3}(\nabla.V)^2 - \frac{2}{3}\rho(\nabla.V)^+ & \text{(off-wall region)} \\ \frac{-\rho\sqrt{k}}{l_D} - C_\mu l_\mu \frac{\rho}{\sqrt{k}}\frac{2}{3}(\nabla.V)^2 - \frac{2}{3}\rho(\nabla.V)^+ & \text{(near-wall region)} \end{cases} \tag{24}$$

for the k-equation and

$$S_0 = C_1 C_\mu \rho k G - C_1 \frac{2}{3} \rho \varepsilon (\nabla . \mathbf{V})^-, \tag{25}$$

$$S_1 = -C_2 C_\mu \rho^2 \frac{k}{\mu_t} - C_1 \frac{\mu_t 2}{k\,3}(\nabla . \mathbf{V})^2 - C_1 \frac{2}{3}\rho (\nabla . \mathbf{V})^+ \tag{26}$$

for the ε-equation (off-wall region only) with

$$(\nabla . \mathbf{V})^- = \min(0, \nabla . \mathbf{V}), \tag{27}$$

$$(\nabla . \mathbf{V})^+ = \max(0, \nabla . \mathbf{V}), \tag{28}$$

$$G = \frac{1}{2}\left(\nabla \mathbf{V} + {}^t\nabla \mathbf{V}\right)^2. \tag{29}$$

The time derivative $\partial(\rho\phi)/\partial t$ is approximated by a first order backward finite difference.

Integrating equation (21) over the hexahedron centered around (I+1/2, J+1/2, K+1/2) point, we obtain:

$$\left(\frac{a_P}{vol_P} - S_1 + \frac{\rho}{\Delta t_P}\right)\phi_P^{n+1} = \frac{1}{vol_P}(a_E \phi_E + a_W \phi_W + a_N \phi_N + a_S \phi_S$$

$$+ a_H \phi_H + a_L \phi_L) + S_0 + \frac{\rho}{\Delta t_P}\phi_P^n. \tag{30}$$

P, E, W, N, S, H and L denote cells centers (I+1/2, J+1/2, K+1/2), (I+3/2, J+1/2, K+1/2), (I-1/2, J+1/2, K+1/2), (I+1/2, J+3/2, K+1/2), (I+1/2, J-1/2, K+1/2), (I+1/2, J+1/2, K+3/2) and (I+1/2, J+1/2, K-1/2) respectively.

Coefficients a_E and a_W are given by

$$a_E = \max\left(-C_e, D_e - \frac{C_e}{2}, 0\right), \tag{31}$$

$$a_W = \max\left(C_w, D_w + \frac{C_w}{2}, 0\right). \tag{32}$$

In the above expressions, the lower-case letters e and w refer to faces centered around points (I+1, J+1/2, K+1/2) and (I, J+1/2, K+1/2) respectively. The convective and diffusive terms C and D are defined by

$$C = \rho \mathbf{V} . \mathbf{S}_\xi \tag{33}$$

and

$$D = \Gamma \nabla \xi \cdot \mathbf{S}_\xi, \tag{34}$$

with ξ the curvilinear coordinate in the I-direction and $\mathbf{S}_\xi$ the area vector directed towards increasing values of I. In the expression for D, the non-orthogonal diffusion terms have been neglected.

On a given cell face, $\rho \mathbf{V}$ and Γ are computed by evaluating the arithmetic average of their adjacent values.

Similar expressions hold for a_N, a_S, a_H and a_L.

Finaly, coefficient a_P is given by:

$$a_P = a_E + a_W + a_N + a_S + a_H + a_L + (C_e - C_w) + (C_n - C_s) + (C_h - C_l). \tag{35}$$

vol_P denotes the volume of the hexahedron centered around point P and Δt_P the local time step for the mean flow at P. For the initialization of k and ε, Δt_P is multiplied by a constant greater than one in order to enhance convergence.

At each time step, the k-equation is solved first and then the ε-equation.

The linear system is solved by Successive-Line-Over-Relaxation (SLOR) with use of a Thomas algorithm for each scalar tridiagonal system.

The code has been run on a CONVEX C220 where it operates at 240 µs per time step and per grid point for the Baldwin-Lomax model and 330 µs per time step and per grid point for the k-ε model. The Baldwin-Lomax model has also been run on a CRAY2 supercomputer with a speed of 60 µs per time step and per grid point.

4.6 Description of Method Used by CFD norway
(I. Øye, E. Ørbekk, H. Nørstrud)

4.6.1 Introduction

The numerical method for solving the Reynolds-averaged Navier-Stokes equations described in this section is based on the time stepping finite volume method which was introduced by Jameson et al. (1981) and later modified by Eriksson (1985, 1987). The method applies cell centered fluxes for spatial discretization, stabilised by a combination of second and fourth order damping terms. Time integration is performed by a three-stage Runge-Kutta method. This explicit approach yields a high performance on vector or parallel computers, is simple to program and can easily be extended to multi-block grids. This algorithm is currently implemented into the codes TwoFlow and ThreeFlow, multi-block flow solvers for two- and three-dimensional flows, respectively.

4.6.2 Governing Equations

The Navier-Stokes equations for general three-dimensional flows are given in cartesian coordinates by

$$\frac{\partial \mathbf{U}}{\partial t} + \frac{\partial \mathbf{F}}{\partial x} + \frac{\partial \mathbf{G}}{\partial y} + \frac{\partial \mathbf{H}}{\partial z} = 0, \tag{1}$$

where **U** is the vector containing the conservative variables

$$\mathbf{U} = \begin{bmatrix} \rho & \rho u & \rho v & \rho w & E_t \end{bmatrix}^T,$$

and **F**, **G** and **H** are the flux vectors containing both inviscid (Euler) terms and viscous terms. The total energy is given by the relation

$$E_t = \frac{p}{\gamma - 1} + \tfrac{1}{2}\rho(u^2 + v^2 + w^2), \quad \gamma = \frac{c_p}{c_v},$$

where γ is the ratio of specific heats. For the viscous stress tensor Stokes' hypothesis, equation (3.3), is assumed. The conductive heat flux terms are expressed in terms of the specific enthalpy (h) and the Prandtl number (Pr):

$$k\nabla T = \frac{\mu}{\Pr} \nabla h .$$

For turbulent flows the viscosity and conductivity coefficients (i.e. μ and k) are interpreted according to equations (3.1–3.2), with Prandtl numbers of 0.72 and 0.9 for laminar and turbulent flows, respectively.

4.6.3 Numerical Method

In this section the numerical solution procedure is briefly described. For simplicity, we assume two-dimensional coordinates. The extension to three dimensions should though be obvious.

Equation (1) transformed to body-fitted coordinates (ξ,η) reads

$$\frac{dU}{dt} = -\frac{1}{Vol}\left(\frac{\partial f}{\partial \xi} + \frac{\partial g}{\partial \eta}\right) + D_2 + D_4, \qquad (2)$$

where the rotated fluxes are defined by

$$f = \frac{\partial y}{\partial \eta}F - \frac{\partial x}{\partial \eta}G$$

$$g = -\frac{\partial y}{\partial \xi}F + \frac{\partial x}{\partial \xi}G \;.$$

The second-order numerical damping term, often also denoted as the shock capturing term, may be written

$$D_2 = d_2\left(\delta_\xi \alpha_1 \delta_\xi + \delta_\eta \alpha_2 \delta_\eta\right)U,$$

where δ is the first-order difference operator, d_2 is a user specified constant and the α's are normalised pressure sensors based on a nonlinear second derivative of the pressure which detects discontinuities as shocks.

The fourth-order artificial dissipation is given by

$$D_4 = -d_4(\delta_\xi^4 + \delta_\eta^4)U,$$

where δ^4 is the fourth-order difference operator and d_4 is a user specified constant.

Time integration of equation (2) is performed by the second order accurate 3-stage Runge-Kutta scheme

$$U^0 = U^n$$
$$U^1 = U^0 + \Delta t \cdot R(U^0)$$
$$U^2 = U^0 + \tfrac{1}{2}\Delta t \cdot R(U^0) + \tfrac{1}{2}\Delta t \cdot R(U^1)$$
$$U^{n+1} = U^0 + \tfrac{1}{2}\Delta t \cdot R(U^0) + \tfrac{1}{2}\Delta t \cdot R(U^2)$$

where the residual **R** equals the right hand side of equation (2). Local time-stepping is applied for rapid convergence towards the steady-state solution. A stability analysis of the scheme reveals a maximum allowed CFL number of 1.8.

For thin shear layers the velocity gradients normal to the walls are dominant. Hence the assumption

$$\frac{\partial}{\partial \eta} \gg \frac{\partial}{\partial \xi}$$

is valid for the viscous stresses, where ξ are the coordinate along the wall and η is the coordinate normal to it. Depending on the characteristics of the flow either the thin-layer equations or the complete Navier-Stokes equations are solved.

4.6.4 Modifications to 'Standard' Turbulence Models

The Baldwin-Lomax model in its original form (Section 3.1.2) has been employed as the standard turbulence model for both two- and three-dimensional flows. Its numerical implementation is simple, and the performance is verified for attached flows. For more complex flows the Chien k-ε model (Section 3.4.4) has been incorporated into the two-dimensional flow solver.

The transport equations for (ρk) and ($\rho \varepsilon$) are slightly modified from equations (C1–C2). On the left-hand side a time derivative term has been added to obtain the same form as equation (2), with the exception of the new source terms on the right-hand side. The solution procedure for the turbulent transport equations is the same as outlined above for the Navier-Stokes equations, except for the artificial viscosity terms which have been replaced with the second order term:

$$\mathbf{D}_2 = e_2 \left(\delta_\xi \lambda_1 \delta_\xi + \delta_\eta \lambda_2 \delta_\eta \right) \mathbf{U}, \quad \mathbf{U} = [\rho k, \rho \varepsilon]^T.$$

The λ's are scaled with the contravariant velocities and e_2 is a small number. For stability reasons, this damping term is added in regions where the local Reynolds number based on the mesh size is less than 2.

4.6.5 Mesh Generation

The computational grids have been constructed with the aid of the mesh generation packages TwoMesh and ThreeMesh. These are of an algebraic type which applies transfinite interpolation technique (Eriksson, 1982) to mapping between user specified boundaries. For improved grid resolution control, internal control curves or surfaces may be specified based on a macro-block concept by Eriksson (1990). In addition, both algebraic and elliptic type smoothing algorithms are available for the smoothing of discontinuities.

4.7 Description of Methods Used by Deutsche Airbus (DA) (E.Elsholz)

4.7.1 Introduction

Work within several tasks of the EUROVAL-project had been carried out using different computational methods which include 2D Navier-Stokes codes as well as 'Locally Infinite Swept Wing' and true 3D boundary layer methods.

4.7.2 Navier-Stokes Approach

The method in use within EUROVAL task 1.1 (RAE 2822 test cases) had been developed by Radespiel/Rossow at DLR-Braunschweig. The 3D version of the CEV-CATS-code is described in Radespiel (1989) and Radespiel, Rossow, Swanson (1989). The CEVCATS-2D method is a cell-vertex finite-volume multigrid code which here is used for transonic airfoil flow computations within C-type meshes.

The main characteristics of this code are:
- Cell-vertex finite-volume scheme based on central difference approximations with artificial dissipation model.
- Thin shear layer approximation
- 5-stage Runge-Kutta explicit (pseudo-) time stepping with implicit residual smoothing and multigrid
- Far-field boundary conditions are vortex-corrected
- Turbulence models in use: Baldwin-Lomax and Johnson-King model, modified.

Numerical Scheme

In the integral form of the mass-averaged Navier-Stokes equations,

$$\partial/\partial t \iint_s W \, dx \, dy + \int_{ds} F \, dy - G \, dx = 0 \qquad (1)$$

where W denotes the vector of conserved quantities, i.e. density, velocity components and specific total energy

$$W = (\rho, \rho u, \rho v, \rho E)^T \qquad (2)$$

and F, G are the total flux vectors. The discretization of the integral equation in space and time are performed separately, the latter of which mainly takes advantage of special techniques - such as residual smoothing, local time stepping and multigrid - to increase the overall computational efficiency of the code. The spatial discretization of equ.(1) is performed with respect to the cell vertices, i.e. the flow variables are located at the nodal points i,j of the mesh. - The line integral of equ.(1) is evaluated by applying the trapezoidal rule along the cell borders. For the cell midpoint location, averaged values of the quantities are taken from the adjacent vertices. The resultant convective inflow of mass, momentum and energy is associated to the vertex by summing up the contributions of the neighbouring cells, understanding the vertex i,j itself as the midpoint location of a 'super-cell', consisting of four component cells. The viscous fluxes that include derivatives of the variables are determined at the midpoints of these component cells by introducing local curvilinear coordinates and central differencing technique. Then, they are averaged with respect to the cell vertex i,j.

Computing the viscous fluxes, a simplifying 'thin shear layer' approximation is introduced, i.e. only the viscous terms associated with the direction normal to the shear layers are taken into account.

An artificial dissipation model, which is a blend of 2nd and 4th differences (Jameson, Schmidt, Turkel, 1981), is introduced in order to prevent oscillations due to odd-even point decoupling as well as due to strong gradients in the shock region. So a background dissipation of third order is maintained throughout 'smooth' regions of the flow field and, driven by the local pressure gradient, first order terms appear in the vicinity of the shock.

A 5-stage Runge-Kutta explicit time stepping procedure with local time stepping and implicit residual averaging is implemented, typically yielding CFL=7...9. Within the basic Runge-Kutta procedure, the physical viscosity terms are computed at the first stage only and are kept frozen for the remaining stages. This is introduced with respect to increasing the computational efficiency. In contrast, the dissipative terms are evaluated for m=1,3,5. As had been shown by Martinelli (1987), this technique exhibits increased parabolic stability limits. - At each time stage, implicit residual smoothing takes place as had been devised by Jameson (1983) for Runge-Kutta type schemes. This technique also increases the stability range of the time stepping routine. In order to avoid the need for enthalpy damping, this smoothing incorporates variable coefficients that account for large variations of the local cell aspect ratio - for further details on this technique, see Radespiel, Rossow, Swanson (1989).

In order to increase the overall efficiency of the method, local time stepping is introduced which is based on the local maximum time step due to stability reasons. By doing this, however, the code does not integrate for truely time dependent solutions but the steady state solution is approached as fast as possible.

The method includes a multigrid acceleration technique (Jameson, 1985), where each coarser grid level is obtained by doubling the mesh spacing of the next finer grid in each coordinate direction. - Moving from one grid level to another, the solution and certain corrections have to be transferred from the current grid to the next. So transferring the solution down to coarser meshes is done directly by 'injection', while at the same time, residuals are weighted. The solution on the coarse mesh is driven by a forcing function constructed from the residuals within the finer mesh level. This procedure is repeated until the coarsest mesh level is reached. Then, the corrections

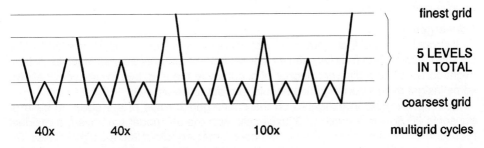

Fig. 1 Typical procedure: Full multigrid

are transferred up to the next finer level by bi-linear interpolation. For increasing the robustness of the whole procedure, the total corrections are smoothed again before finally updating the solution. - All this is done within a series of fixed W-cycles. With respect to full multigrid strategy, the solution may be obtained on several subsequently refining mesh systems, see Fig. 1.

Boundary Conditions

At the surface of the airfoil the non-slip condition is realized by setting $u=v=0$. To solve for adiabatic walls, additional control volumes are required inside the airfoil specifying the appropriate symmetric or antisymmetric mirror images of the flow properties of the nearest cells outside. - The wake closure along the C-type mesh cutting line is ensured due to overlapping the opposite mesh portions by a single row of cells on each side of the cutting line.

Far-field boundary conditions are derived by superimposing the flow of a single vortex to the free stream flow. The type of the far-field boundary condition, i.e. free-stream or vortex corrected velocity prescribed, may be selected by the user.

Turbulence Models

From the different turbulence models incorporated in the code, two models are tested here: the Baldwin-Lomax model (BL, Baldwin, Lomax 1978) for equilibrium flow and the Johnson-King model (JK, Johnson,King 1985) for non-equilibrium flow situations. The BL model is introduced in the standard formulation while the JK model had been modified according to Abid, Vatsa, Johnson, Wedan (1989). The main modifications introduced to the original Johnson-King model, as is described in sect. 3, are:

- a function of vorticity is used instead of the maximum turbulent shear stress and
- through the outer layer, the relation of Baldwin-Lomax that is multiplied by the non-equilibrium factor σ is used rather than the displacement thickness.

The simplified transport equation equ. (JK5) which was originally introduced by Johnson, King (1985), here becomes a partial but still linear differential equation. This equation is solved for by cell-centered finite-volume discretization. For a more detailed description of this model variant, the reader may be referred to Radespiel (1989). - With respect to the model descriptions of sect. 3, the coefficients of the models used here are:
- BL model: $C_{CP}=1.6$, $C_{WK}=1.0$, $C_{KLEB}=0.3$, $A^+=26$.
- JK model: $a_1 = 0.25$, $C_{dif} = 0.5$, $A^+=17$.

A special treatment of the transition region, as is mentioned in equ. (BL11), has not been implemented here.

Convergence

It should be noted that there is no explicit convergence criterion implemented in the CEVCATS code. This may be seen as a drawback as computed series, such as polars, might converge to a different degree of accuracy at the end, but, in practice, it allows some insight even if convergence should turn out to be difficult.

Radespiel (1989) investigated on the convergence rates of the code. In case of transonic 2D flow in a mesh of 320x64 cells with the JK model employed, a residual reduction of 4 orders of magnitude had been obtained within 200 multigrid cycles. So from an engeneering point of view, the standard solution procedure has been defined to be 40/40/100 full multigrid cycles, covering subsequently refined meshes over 5 levels totally and reaching approx. 4 digits relative accuracy for both, C_L and C_D in general. In Fig. 2, a typical convergence history plot is given (RAE 2822 airfoil, case 10, BL model applied).

The computational effort required according to this standard procedure is about 120 CPU-seconds on a VP-200 vector computer.

4.7.3 Boundary Layer Approach

Two available boundary layer codes have been used within EUROVAL task 5.1 (F5 wing and 3D boundary layer test cases). The methods are a 'Locally Infinite Swept Wing' (BLLISW, Elsholz 1988) and a true 3D code (3D-INV, Xue, Lesch, Thiele 1989). In the context of task 3.1 (2D boundary layers) the BLLISW code has been used.

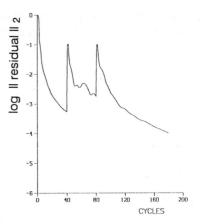

Fig. 2
Typical convergence history plot

4.7.3.1 'Locally Infinite Swept Wing' (LISW) Boundary Layer Method
Numerical Scheme

The LISW- boundary layer code is an implicit finite-difference code of cell-vertex type, solving for the x_1, x_2 momentum and total energy equations within a 3D surface-aligned coordinate system - but with the spanwise (x_2-direction) derivatives of flow properties - not the metric properties - dropped. This formulation allows locally infinite swept wing and true conical wing section boundary layer solutions.

An implicit marching procedure along the main direction x_1 is implemented that enables switching into different computational modes: standard mode, inverse or semi-inverse mode (for definitions, see below). Separating flows should be computed using the inverse or semi-inverse mode. - The discretization along the main direction is of second order Crank-Nicolson or first order upwind type, respectively, with a possible blend between both techniques. Here, the upwind discretization is used. In the separated flow regime, no FLARE-approximation is adopted. - Normal direction is approximated by 4th order Hermitian polynomials. At each x_1 station, the solution across the boundary layer is obtained by a direct solver for each of the momentum equations separately from 2x2 block tridiagonal matrices which include, however, an extra column accounting for the inverse boundary condition at the outer edge. A standard recursive algorithm is used to solve the energy equation. Hence, the coupled system of boundary layer equations is solved iteratively by updating the elements of matrices.

To ensure adequate normal resolution across the boundary layer, progressive mesh spacing is introduced with $\Delta\eta_j / \Delta\eta_{j-1} =$ VGP=const. with VGP of order 1.03...1.06. Along the streamwise direction, the grid stations utilized correspond to the input geometry points.

Within the marching procedure the number of normal grid points required is actually increased (or decreased) in order to adjust the computational domain securely to the local boundary layer thickness. This is done by adjusting the maximum normal distance y_{max} approximately to 15 displacement thicknesses.

Depending on the computational mode selected, the method requires the boundary layer edge velocity components u_{e1}, u_{e2} or the pressure distribution in the standard mode but the displacement thicknesses δ^*_1, δ^*_2 when computing in full inverse mode. Consequently, a combination of both, δ^*_1, u_{e2} is required for semi-inverse mode selection. - Computations are started at the attachment line of a swept wing section or from a given 3-D velocity profile at an arbitrary starting position from similar solutions.

Turbulence Models

Present state of the LISW method allows the choice of two models: Cebeci-Smith equilibrium model (CS) and Johnson-King non-equilibrium model (JK).

Cebeci-Smith model (CS): The Cebeci-Smith turbulence model is mainly based on the description of sect. 3, except that in the Prandtl-van Driest formulation, equ. (CS2), here the normal derivatives of the resulting velocity profile are used (Cebeci, Kaups, Ramsey 1977) rather than the vorticity as denoted in equ. (CS4).

In addition to this standard formulation, a simple but useful modification concerning the inner eddy viscosity model has been introduced which does not allow any decrease of (inner) eddy viscosity when propagating across the boundary layer. This modification already provides stable and satisfying results even throughout moderate reverse flow regions.

Johnson-King model (JK): This model implementation is the 3D extension of Abid (1988). This formulation differs from equ. (JK3) (inner layer) in using the resulting normal derivative of the velocity for the inner viscosity. The outer viscosity, equ. (JK2), still holds except that δ^*_i is computed from the resulting velocity profile, too. Due to these modifications, the simplified transport equation corresponding to equ. (JK5) now contains partial derivatives with respect to x_1, x_2, but it remains linear, see Abid (1988). However, due to the present 'locally infinite swept wing' assumptions, the spanwise component of the resulting partial differential equation has been dropped.

The model coefficients adopted here - see sect. 3 - are $a_1 = 0.25$, $C_{dif} = 0.5$, $A^+ = 15$ but the non-equilibrium factor σ has been restricted not to exceed a value of two.

4.7.3.2 The 3D Boundary Layer Method

This method was originally developed at TU-Berlin (Xue, Lesch, Thiele 1989). After the basic version of this code had been tested on the DA facilities, the code was extended by a pressure to velocity conversion module as well as additional post-processing options.

Numerical Scheme

The method is a finite-difference implicit true 3D boundary layer method (cell vertex type) solving for the x_1, x_2 momentum and total energy equations within a surface-aligned coordinate system. An implicit marching procedure is implemented that enables Δx_1 step size control as well as switching the computational modes: from standard mode to semi-inverse (= standard in x_2-direction, inverse in x_1-direction) mode. Separating flow regions should be computed within this semi-inverse mode. - The discretization of x_1, x_2-derivatives is of Crank-Nicolson type ZIG-ZAG schemes directed by the local domain of influence. At the spanwise (x_2) boundaries an implicit BOX discretization applies. Propagation from the current to the next line $x_1 = $ const is initiated at appropriate diverging streamline locations by an implicit double ZIG-ZAG scheme. If there is no such location detected, the solution at this line is started from a similar solution at the appropriate spanwise (x_2) boundary.

Normal direction is approximated by 4th order Hermitian polynomials. At each x_1,x_2 location, the solution across the boundary layer is obtained by directly solving for both, x_1- and x_2-momentum equations simultaneously from 4x4 block tridiagonal matrices. According to the computational semi-inverse mode, an additional column appears within these matrices to account for the outer boundary condition defined by a given δ^*_1. This solution is updated by the density changes from the subsequent solution of the energy equation.

A surface oriented grid is considered in which the boundary conditions are interpolated for constant step sizes Δx_1 and Δx_2. Marching along the main direction x_1, the internal Δx_1 step size is controlled according to the asymptotic quality of the outer portion of resulting boundary layer profiles. Satisfying asymptotic quality indicates a possible increase in step size while poor quality forces the step size to be decreased.

Normal step size progression is defined by $\Delta \eta_j / \Delta \eta_{j-1} = \text{VGP} = \text{const}$ throughout 85% of the normal distance ($\text{VGP} = 1.03 \ldots 1.06$) and by $\text{VGP} = 1.0$ for the outer part (constant step size). The overall extension of the computational domain across the boundary layer is controlled via an x_1-dependent proportionality factor $\alpha(x_1)$. By incorporating this factor into the basic transformation of the boundary layer equations, the computational domain is adjusted by updating $\alpha(x_1)$. As a consequence, there is no change of the normal total grid point number in the 3-D method.

Starting and Boundary Conditions

According to standard or semi-inverse computational mode, the method requires boundary layer edge velocity components u_{e1}, u_{e2} in the standard mode or displacement thickness/velocity combination δ^*_1, u_{e2} in the semi-inverse mode. At the spanwise boundaries similar solutions are used in case the mean flow enters the computational domain, otherwise no additional conditions are necessary. - Computation is started along an arbitrary starting line using similar solutions again and - in case of turbulent starting conditions - an appropriate guess of Re_x for the Cebeci-Smith turbulence model to fit the profile parameters required at the starting line.

Turbulence Models

Cebeci-Smith model (CS): The Cebeci-Smith turbulence model is mainly based on the description of sect. 3, except that in the Prandtl-van Driest formulation, equ. (CS2), here the normal derivatives of the resulting velocity profile are used (Cebeci, Kaups, Ramsey 1977) rather than the vorticity as denoted in equ. (CS4).

Johnson-King model (JK): The model implemented here is the 3D version of Abid (1988). This formulation differs from equ. (JK3) (inner layer) in using the resulting normal derivative of the velocity for the inner viscosity. The outer viscosity, equ. (JK2), still holds but δ^*_i is computed from the resulting velocity profile again. Due to these modifications, the simplified transport equation corresponding to equ. (JK5) now contains partial derivatives with respect to x_1, x_2, but it remains linear, see Abid (1988).

In terms of sect. 3, the model coefficients adopted here are $a_1 = 0.25$, $C_{dif} = 0.5$, $A^+ = 15$ and the non-equilibrium factor σ has been restricted not to exceed a value of two.

4.8 Description of Method Used by DLR (D. Schwamborn)

4.8.1 Introduction

This chapter gives a brief review of the Navier-Stokes method used by the DLR in the two dimensional applications of chapters 5.1 (RAE2822) and 5.2 (bump flow) as well as in the three dimensional F5 wing application (chapter 5.7). Furthermore the boundary conditions and the turbulence models employed are briefly discussed.

4.8.2 The Block Structured Navier-Stokes Method

Both the two and three dimensional version of the so-called *ViB-code* have essentially the same features as will be described in the following. These codes solve the integral form of the time-dependent compressible full Navier-Stokes equations in Cartesian coordinates:

$$\frac{\partial}{\partial t} \iiint_{Vol} \mathbf{U} \, dVol + \iint_S \mathbf{H} \cdot \mathbf{n} dS = 0, \qquad (1)$$

where $\mathbf{U} = (\rho, \rho u, \rho v, \rho w, \rho E)^T$ is the solution vector of volume-averaged mass, momentum and total energy. The unit normal vector of the surface S of the volume Vol is denoted by $\mathbf{n}$ and $\mathbf{H}$ represents the tensor of the inviscid and viscous fluxes.

This set of equations is closed by the equation of state and by Sutherland's law for the dependence of viscosity and heat conductivity on the temperature. For turbulent flow equation (1) is equivalent to the Reynolds averaged Navier-Stokes equations if the coefficients μ, k are supplemented by the corresponding effective turbulent quantities as described in chapter 3.1.1.

Equation(1) is solved using a cell-centered finite-volume Runge-Kutta approach, where the variables in fluxes across the cell faces are averaged from the neighbouring cell centers which is equivalent to central-differencing and thus makes the use of articial damping necessary. These damping terms are the usual blend of second and fourth order differences employing a pressure switch for the second order part. Up to here the method is very similar to the approach as it was first developed by Jameson et al.(1981). For all applications in this book we used the same standard switches $\varepsilon^{(2)}$ and $\varepsilon^{(4)}$ for the second and fourth order terms with

$$\varepsilon^{(2)}_{i+1/2} = c_2 \max(v_i, v_{i+1}) \qquad (2)$$

and

$$\varepsilon^{(4)}_{i+1/2} = \max(0, c_4 - \varepsilon^2_{i+1/2}) \qquad (3)$$

where the standard second difference pressure sensor

$$v_i = \frac{|p_{i+1} - 2p_i + p_{i-1}|}{|p_{i+1}| + 2|p_i| + |p_{i-1}|} \qquad (4)$$

is employed. The coefficients used in equations (2,3) are $c_2 = 1/4$ and $c_4 = 1/256$. Only for the F5 wing (chapter 5.7) an additional calculation was made with these coefficients doubled in order to see the influence of the artificial dissipation.

The first derivatives at cell faces needed for the calcution of the viscous terms are calculated via finite difference formulas using a local coordinate system (Deiwert, 1980).

For the time discretization of the system of ordinary differential equations resulting from the finite volume approach a linearized explicit four-stage Runge Kutta with the classical coefficients is employed. This is formally second order accurate in time for non-linear equations. Since we are, however, only interested in steady state results we freeze the numerical and physical dissipation terms after the first stage to save computation time and employ locally the largest possible time step obtained from a stability analysis. This local time stepping is the only standard method we use for convergence acceleration.

Some of the two dimensional results were, however, obtained using an full approximation storage (FAS) multigrid (MG) approach which decreased the computation time by a factor of five to six. In connection with this approach implicit residual smoothing was used additionally.

Both the two and three dimensional version of the *ViB-code* utilize a block structure thus allowing for meshes of arbitrary size and for configurations of high geometric complexity where the grid is no longer a cube in the computational space. To fully exploit the block structured approach an arbitrary number of neighbouring blocks and/or boundary conditions per block face is allowed through the segmentation of those faces. This can strongly reduce the necessary number of blocks within a given computational domain thus resulting in larger blocks, i.e. larger vector length on vector computers.

The information transport between blocks and the fulfillment of boundary conditions is enabled via a layer of dummy cells around each block, such that the algorithm for all interior cells is not influenced by the number of segments of the block. The data in the dummy cells are updated at each stage of the Runge-Kutta if this is possible from data within the same block, i.e. for standard boundary conditions and faces or segments where the block is connected to a different segment of itself (e.g. along the wake cut of a single block mesh). At faces or segments where a neighbouring block exists the update of the dummy cells is only done at the beginning of each time step since the data in the corresponding neighbour block will only change when that block is treated in core. This procedure results in a time lag between the different blocks and thus even between the information in the dummy cells of one block if this stems from different neighbours. This time lag, however, seems not to be harmful for steady state calculations considering only the final result, and at least in some test cases the convergence was not delayed with respect to a single block calculation. This may, however, be different if the blocks are very small, especially if the multigrid approach is used.

4.8.3 Boundary Conditions

In order to arrive at a well-posed problem we have to supplement the system of differential equations by a set of boundary conditions. At walls we use the no-slip and the adiabatic wall condition. The pressure is derived from the assumption of zero pressure gradient normal to the wall which is justified for the very small step sizes normal to the wall that are used in Navier-Stokes calculations for turbulent flows.

At the far field boundary we use one-dimensional Riemann invariants normal to the boundary in order to obtain boundary conditions. In case of the two dimensional flow about the RAE2822 airfoil we employ an additional compressible farfield vortex to modify the outer flow conditions at this boundary due to the circulation about the airfoil.

For internal flows, i.e. the bump flows cases and DLR F5 wing, we use outflow conditions in the subsonic case where four variables (namely $\rho, \rho u, \rho v, \rho w$) are extrapolated by a second order formula while the pressure is prescribed via a non-reflecting boundary condition (Rudy-Strikwerda, 1985), which reads

$$p_t = \rho c u_t + \beta(p - p_{out}), \qquad (5)$$

where p_{out} is the prescribed pressure, c is the speed of sound, and the index t denotes partial differentiation with respect to time. This formulation allows the pressure waves to leave the computational domain, but ensures that the pressure approaches the prescribed value if the solution converges to the steady state, as long as β is not zero. In the case of the DLR F5 wing we also empoyed the farfield condition at the outlet of the tunnel for one of the calculations.

For the bump flow cases we use inlet conditions based on the assumption of isentropic isenthalpic parallel flow, i.e. we assume the total temperature and the total pressure to be constant at the inlet, set the vertical velocity to zero and extrapolate the horizontal velocity linearly.

At the inflow boundary for the DLR F5 wing we prescribe the given inflow Mach number, temperature and flow direction while the pressure is linearly extrapolated.

4.8.4 Performance of the code

The code implemented on a CRAY - YMP 232 needs $1.1 \cdot 10^{-5}$ and $2.4 \cdot 10^{-5}$ CPU seconds per time step and grid point on one processor for the two and three dimensional version, respectively.

Typical numbers of time steps to drop the residual by four orders of magnitude are between 8000 and 10000 for two dimensional flow without multigrid, while with multigrid 1000 to 2000 work units are necessary. In the three dimensional case between 5000 and 8000 time steps are needed to drop L_2-norm of the residual at least three orders of magnitude.

4.8.5 Turbulence Models

4.8.5.1 The Baldwin-Lomax Model (BL)

This model is used with a number of modifications to the original model as described in chapter 3.1.2. First, we use $C_{WK}=1.0$ instead $C_{WK}=0.25$ in equations (BL7) and second we base C_{Kleb} on equation (3.7) of chapter 3.6 which results in $C_{Kleb}=0.52$ instead of $C_{Kleb}=0.3$. Only in the bump flow application of chapter 5.2 we use also the original coefficients of the BL model to examine the influence of these. There it turns out that only the first modification has a greater influence on the results since only the first formulation of (BL&) is used when $C_{WK}=1$ is used. All further modifications are believed have only marginal influence on the solution:

The transition switch of the model (BL11) is not used, i.e. the flow is turbulent everywhere in the bump flow cases while the model is switched on at the prescribed transition location in the airfoil test cases.

In the van-Driest damping the maximum of the laminar shear-stress in the local profile is used instead of the wall value τ_{wall} in order to avoid numerical difficulties near separation and reattachment points.

Finally the inner formulation of the eddy viscosity is not switched of immediately behind the trailing edge in the airfoil cases in order to avoid a sudden jump at the trailing edge.

4.8.5.2 The Granville Modification of the Baldwin-Lomax Model (GR)

The implementation of this model is as described in chapter 3.1.2 except that again the modifications of the BL model discused above are also used here, except of course the modification with respect to C_{Kleb}.

4.8.5.3 The Johnson-King Model (JK)

This model is different from that described in chapter 3.2.2 in two aspects following an approach by Abid et al. (1989) for three dimensional flow. First the outer eddy viscosity formulation of the JK model is based on the BL model approach (equ. BL6) instead on the CS model approach (equ. CS6) of Johnson and King such that our equivalent to equation (JK2) reads

$$v_{to} = (kC_{CP}F_{WAKE}F_{Kleb}) \cdot \sigma(s). \tag{6}$$

The other difference is with respect to the form of the differential equation (JK10) and is more a procedural one, i.e. equation (JK10) is multiplied by u_m and supplemented by a time derivative of g to result in the following equation:

$$\frac{\partial g}{\partial t} + u_m \frac{\partial g}{\partial x} = \frac{a_1}{2L_m}\left[(1-\frac{g}{g_{eq}}) + \frac{C_{dif} L_m}{a_1(0.7\delta - y_m)}|1-\sigma^{1/2}|\right]. \tag{7}$$

The advantage of this formulation is that it can be solved by an unsteady approach for all locations x along the surface without considering the local flow direction, which is not the case for a space-marching solution of equation (JK10). This is especially valuable in three dimensional flow where just a second surface tangential term $v_m \partial g/\partial y$ is added.

Though the basic equations have been changed somewhat and the numerical approach to equation (7) is made with a Runge-Kutta time-integration which needs an additional fourth order damping term for stability it is believed that the basic features of the JK model are retained.

For the airfoil cases we use the BL model the wake of the airfoil and allow for a smooth transition from the boundary layer to the wake formulation in order to avoid a jump in eddy viscosity at the trailing edge.

4.8.5.4 The Johnson-Coakley Model (JC)

For this model we introduce the same modifications as presented in chapter 3.2.3 to our JK model. For the new velocity scale according to equation (JC3) we use the mixing length approach instead of the Clauser formulation resulting in

$$u_s = ky|\frac{du}{dy}|[1-\tanh(y/L_c)] + \sqrt{\rho_m/\rho}\ u_m\tanh(y/L_c) \tag{8}$$

but using $A^+ = 17$ in the van Driest damping for the complete formulation.

Note that for the airfoil test cases the same approach as described above is employed for the wake part of the flow.

4.9 Description of Method Used by Dornier (W. Haase, W. Fritz)

4.9.1 Introduction

In the context of the EUROVAL project, Dornier took part in the work on the validation of CFD codes with respect to turbulence model investigations using a boundary layer method as well as Navier-Stokes methods for solving two-dimensional, turbulent and transonic flows.

The single-block and multi-block Navier-Stokes methods were in use for single- and multi-element airfoil flows and for a channel bump flow case. Both Navier-Stokes methods are based on the same fundamentals and, therefore, will be described together in section 4.9.3.

The quasi-two-dimensional boundary layer method has been applied to equilibrium and non-equilibrium boundary layers comparing different types of turbulence models on a free-of-numerical-dissipation basis, refer to section 5.5.

4.9.2 The Boundary Layer Method

The code for computing quasi-two-dimensional, compressible, laminar and turbulent boundary layers was used in its 1990/91 version and has been developed by H. Horton, Queen Mary and Westfield College, London. Acknowledgements are due to H. Horton's valuable help and support during the context of the work.

4.9.2.1 Numerical Approach

For both, laminar and turbulent flows, the Levy transformation is applied to the Reynolds-averaged boundary layer equations yielding a set of partial differential equations for momentum and energy, respectively. Approximation of the right-hand sides of these equations, containing the streamwise derivatives, with three-point backward differences (except at the first station downstream of the origin, where two-point differencing has to be used), leads to a coupled, non-linear set of ordinary differential equations to be solved at each streamwise ξ-station.

At each streamwise station, the system of ordinary differential equations is solved by applying a Newton iteration to the non-dimensional stream function $f(\eta)$ and its derivatives, thus achieving a quasi-linear form of the x-momentum equation with unknown properties assumed to be known from the previous streamwise station. The method is fully implicit, second-order accurate in streamwise and fourth-order accurate in the wall-normal direction.

An iteration at each streamwise station is required, using the mentioned quasi-linearization together with invariant embedding algorithms (generalized Ricatti transformation) for the two-point boundary value problem. The pair of initial value problems is solved in sequence, starting with an outward integration pass from $\eta=0$ to an edge value η_∞ followed by an inward pass with backward integration from $\eta=\eta_\infty$ to 0. An implicit 4th-order integration scheme (due to Gear) is used, with a 4th-order Runge-Kutta starter for the outward integration and Gear low-order starter for the inward pass. After solving the momentum and energy equations and, in the case of turbulent flows, applying an eddy viscosity turbulence model, an iteration is performed by use of an underrelaxation process on $f(\eta)$ in order to achieve a prescribed accuracy for the profiles at the current streamwise station.

Compared to shooting methods, for which the streamwise intervals must be large, the present procedure offers the advantage that, whilst marching downstream, mesh intervals of any required fineness can be accommodated by the use of sufficiently small integration steps in wall-normal direction.

If the derivatives of the profiles at the boundary layer edge exceed a given value, additional points are added at the end of the boundary layer. In the (wall-normal) η-direction, a geometrical stretching is used on a basis of a prescribed stretching factor, i.e. a fixed ratio between two successive η-step sizes, a certain number of intervals between $\eta=0$ and $\eta=\eta_{max}$. In most cases, 98 intervals with a stretching factor of 1.02 between $0<\eta<6.4$ have been taken as an initial guess for the start-up phase of the computations and are found to yield a good accuracy over a wide range of different flow conditions. In the case that the mesh has to be extended in the wall-normal direction, the total number of mesh intervals have been increased, the geometrical stretching factor, however, was kept constant. In many circumstances, the present method will be more accurate than a finite difference method for a given number of mesh points.

The mean velocity profile at the initial streamwise station was generated in each case by computing the Coles velocity profile corresponding to the measured values of shape factor, H, and momentum thickness Reynolds number R_Θ.

4.9.2.2 Turbulence Model Implementation

Because the numerical method is fully implicit, initial shear stress or eddy viscosity profiles are not required when algebraic models are used, by contrast with Keller box methods such as that of Cebeci.

The Johnson-King, 1/2-equation, model is implemented in a fully-implicit manner, using an iterative procedure contained within the loop already used for algebraic models. This does not have any appreciable adverse effect on the rate of convergence of the iteration by comparison with purely algebraic models.

For those calculations employing the Hassid-Poreh model, the equation for the turbulent energy, k, is discretised and solved by procedures similar to those used for the mean momentum equation. The momentum and k-equations are solved in succession inside an iterative loop, the eddy viscosity being updated after each iteration. In this case an initial profile of the turbulent energy, k, is required, which is generated by computing an eddy viscosity profile from the van Driest relation in the inner region and a constant mixing length in the outer region. This, in conjunction with the relations given by Hassid and Poreh, provides an implicit expression for k which is solved by iteration at each mesh point in the initial profile.

4.9.3 The Navier-Stokes Method

4.9.3.1 Governing Equations

The integral form of the Navier-Stokes equations in cartesian coordinates, describing two-dimensional, unsteady and compressible flows,

$$\frac{\partial}{\partial t}\iint_{Vol} U \, dVol + \int_S H \cdot n dS = 0 \quad , \tag{1}$$

with $\mathbf{U}=(\rho, \rho u, \rho v, E)^T$ being the vector of the dependent variables, density, momentum and total energy per unit volume. **n** denotes the unit normal vector of the cell surface S and **H** represents the tensor of viscous and inviscid fluxes.

The perfect gas equation of state is used to define the mean static pressure via the internal energy.

$$p = (\gamma - 1)\rho e \qquad (2)$$

For viscous laminar flows, the set of equations is closed by Sutherland's law for the molecular viscosity. For turbulent flows, molecular viscosity, μ, and heat conductivity, k, are replaced by $\mu=\mu+\mu_t$ and $k/c_p=\mu/Pr$ by $\mu/Pr+\mu_t/Pr_t$, respectively.

The Prandtl numbers are chosen to be Pr=0.72 for laminar and Pr_t=0.9 for turbulent flows. The second coefficient of viscosity, λ=-2/3μ, is kept unchanged in the occurrence of turbulent flows. The ratio for specific heat, γ, is maintained constant at 1.4.

4.9.3.2 Finite Volume Method

Applying the integral form of the Navier-Stokes equations, equ. (1), to each cell of a computational domain separately, where all quantities are defined to be constant, results in a system of ordinary differential equations in time and is solved by means of a finite volume approach (Jameson et al, 1981, and Haase et al, 1983) with multigrid acceleration.

Particularly, the system of ordinary differential equations is solved explicitly by an m-stage Runge-Kutta type time-stepping method:

$$\begin{array}{c} u^{(0)} = u^{(n)} \\ u^{(\nu)} = u^{(0)} - \alpha_\nu \, P u^{(\nu-1)}, \quad \nu = 1, m \\ u^{(n+1)} = u^{(m)} \end{array} \qquad (3)$$

where n denotes the previous time-level and P represents a spatial (central and therefore second order) difference operator.

For all applications, apart from the two-element airfoil computations, a 3-stage scheme was in use with coefficients

$$\alpha_1=0.8, \, \alpha_2=0.8 \text{ and } \alpha_3=1.0.$$

This scheme is stable up to a Courant number of approximately 1.5. An advantage of this scheme, i.e. using the coefficients 0.8/0.8 instead of the more commonly used values of 0.6/0.6, is an increased damping property in the low(er) frequency range. The slight disadvantage, however, i.e. the reduction of the maximum CFL number, which is for 0.6/0.6 CFL_{max}=1.8, can be easily overcome by use of acceleration techniques as there are residual averaging and/or multigrid.

For the two-element airfoil flow cases, again a three-stage scheme was in use, however, the coefficients were taken in their standard version, i.e.

$$\alpha_1=0.6, \, \alpha_2=0.6, \, \alpha_3=1.0,$$

allowing, as just mentioned, for a maximum CFL number of 1.8.

4.9.3.3 Filtering Technique

To prevent an odd-even decoupling, blended second and fourth order artificial dissipation is used, Jameson et al (1981). If the filtering technique is applied only once for the 3-stage and twice for the 5-stage scheme, stability analysis indicates the best damping property as well as the largest extension of the stability region to the left of the real axis giving latitude in the introduction of dissipative terms. In practice, the fourth order filter is active throughout the computational domain except in areas with larger pressure gradients where the second order filter takes over.

In order to minimize the numerical dissipation, especially in areas with large (geometrical) aspect ratios, filtering is applied taking the eigenvalues in the x,(i)- and y,(j)-direction independently, instead of using the sum of these eigenvalues.

Additionally, the filter terms (filter fluxes) in the wall normal direction are scaled with the ratio of the local Mach number and the Mach number at infinity (local Mach number and local isentropic Mach number for the 2-element airfoil cases). This procedure reduces the filter fluxes in the boundary layers significantly and, consequently, the amount of artificial total pressure losses.

4.9.3.4 Convergence Acceleration

Introducing the residual averaging approach, i.e. collecting the information from residuals implicitly, permits stable calculations beyond the ordinary Courant number limit of the explicit scheme. In the present work, the Courant number is chosen to be 3.5 for all single-block calculations and CFL=3.0 for the two-element airfoil flow cases.

Furthermore, since only the steady state is of interest, a variable timestep approach is used accelerating convergence drastically.

To increase convergence speed even more, a multigrid approach is used in all present calculations. In case that more than two grid levels are involved, a W-cycle is applied.

Additional reduction of the computation time is achieved by a multi-level technique, i.e calculations are started on a coarse mesh, are then transferred to a medium mesh and are finalized in the finest mesh. In all subsequent meshes, a multigrid approach may be applied and computations in each mesh are pursued until a predefined accuracy level is reached. Compared to a computation using a multigrid technique only on the finest mesh, i.e without multi-level strategy, a speed-up factor of almost 3 was reached for the single-airfoil cases.

4.9.3.5 Steady State

The steady state is defined, either by reaching a maximum error norm (L2 norm of $d\rho/dt$) reduction of 3.5 decades, or by reaching a variation of all force coefficients within a certain monitoring sequence (typically between 10 - 30 multigrid cycles) of less than 0.05% in accordance with the number of supersonic points staying constant.

4.9.3.6 Boundary Conditions

Airfoil Applications

At the solid wall boundary no-slip conditions are implemented and the flow is assumed to be adiabatic with a zero pressure gradient in the wall-normal direction.

A wake boundary is defined by overlapping lower and upper wake cells. For the 2-element airfoil cases, the block-structured solver allows for arbitrary boundary conditions along all block faces and uses an overlapping cell strategy.

At the outflow boundary linear extrapolation is used for density and mass fluxes, and the static pressure is fixed to the static pressure at infinity.

Assuming the flow is subsonic at infinity, fixed and extrapolated Riemann invariants are introduced as farfield boundary conditions. For the single-element airfoil calculations, a farfield vortex has been applied; for the multi-element airfoil cases, however, linear extrapolation was used where the flow leaves the domain, and freestream values were set in cases where the flow approaches the domain.

Channel Flow Applications

At the solid wall boundary no-slip conditions are implemented and the flow is assumed to be adiabatic with a zero normal pressure gradient.

Symmetry conditions are imposed at the channel centerline for the ONERA bump, case A.

At the outflow boundary linear extrapolation is used for density and mass fluxes, and the static pressure is fixed to a prescribed exit pressure.

The inflow boundary is treated in two different ways: Firstly, the isentropic conditions proposed by M. Leschziner (UMIST) are used, apart from the situation that the streamwise u-velocity component is extrapolated and the density is calculated on isentropic relations. This results in a "well defined" inlet velocity profile which is no more "rectangular", nevertheless, the lateral v-velocity component remains "unphysically" zero.

Therefore, as an interesting alternative to the latter approach, both u- and v-velocities are extrapolated under the constraint that v never hits the "kinematic limit", i.e. the computed density remains positive and follows:

$$\rho = \rho_0 \left[1.-min\left\{ 1., 0.5\,(u^2 + v^2)\frac{\gamma-1}{\gamma R T_0} \right\} \right]^{1/(\gamma-1)} \quad (4)$$

4.9.3.7 Mesh Generation

For all channel flow simulations, the mesh provided by UMIST (M. Leschziner) is used without any change. This mesh uses 192 x 64 volumes for discretizing the computational domain with a first stepsize at the surface ensuring y^+-values in the order of 1 and below.

For the single-element airfoil, the BAe-provided mesh (by G. Page) has been used for all final calculations. Results of other investigations, using own meshes and different farfield boundary conditions may be taken from intermediate reports of the EUROVAL project and are available on special request.

As the above mentioned flow cases, computations for the 2-element airfoil were started in a mandatory grid which was initially supplied by BAe and, furthermore, in a refined mesh version delivered by CFD-norway to all partners involved. It turned out, however, that the high-angle-of-attack test case computations failed to converge. An identification of the reasons for that failure led to the result that the (Dornier) Navier-Stokes method is rather sensitive to "inappropriate" meshes, e.g. to (fine) meshes which are not, at least slightly, adapted to the flow.

Due to these circumstances, for the two-element airfoil flow at high angle of attack, a third version of the initial BAe mesh was performed at Dornier, taking into account a solution adaptive process using a multi-level technique with three mesh levels.

To achieve the final solution adapted (fine) mesh, a first solution in a coarse mesh is performed. From the solution obtained in the coarse mesh, the total pressure loss distribution is taken as a weighting function for the source terms of the elliptic grid generation process (Fritz, 1987). Particularly, a blend of the first and second derivatives of the pressure is used to determine the source terms in the streamwise ξ-direction, and the total pressure loss $(1-p_t/p_{t\infty})$ is taken for the corresponding term in the (lateral) η-direction. From that, the next finer mesh is adapted. The same process is repeated in the medium mesh, i.e. the medium mesh results are used for an adaptation of the finest mesh. It should be mentioned at this point that another, possible, adaptation loop in the finest mesh did not prove to be necessary.

4.9.3.8 Modifications to Original Versions of Turbulence Models

A description of the original version of all turbulence models which have been used in the context of the EUROVAL work can be taken from chapter 3. In the following, modifications to the original versions of the turbulence models which have been applied by Dornier, are presented; equation numbers, put into parentheses, refer to the equation numbers of chapter 3.

Baldwin-Lomax and Cebeci-Smith Model

Both the Baldwin-Lomax and the Cebeci-Smith (algebraic) turbulence models are based on their original descriptions apart from the following modifications:

The maximum value of the "moment of vorticity" F_{max} (BL8) at y_{max} is calculated via a second order interpolation.

To derive the "correct" F_{max}-value (BL8), at a certain streamwise position (station), all maxima of the corresponding F-function are calculated. The absolute F_{max} is then chosen in a way that only a minimum variation between the current y_{max} and the y_{max} of the previous station is allowed.

The "transition switch" (BL11) is not in use, i.e. the C_{MUTM} parameter is set identically to zero and not to the originally given value of 14.

To overcome difficulties for both algebraic models at (or near) separation and reattachment, where τ_{wall} tends to zero, τ_{wall} in equations (BL5 and CS5) is replaced by the maximum value of the laminar shear stress profile, τ_{max}. Although this is only a minor modification, due to the fact that it is only active in the vicinity of du/dy=0, it is stabilizing the iterative process and, thus, improving convergence.

The constant C_{Wk} in the Baldwin-Lomax model was taken to be 0.25, i.e. to the original value, in all calculations. For the RAE2822 airfoil, case 10, however, a flow which exhibits a strong shock induced separation, $C_{WK}=1.0$ was taken in order to achieve a stable and steady solution. Unfortunately, all attempts to use the original value of 0.25, failed.

Johnson-King Model

The Johnson-King model is used in the way proposed by Johnson (1987) in his AIAA publication with some minor modifications:

For both equilibrium and non-equilibrium approaches, an Aitken iteration is chosen to solve the implicit formulation for v_t (JK7), i.e. the value for the wall friction velocity, $u_{\tau m}$, is not taken from the last iteration, as proposed by Johnson-King, but is iterated at the current timestep down to approximately 1.5% accuracy. Depending on the streamwise station, 1 to 5 iterations are needed. Consequently, for the Johnson-King model in non-equilibrium formulation, both equilibrium and non-equilibrium values for $u_{\tau m}$ are iterated in the same manner.

Calculations for the Johnson-King model are always started by choosing the Cebeci-Smith model until a certain convergence (1.5 decades reduction in the L2-norm) is achieved. After switching to the Johnson-King equilibrium model, the computation is pursued until the final convergence criterion is reached.

In case of applying the Johnson-King model in non-equilibrium form, the next switch (from equilibrium Johnson-King) is performed after another half decade of error norm reduction, and then the calculation is finalized until the steady state convergence criterion (see section 4.9.3.5: 'Steady State') is fulfilled.

Boundary Layer Length Scales

The boundary layer length scales, particularly the displacement thickness, being an input parameter for both the Cebeci-Smith and the Johnson-King model, has been calculated numerically by using the approach of Stock & Haase (1989), which is briefly presented in section 3.6.

4.10 Description of Method Used by NLR (F.J. Brandsma)

4.10.1 Introduction

In the framework of the Dutch ISNaS project, in which several research institutes and universities work together on an information system for flow simulations based on the Navier-Stokes equations, NLR and the University of Twente coorporate on the development of a multi-block flow solver for 3D compressible flows based on the Reynolds-averaged Navier-Stokes equations. As a first pilot study a 2D single block Navier-Stokes code has been developed, TURB2D, which can be used for analysis of the compressible turbulent flows around (single-element) airfoils (see Brandsma & Kuerten, 1990). This solver has been initially used in of the EUROVAL project. In order to investigate convergence acceleration methods for the final 3D flow solver, first the existing 2D single-block code has been supplied with implicit residual smoothing and a multigrid strategy. Furthermore, by introducing more types of boundary conditions and some modifications to the turbulence models implemented, the range of applicability has been enlarged. The resulting 2D single-block multigrid flow solver is called MUTU2D and has extensively been validated in the present EUROVAL project with respect to its applicability to transonic airfoil flow (see section 5.1) as well as with respect to its capability to predict maximum lift for (single-element) airfoils (see section 5.3). The MUTU2D code is described below where it should be mentioned that as far as the discretization in space is concerned, the description is also valid for the TURB2D code (initially used for the A-airfoil testcases described in section 5.3).

4.10.2 Governing Equations

The 2D time dependent Reynolds-averaged Navier-Stokes equations for compressible flow are integrated towards a steady state, where the equations are written in a fully conservative integral form in a cartesian coordinate system:

$$\frac{\partial}{\partial t} \iint_V \boldsymbol{W} \, dt + \int_S (\boldsymbol{H}_c - \boldsymbol{H}_v) \cdot \boldsymbol{n} \, dS = 0 \ . \tag{1}$$

The dependent variables are the density, the components of the momentum vector per unit of volume, and the total energy per unit of volume: $\boldsymbol{W} = (\rho, \rho u, \rho v, \rho E)^T$. The tensors $\boldsymbol{H}_c$ and $\boldsymbol{H}_v$ contains the contribution of respectively the convective fluxes and the viscous fluxes at the 'surface' S of the control 'volume' V where $\boldsymbol{n}$ is the unit outward normal vector. The set of equations is closed by the equation of state for an ideal perfect gas (γ=1.4), whereas Sutherland's law is used to relate the dynamic viscosity, μ, to the temperature. The thermal conductivity coefficient is related to the viscosity coefficient by assuming a constant Prandtl number, Pr=0.72, throughout the flow. Furthermore the Reynolds stress terms included in the viscous stress tensor, are modelled using the eddy viscosity concept as described in section 3.1.1, replacing μ by $\mu+\mu_t$, and the thermal conductivity, k, by $k+k_t$. The effective turbulent thermal conductivity, k_t, is related to the effective turbulent viscosity by assuming a constant value for the turbulent Prandtl number, Pr_t=0.90.

4.10.3 Discretization in Space

For the discretization in space, a finite volume technique is used with the unknowns located at the vertices of the the grid cells of a curvilinear (boundary conforming) computational grid. For the convective contributions to the equations in grid point (i,j), a control volume consisting of four adjacent grid cells is formed, with (i,j) as a common vertex (see Fig. 1). The convective fluxes (see equation 1) are integrated over the boundary of this control volume, V, using the trapezoidal rule. For the viscous flux contributions, the necessary derivatives of the dependent variables are calculated in cell centers by applying a 2D formulation of Gauss' theorema. The viscous flux contributions are integrated over the boundary of a secondary cell around point (i,j) consisting of the four cell centers in the primary control volume. The net viscous contribution is then multiplied by four to obtain an approximation for the viscous flux contribution of the primary control volume. On a cartesian grid this method is equivalent with a 2nd-order finite difference method using central differences. The stencil used for the interior grid points, depicted in Fig. 1, uses nine point for the discretization of the convective and viscous terms in the equations.

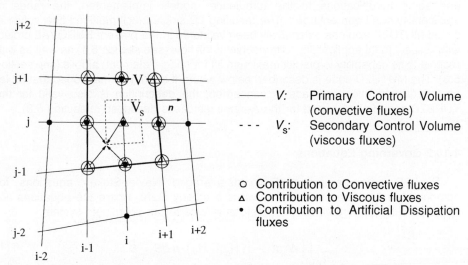

Fig. 1 Stencil used for the spatial discretization in the cell-vertex scheme employed by the MUTU2D code

In order to obtain interior stability of the central difference scheme described above, 4th-order artifical dissipation terms are introduced. Furthermore, in order to be able to capture shock waves, which manifests themselves as discontinuities on the scale that can practically be resolved, 2nd-order artifical dissipation terms are added making the scheme locally 'upwind' according to the local physical flow properties. The 'blended' 2nd- and 4th-order artifical dissipation terms are formulated in the same way as Jameson et al. (1981), but applied in the context of the cell vertex formulation employed here. Some details ar presented below.

The artificial dissipation term in i-direction, $D_i(W)$, takes the form

$$D_i(W) = d_{i+1/2,j} - d_{i-1/2,j} \; , \qquad (2)$$

where

$$d_{i+1/2,j} = R_{i+1/2,j} [\varepsilon^{(2)}_{i+1/2,j}(W_{i+1,j} - W_{i,j}) - \varepsilon^{(4)}_{i+1/2,j}(W_{i+2,j} - 3W_{i+1,j} + 3W_{i,j} - W_{i-1,j})] \; . \qquad (3)$$

In order to avoid excessively large dissipation levels in the large aspect ratio cells and to obtain optimal smoothing properties, a variable scaling coefficient is used, similar to the work of Martinelli (1987),

$$R_{i+1/2,j} = (\tilde{\lambda}^i_{i,j} + \tilde{\lambda}^i_{i+1,j})/2 \; , \qquad (4)$$

with

$$\tilde{\lambda}^i_{i,j} = \lambda^i_{i,j}[1 + (\lambda^j_{i,j}/\lambda^i_{i,j})^{2/3}] \; , \qquad (5)$$

in which λ^i and λ^j are the spectral radii of the flux Jacobian matrices in i- and j-direction respectively, which gives for the i-direction,

$$\lambda^i_{i,j} = |u \cdot S_i| + a|S_i| + |4\alpha S_i \cdot S_i|/V \; . \qquad (6)$$

Here, a is the local speed of sound, u is the velocity vector, V is the 'volume' of the primary control volume, and

$$S_i = 2 \int_{\partial V_i} n \, dS \; , \qquad (7)$$

with ∂V_i the conjunction of the two cell surfaces with constant i meeting in grid point (i,j). The parameter α in equation (6) is given by

$$\alpha = \frac{\max\{4(\mu + \mu_t)/3, \gamma(\mu/Pr + \mu_t/Pr_t)\}}{\rho \, Re} \; , \qquad (8)$$

with γ the specific heat ratio, and Re the Reynolds number. For the parameters $\varepsilon^{(2)}$ and $\varepsilon^{(4)}$ the standard formulation is used,

$$\varepsilon^{(2)}_{i+1/2,j} = \kappa^{(2)} \max(v_{i+1,j}, v_{i,j}) \; , \quad \varepsilon^{(4)}_{i+1/2,j} = \max(0, \kappa^{(4)} - \varepsilon^{(2)}_{i+1/2,j}) \; , \qquad (9)$$

where the standard pressure 'shock' sensor is employed

$$v_{i,j} = \left| \frac{p_{i+1,j} - 2p_{i,j} + p_{i-1,j}}{p_{i+1,j} + 2p_{i,j} + p_{i-1,j}} \right| \; . \qquad (10)$$

The coefficients $\kappa^{(2)}$ and $\kappa^{(4)}$ should be supplied by the user of the code (see section 4.10.9 for the values used for the calculations).

Furthermore, in order to reduce the amount of artificial dissipation in the boundary layer, the artificial dissipation can optionally be scaled with the ratio of the local- and the free stream Mach number (for both the *i*- and *j*-direction separately).

4.10.4 Time Integration

When the system of equations (1) is discretized in the way described above, a system of ordinary differential equations result,

$$V_{i,j} \frac{dW_{i,j}}{dt} + Q_{c_{i,j}} - 4Q_{v_{i,j}} + D_{i,j} = 0 , \qquad (11)$$

where $V_{i,j}$ is the volume of the primary conservation cell around point *(i,j)* (see Fig. 1), $Q_{c_{i,j}}$ is the contribution of $H_c \cdot n$ (see equation 1) along the boundary of $V_{i,j}$, $Q_{v_{i,j}}$ is the contribution of $H_v \cdot n$ (see equation 1) along the boundary of the secondary conservation cell around point *(i,j)* (V_s in Fig. 1), and $D_{i,j}$ represents the total of the artificial dissipation terms as described above. This set of equations is integrated in time towards a steady state using a Runge-Kutta scheme (first order in time) of which the coefficients have been set such that a maximum stability region is obtained. In the present method a five-stage scheme is used. The viscous terms are only evaluated at the first stage and are frozen on the remaining stages, whereas the artificial dissipation terms are only calculated at the stages 1, 3, and 5, and at each stage a linear combination of previously calculated contributions is used. For the mth stage, the scheme is written as (omitting the subscripts *i,j*)

$$W^{(m)} = W^{(0)} - \alpha_m \Delta t [Q_c^{(m-1)} - 4Q_v^{(0)} - \sum_{n=1}^{m} \beta_{mn} D^{(n-1)}] , \qquad (12)$$

where (0) denotes the old time level, Δt is the time step, and $W^{(5)}$ is the solution at the new time level. The coefficients α_m are taken to be

$$\alpha_1 = 1/4 , \quad \alpha_2 = 1/6 , \quad \alpha_3 = 3/8 , \quad \alpha_4 = 1/2 , \text{ and } \alpha_5 = 1 . \qquad (13)$$

The non-zero β_{mn}'s ($n \leq m$ with n=1, 2, or 3) are given by

$$\beta_{11} = 1 , \quad \beta_{22} = 0.56 , \quad \beta_{55} = 0.44$$

and

$$\beta_{21} = \beta_{11}, \quad \beta_{31} = 1 - \beta_{33} , \quad \beta_{41} = 1 - \beta_{33} , \quad \beta_{43} = \beta_{33} , \qquad (14)$$

$$\beta_{51} = (1 - \beta_{33})(1 - \beta_{55}) , \quad \beta_{53} = \beta_{33}(1 - \beta_{55}) .$$

The choice of these parameters is motivated in the work of Martinelli (1987).

The scheme employs local time stepping where the time step is taken according to linear stability analysis $\Delta t_{i,j} = \min(\Delta t_{c_{i,j}}, \Delta t_{v_{i,j}})$, with the allowable 'convective' time step as

$$\Delta t_{c_{i,j}} = CFL\, V_{i,j}/(|u \cdot S_i| + |u \cdot S_j| + a|S_i| + a|S_j|) , \qquad (15)$$

and the allowable 'viscous' time step as

$$\Delta t_{v_{i,j}} = RK \, V_{i,j}^2 / (4\alpha \, |\boldsymbol{S}_i \cdot \boldsymbol{S}_j|) \quad , \tag{16}$$

where it should be noticed that with a specific choice for RK, also the allowable values of the artificial dissipation coefficients are bounded. The scheme described above is (linearly) stable for Courant numbers up to CFL=3.25, whereas also RK=3.25 can be employed.

4.10.5 Convergence Acceleration

Besides local time stepping, convergence to steady state has been accelerated by by implicit residual averaging. After each Runge-Kutta time step the residuals, ΔW, are smoothed using

$$(1 - \varepsilon^i \nabla_i \Delta_i)(1 - \varepsilon^j \nabla_j \Delta_j) \overline{\Delta W} = \Delta W \quad , \tag{17}$$

with $\overline{\Delta W}$ the smoothed residuals, and ∇_i and Δ_i forward and backward differences in the i-direction (the same notation is used for the j-direction). In order to obtain optimal smoothing properties, the coefficients ε^i and ε^j are variable throughout the flow field as suggested by Martinelli (1987), and also Radespiel (1989). For the i-direction the following expresion is used,

$$\varepsilon_{i,j}^i = \max[0, \frac{1}{4} \{ \left(\frac{CFL \, \tilde{\lambda}_{i,j}}{CFL_{ex}(\lambda_{i,j}^i + \lambda_{i,j}^j)} \right)^2 - 1 \}] \quad , \tag{18}$$

where the coefficients λ are given in equations (5-6), and CFL_{ex} denotes the CFL number of the explicit time stepping scheme described in the previous section. With this formulation for the residual averaging, the CFL number can in practical situations be increased with a factor two to three.

Finally, a multigrid strategy is employed which follows closely the work of Martinelli (1987), employing a Full Approximation Scheme (FAS). Optionally use can be made of a fixed V- or W-cycle up to five grid levels. On each grid level the number of pre-relaxations (with the Runge-Kutta scheme described above), v_1, and the number of post-relaxations, v_2, within each multigrid cycle can be specified by the user of the MUTU2D code, whereas also the number of relaxations on the coarsest level, v_0, can be controlled. The iteration process is initialized by a number of Full Multigrid (FMG) cycles in order to obtain a suitable initial fine grid solution.

4.10.6 Boundary Conditions

At the airfoil surface the no-slip condition is imposed,

$$\boldsymbol{u} = \boldsymbol{0} \quad , \tag{19}$$

while at each boundary point the continuity equation as well as the energy equation are solved, for which an artificial (four cell) control volume is used consisting of the two cells meeting at the boundary point and their mirror images. In the energy equation, the adiabatic wall condition is applied,

$$\frac{\partial T}{\partial n} = 0 \quad . \tag{20}$$

For the convective fluxes at the faces of the dummy cells, the velocity components are taken to be the symmetric images of their values at the first cell face inside the computational domain, whereas for the viscous fluxes the anti-symmetric images of the velocity components are used.

At the far field boundaries approximate Riemann-invariants are used which follow from the linearized equations for a locally one dimensional flow normal to the boundary. The characteristic variables corresponding to outgoing characteristics are linearly extrapolated from the interior of the computational domain (except on the coarsest grid levels for which 0th-order extrapolation is used), whereas characteristic variables corresponding to incoming characteristics are obtained using (basically) free stream values for the flow variables. However, as demonstrated for the RAE2822 calulations (see section 5.1), merely taking free stream flow variables for the incoming characteristics makes the solution rather sensitive to the location of the far field boundary. In order to be able to keep the far field boundaries at a reasonable distance (e.g. less than 20 chords), at the boundary points the free stream values of the flow variables used in the boundary conditions are corrected by adding the leading term in the far-field small-disturbance potential approximation, being a point vortex representing the lift (circulation). The formulation of this vortex correction is taken from Thomas&Salas, (1985).

4.10.7 Turbulence Models

As mentioned before, the Reynolds stress terms are modelled using the eddy viscosity concept. The algebraic turbulence model of Baldwin&Lomax (BL) as described in section 3.1.2 has been implemented. With respect to the standard version of this model, the following modification are introduced. First of all, the parameter C_{wk} (equation BL7) is taken to be $C_{wk} = 1.0$ rather than the standard value of 0.25. This is usually done for reasons of stability of the calculation process for separated flows. Furthermore, in the near wake an intermittency function is used in order to obtain a smooth eddy viscosity distribution in the trailing edge region. The formulation used here is

$$\tilde{v}_t(s) = v_t(s) + [v_t(s_{te}) - v_t(s)] \exp[-\frac{(s-s_{te})}{B}] \quad , \tag{21}$$

in which s is the streamwise coordinate with a value s_{te} at the trailing edge. The constant B is taken to be $B = 8Re^{-0.2}$ which corresponds (at least for the flow over a flat plate) to approximately 20 times the boundary layer thickness at the trailing edge. The last modification is that in (equation BL5) $|\tau_w|$ is replaced by the maximum of the (laminar) shear stress in a profile, $|\tau|_{max}$, in order to avoid difficulties near separation points. It should also be stated here that in the wake formulation the value of U_{min} is taken to be the velocity at the wake-cut of the (C-type) computational grids used. This implies that the wake-cut of such a grid is required to be a good initial guess for the wake centerline.

In order to improve the predictive capabilities of the MUTU2D code with respect to separated flows, a second turbulence model has been implemented wich consists of a combination of the BL model (with the same modifications as described above) in regions where the flow is attached and a backflow model of Goldberg (GB), described in section 3.1.5) in regions of separated flow. The combination is implemented in such a way that in the region above and behind a separation bubble, the turbulent viscosity is taken to be the average of the value which follows from the GB model at the edge of the separation bubble (eqation GB6 with $G \equiv 1.0$ and $n = n_b$), and the value which follows from the BL model outside the separation bubble. This combination has proven to be rather successfull for the A-airfoil testcases in predicting realistic velocity profiles inside separation bubbles.

With respect to transition, only fixed transition can be handled by MUTU2D. At the prescribed transition location, the turbulence model is smoothly 'turned' on using

$$\tilde{v}_t(s) = v_t(s)(1 - \exp[-\frac{(s-s_{tr})^2}{gn_{\Delta s}}]) , \qquad (22)$$

where s is again the streamwise coordinate with a value of s_{tr} at the transition location, the constant g is taken to be $g = 0.36$, and $n_{\Delta s}$ is a fixed number if grid cells which is usually is taken to be 3 or 4.

4.10.8 Parameter Settings and Code Performance

In the EUROVAL project the MUTU2D code, as described above, has been applied to the transonic RAE2822 airfoil cases (section 5.1) and the low subsonic A-airfoil cases (section 5.3). For all calculations C-type grids have been used. The coefficients for the artificial dissipation have been set to the values of $\kappa^{(2)}$=0.5 and $\kappa^{(4)}$=0.0625 for the RAE2822 airfoil cases, whereas for the A-airfoil cases $\kappa^{(2)}$ has been set equal to zero (no shock waves present) and for $\kappa^{(4)}$ a slightly increased value of $\kappa^{(4)}$=0.075 has been used. The option to reduce the artificial dissipation in the boundary layer by scaling with the local Mach number is applied for the surface normal (*j*-) direction. In all cases residual averaging has been employed allowing the *CFL* number to be *CFL*=6.5 and also for *RK* a value of 6.5 has been used. All calculations have been performed with a fixed 4-level W-cycle multigrid strategy using v_1=5 pre-relaxations and v_2=6 post-relaxations on each grid level, while on the coarsest level v_0=10 relaxations has been employed. Each time the calculations have been initialized using 4 FMG cycles. For the airfoil calculations reported in this work the scheme proved to be robust and have shown good convergence properties. With the settings described above, typically 50 to 60 FAS cycles have turned out to been sufficient to obtain well converged solutions, meaning that the L_2-norms of the residuals in all four equation have been decreased by more than four orders of magnitude. As an example, in Fig. 2 the convergence histories for the L_2-norm and the maximum norm of the residual in the continuity equation (both normalized with their initial values) as well as for the lift- and drag coefficients are presented, for one of the RAE2822 cases (calculated on a 264x64 grid). For that particular case, the 90 FAS cycles presented in Fig. 2 (nearly six orders of magnitude reduction of the residual norms shown there) took about 300 CPU secs. on the NEC SX-3 supercomputer of NLR.

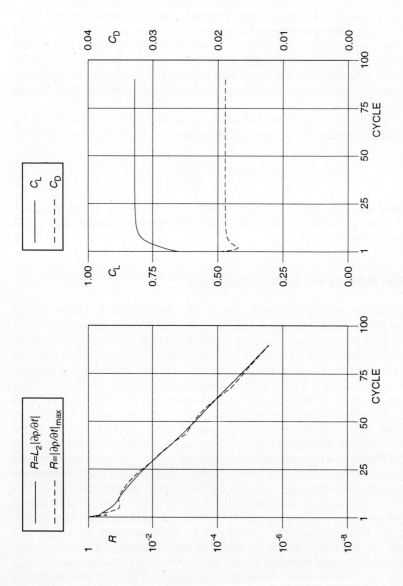

Fig. 2 Convergence histories of L_2- and maximum norm of the residuals in the continuity eqation, and of the lift and drag coefficients, for a calculation with the MUTU2D code for RAE2822 case 9 ($M_\infty=0.734$, $\alpha=2.54°$, $Re_c=6.5\cdot10^6$) on the RAE2822G1 grid (264x64 cells)

4.11 Description of Method Used by RAE (P.D.Smith)

4.11.1 Governing Equations

The method used is a time-marching boundary-layer method based upon the work of Van Dalsem and Steger (1985). The method takes the three-dimensional Navier-Stokes equations in Cartesian coordinates (x,y,z) with velocity components (u,v,w), pressure p, density ρ and specific enthalpy H. Transforming the independent variables x, y and z to body conforming coordinates ξ, η and ζ gives

$$(J^{-1}\rho)_t + (J^{-1}\rho U)_\xi + (J^{-1}\rho V)_\eta + (J^{-1}\rho W)_\zeta = 0 \tag{1}$$

$$\rho u_t + \rho U u_\xi + \rho V u_\eta + \rho W u_\zeta + (\xi_x p_\xi + \eta_x p_\eta + \zeta_x p_\zeta) = R_{xmom} \tag{2}$$

$$\rho v_t + \rho U v_\xi + \rho V v_\eta + \rho W v_\zeta + (\xi_y p_\xi + \eta_y p_\eta + \zeta_y p_\zeta) = R_{ymom} \tag{3}$$

$$\rho w_t + \rho U w_\xi + \rho V w_\eta + \rho W w_\zeta + (\xi_z p_\xi + \eta_z p_\eta + \zeta_z p_\zeta) = R_{zmom} \tag{4}$$

$$\rho H_t + \rho U H_\xi + \rho V H_\eta + \rho W H_\zeta - p_t = R_{ener} \tag{5}$$

where the terms on the right hand side represent the viscous terms and where U, V and W are contravariant velocities given by

$$U = \xi_x u + \xi_y v + \xi_z w \tag{6}$$

$$V = \eta_x u + \eta_y v + \eta_z w \tag{7}$$

$$W = \zeta_x u + \zeta_y v + \zeta_z w \tag{8}$$

and J is the transformation Jacobian

$$J = 1/(x_\xi y_\eta z_\zeta + x_\eta y_\zeta z_\xi + x_\zeta y_\xi z_\eta - x_\xi y_\zeta z_\eta - x_\eta y_\xi z_\zeta - x_\zeta y_\eta z_\xi).$$

Note that the Cartesian velocity components u, v, w are still retained as dependent variables. This avoids the introduction of destabilising coordinate source terms.

Applying the boundary layer approximation and taking ζ to be the direction normal to the wall we find $p_\zeta = 0$ and the viscous terms simplify to

$$R_{xmom} = \zeta_x \frac{\partial}{\partial \zeta}\left[2\mu\zeta_x u_\zeta + \lambda(\zeta_x u_\zeta + \zeta_y v_\zeta + \zeta_z w_\zeta) \right]$$

$$+ \zeta_y \frac{\partial}{\partial \zeta}\left[\mu(\zeta_y u_\zeta + \zeta_x v_\zeta) \right] + \zeta_z \frac{\partial}{\partial \zeta}\left[\mu(\zeta_z u_\zeta + \zeta_x w_\zeta) \right]$$

$$R_{ymom} = \zeta_y \frac{\partial}{\partial \zeta}\left[2\mu\zeta_y v_\zeta + \lambda(\zeta_x u_\zeta + \zeta_y v_\zeta + \zeta_z w_\zeta) \right]$$

$$+ \zeta_x \frac{\partial}{\partial \zeta}\left[\mu(\zeta_y u_\zeta + \zeta_x v_\zeta) \right] + \zeta_z \frac{\partial}{\partial \zeta}\left[\mu(\zeta_z v_\zeta + \zeta_x w_\zeta) \right]$$

$$R_{zmom} = \zeta_z \frac{\partial}{\partial \zeta}\left[2\mu\zeta_z w_\zeta + \lambda(\zeta_x u_\zeta + \zeta_y v_\zeta + \zeta_z w_\zeta) \right]$$

$$+ \zeta_y \frac{\partial}{\partial \zeta}\left[\mu(\zeta_z v_\zeta + \zeta_y w_\zeta) \right] + \zeta_x \frac{\partial}{\partial \zeta}\left[\mu(\zeta_z u_\zeta + \zeta_x w_\zeta) \right]$$

where μ and λ are the molecular and bulk viscosity coefficients respectively. For turbulence models of the eddy viscosity type μ is replaced by $\mu + \mu_T$, where μ_T is the eddy viscosity. Usually for flows over adiabatic walls (and for all the work on this project), rather than solve the energy equation, the Crocco relation

$$H = H_e + \frac{r-1}{2}(Q_e^2 - Q^2)$$

is assumed. Where suffix e denotes conditions at the boundary-layer edge, r is the recovery factor and

$$Q^2 = u^2 + v^2 + w^2 .$$

4.11.2 Numerical Method

Spatial derivatives in the momentum equations are approximated with implicit second order accurate central difference operators in the ζ direction and upwind second order accurate operators in the other two directions. In the continuity equation central differences are used in directions parallel to the body and trapezoidal integration in the normal direction. The algorithm is second order accurate in space and first order accurate in time.

The advantage of the time marching technique is that stagnation points and attachment lines arise naturally in the calculation and, unlike space marching methods, no special marching techniques are required.

Equations (2) to (5) are used to update u, v, w and H. New values of U and V are obtained from equations (6) and (7) and ρ is obtained from the equation of state. Using these updated values a new value for W is obtained from the continuity equation, eq.(1). Finally, updated values for the Cartesian velocity components are obtained from

$$u = x_\xi U + x_\eta V + x_\zeta W$$

$$v = y_\xi U + y_\eta V + y_\zeta W$$

$$w = z_\xi U + z_\eta V + z_\zeta W .$$

This process is then repeated, stepping forward in time, until converged values of the dependent variables are obtained.

For separated flows the method operates in an inverse mode in which the skin friction distribution instead of the pressure distribution is prescribed. At the wall the momentum equations degenerate to

$$p_x \big|_{wall} = (\xi_x p_\xi + \eta_x p_\eta + \zeta_x p_\zeta)_{wall} = R_{xmom}\big|_{wall}$$

$$p_y\bigg|_{wall} = (\xi_y p_\xi + \eta_y p_\eta + \zeta_y p_\zeta)_{wall} = R_{ymom}\bigg|_{wall}$$

$$p_z\bigg|_{wall} = (\xi_z p_\xi + \eta_z p_\eta + \zeta_z p_\zeta)_{wall} = R_{zmom}\bigg|_{wall}$$

so that for a given skin friction distribution the Cartesian pressure gradient at the wall $(p_x, p_y, p_z)_{wall}$ may be evaluated. Then with p_ζ assumed to be zero p_ξ and p_η are constant along a surface normal so that

$$p_\xi = \left(x_\xi p_x + y_\xi p_y + z_\xi p_z\right)_{wall}$$

$$p_\eta = \left(x_\eta p_x + y_\eta p_y + z_\eta p_z\right)_{wall}$$

and the solution procedure then becomes very similar to that for the direct mode.

4.11.3 Initial and Boundary Conditions

The boundary conditions applied are that u=v=w=0 at the surface and, as described above, either the pressure distribution is prescribed at the outer boundary or the skin friction distribution is prescribed at the wall. If either of the boundaries at the first and last stations in the ξ direction is an outflow boundary then upwind differences are used. If the first station in the ξ direction is an inflow boundary then profiles of the dependent variables must be prescribed. For velocity these are generated from Coles and Mager profiles whilst turbulence quantities are given by the Cebeci-Smith algebraic eddy viscosity model. For the side boundaries at the first and last stations in the η direction, upwind differences are used if outflow is occurring and infinite yawed wing conditions are assumed in the presence of inflow.

4.11.4 Turbulence Models

The turbulence models which have been implemented are: the Cebeci-Smith model, a modified Cebeci-Smith model in which the outer eddy viscosity is reduced linearly with increasing limiting streamline angle (up to a maximum reduction of one half at an angle of 30°), the Johnson-King model, the Johnson-Coakley model, a one equation model, which uses the Hassid-Poreh length scale together with the Chien k equation, and the Chien k-ε model.

4.12 Description of Method Used by SAAB (B. Arlinger, T. Larsson)

4.12.1 Introduction

The numerical code used by SAAB originates from Bergman (1991) and has been extensively modified and further developed by SAAB and subcontractors. This concerns mainly the implementation of different turbulence models and multigrid technique.

As a subcontractor of SAAB, the Laboratory of Aerodynamics at the Helsinki University of Technology (HUT) has also performed computations. A short description of their numerical method will also be given below.

4.12.2 Numerical Method (SAAB)

The code solves the two-dimensional time-dependent compressible Reynolds averaged Navier-Stokes equations written in conservative form. The mean flow equations are discretized in space using a cell-centered finite volume approximation.

Central differences are used for the convective fluxes. For the viscous fluxes, the gradients of velocity and temperature are evaluated using the gradient theorem on auxiliary cells. The viscous fluxes are then computed in the same way as the convective fluxes. The molecular viscosity is determined from Sutherland's law.

A blending of adaptive second and fourth order artificial dissipation terms are added to the numerical scheme to prevent oscillations in the vicinity of shock-waves and suppress odd/even decoupling in the solution. The second order terms are directly related to the local second differences of pressure which implies that these terms are small except in regions of large pressure gradients such as in the neighbourhood of shock-waves. Fourth order dissipation is added everywhere except where the second order dissipation is large. In boundary layers the influence of artificial dissipation can be decreased through a local Mach number scaling.

To reach a steady state solution the mean flow equations are integrated in time using an explicit five-step Runge-Kutta scheme where the contribution from the dissipation (artificial and physical) stays frozen after the second stage. Local time steps as well as multigrid technique are available for convergence acceleration. The multigrid technique is based on a FAS scheme and has been implemented by Weinerfelt (1992).

At the airfoil surface no-slip and adiabatic wall conditions are assumed while the far-field boundary conditions are based on the one-dimensional Riemann invariants.

The equations are solved on a structured grid which can be decomposed in several patched blocks. In each block the user can specify whether the Euler or the Navier-Stokes equations should be used. Navier-Stokes blocks can also be divided in several laminar and turbulent regions.

4.12.3 Turbulence Modelling

To close the Reynolds averaged Navier-Stokes equations the Reynolds stresses and the turbulent heat fluxes are related to the mean flow quantities using the Boussinesq eddy viscosity concept applying either the algebraic Baldwin-Lomax model or a two-layer k-ε model.

4.12.3.1 The Baldwin-Lomax (BL) Model

The Baldwin-Lomax model used by SAAB is more or less the standard one described in chapter 3.1.2. In the wake region however, the original outer formulation has been supplemented with a wake model proposed by Visbal and Shang (1985).

This includes a smooth transition zone near the airfoil trailing edge between profile and wake region. The eddy viscosity in the near-wake is computed by allowing the trailing edge eddy viscosity profile to exponentially reach its far-wake value. The distance from the trailing edge to the far-wake region is chosen to be of order 10δ, where δ denotes the average boundary layer thickness at the trailing edge. This boundary layer thickness is approximated using the empirical formula $\delta = 0.37Re^{-0.2}$, where Re is the Reynolds number based on the free-stream velocity and the airfoil chord (Schlichting, 1955).

The location of the wake center line, defined by the point of minimum velocity in the profile, is found at each time step using an under-relaxation procedure in order to further stabilize the calculation of the eddy viscosity distribution in the wake region.

The constant C_{WK} appearing in the formulation of F_{WAKE} (BL7) is set equal to 1.0 instead of the standard value 0.25.

Transition from laminar to turbulent flow is simulated by simply setting the eddy viscosity equal to zero in the laminar regions and starting the full BL procedure at the experimentally determined transition points.

4.12.3.2 The Two-Layer k-ε Model

In the two-layer model the k-ε model is combined with a one-equation model near the wall. The k-ε model used in the bulk of the flow not to close to walls is the "Standard" high-Reynolds-number Jones/Launder (SJL) model described in chapter 3.4.5. Near walls the one-equation model of Wolfshtein (W) (chapter 3.3.2) is used. The matching line between these two models is chosen along a pre-selected grid line.

The k and ε equations are treated implicitly for stability reasons. The diffusive terms are discretized using central differences, while for the convective terms a hybrid upwind/central differencing is used. The discretization results in a tri-diagonal system of linear algebraic equations which are solved using an ADI method (Patankar, 1980). At each time step the steady equations are updated using an under-relaxation factor of 0.5.

Laminar-turbulent transition at low free-stream turbulence levels occurring via shear-layer instability cannot be simulated by this type of turbulence models, hence some artificial triggering is required and the turbulent viscosity is therefore set to a low value in the transition regions, typically 1% of the molecular viscosity. In laminar parts the source terms in the k and ε equations are given some low value close to zero.

The current practice is to start from a solution obtained with the Baldwin-Lomax model and for the first 500 -1000 iterations only the k and ε equations are solved keeping the mean flow variables frozen. The implementation of the k-ε model is described in detail in Davidson (1990).

4.12.4 Grid Generation

Calculations concerning high-Reynolds number turbulent flows make the grid generation an important topic and the location of the grid points can have significant influence on the accuracy of the numerical results. In order to resolve the important physics in boundary layers and wakes, clustering of the cells is necessary in these regions. This clustering should be applied in a way that gives very small stretching close to the airfoil surface.

For the single-element airfoil cases (Airfoil A) the grids of C-type are generated by algebraic means. The distance from the wall to the first cell center is set to satisfy $y^+ \approx 1$. The stretching factor applied in the normal direction of the wall is based on the stretching strategy cited by Blottner (1989) as developed at NASA-Ames. It implies that the stretching factor increases away from the wall as 1 plus a sine function. Thus, the innermost cells have very small stretching. The wake cut extending from the trailing edge is defined as the stream-line given from classical incompressible flow around a flat plate with the prescribed angle of attack. In the circumferential direction 385 grid points are used with 257 grid points around the airfoil and 65 grid points are applied in the normal direction.

The multi-element airfoil cases (NLR 7301) are exclusively calculated using the mandatory multi-block grid generated by CFD Norway.

4.12.5 Numerical Method (HUT)

The basic equations are the compressible thin layer Navier-Stokes equations. The equation system is solved by a cell-centered finite volume scheme utilizing curvilinear structured grids. For the convective fluxes, the splitting of Van Leer is applied. The viscous fluxes are evaluated using conventional central differences.

The Cebeci-Smith (CS) turbulence model is activated in predetermined regions of the flow and the transition zones are modelled by a linear change of viscosity between the fully laminar and fully turbulent regions. The boundary layer thickness δ used in the CS model is derived in the way proposed by Stock & Haase (1989) and described in chapter 3.1.6.

All the boundary conditions on the airfoil are treated explicitly preserving formal second order accuracy. At the farfield boundary of the computational domain, a circulation correction is applied by adding the lifting point vortex induced velocities iteratively to the free-stream values.

The time integration of the discrete flow equations is based on an implicit scheme using approximate factorization which leads to a stability limit. Variable time steps with constant Courant numbers are used to speed up the convergence. The convergence is also accelerated by a multigrid technique employing V-cycling. The solution method is described in detail in Siikonen, Hoffren & Laine (1990).

4.13 Description of Method Used by TU Berlin
(L.Xue, T.Rung & F.Thiele)

An implicit space marching finite-difference method (Xue-Thiele, 1989) is used to calculate the three-dimensional boundary-layer flow for the DLR F5-wing. The solution procedure works either in direct or inverse mode. The external flow field is prescribed by a pressure distribution (Schwamborn, 1991), hence the direct mode is used in connection with several turbulence models. Transition is fixed according to Sobieczky's experimental data (Sobieczky, 1988). For the determination of the velocity components at the boundary-layer edge, a method developed by John (1991) at Deutsche Airbus is used. This method applies locally infinite swept wing assumption or the complete 3-D boundary-layer equations to relate the pressure field to the external velocity component.

4.13.1 Transformation and Discretisation of the Governing Equations

The method solves the x_1, x_2 momentum and total energy equation within 3-D body-fitted curvilinear coordinates.

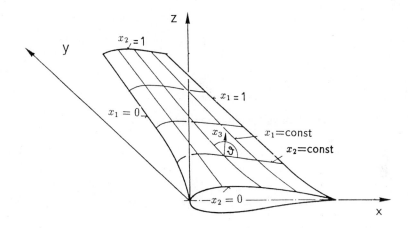

Fig. 1 Surface oriented coordinate system

The continuity equation is satisfied by the use of a two-component vector potential

$$\frac{\partial \Psi}{\partial x_3} = \rho v_1 h_2 \sin\Theta, \quad \frac{\partial \Phi}{\partial x_3} = \rho v_2 h_1 \sin\Theta, \quad -\left[\frac{\partial \Phi}{\partial x_2} + \frac{\partial \Psi}{\partial x_1}\right] = \rho v_3 h_1 h_2 \sin\Theta. \quad (1)$$

Introducing a similarity transformation

$$s_1 = \int_0^x h_1 dx_1, \quad \beta = \alpha(\rho_e v_0 s_1)^{\frac{1}{2}}, \quad d\eta = \left(\frac{\rho v_0}{\beta}\right) dx_3 \quad (2)$$

and two additional functions f and g,

$$\Psi = \beta h_2 \sin \Theta f \quad , \quad \Phi = \beta h_1 \sin \Theta g \tag{3}$$

the momentum equations are reduced to a system of two differential equations of third order.

x_1 momentum

$$(bf'')' + m_1 ff'' - m_2 f'^2 - m_5 f'g' + m_6 f''g - m_8 g'^2$$

$$\begin{cases} +m_{11}c & direct\ mode \\ +c\left(m_2 f_e'^2 + m_5 f_e' g_e' + m_8 g_e'^2 + m_{10} f_e' \dfrac{\partial f_e'}{\partial x_1} + m_7 g_e' \dfrac{\partial f_e'}{\partial x_2}\right) & inverse\ mode \end{cases}$$

$$= m_{10}\left(f'\dfrac{\partial f'}{\partial x_1} - f''\dfrac{\partial f}{\partial x_1}\right) + m_7\left(g'\dfrac{\partial f'}{\partial x_2} - f''\dfrac{\partial g}{\partial x_2}\right) \tag{4}$$

x_2 momentum

$$(bg'')' + m_1 fg'' - m_3 g'^2 - m_4 f'g' + m_6 gg'' - m_9 f'^2$$

$$\begin{cases} +m_{12}c & direct\ mode \\ +c\left(m_3 g_e'^2 + m_4 f_e' g_e' + m_9 f_e'^2 + m_{10} f_e' \dfrac{\partial g_e'}{\partial x_1} + m_7 g_e' \dfrac{\partial g_e'}{\partial x_2}\right) & inverse\ mode \end{cases}$$

$$= m_{10}\left(f'\dfrac{\partial g'}{\partial x_1} - g''\dfrac{\partial f}{\partial x_1}\right) + m_7\left(g'\dfrac{\partial g'}{\partial x_2} - g''\dfrac{\partial g}{\partial x_2}\right). \tag{5}$$

The boundary conditions of the physical plane result in

$$\eta = 0 : f_w' = 0, \quad g_w' = 0 \tag{6}$$

$$m_1 f_w + m_6 g_w + m_{10}\left(\dfrac{\partial f}{\partial x_1}\right)_w + m_7 \left(\dfrac{\partial g}{\partial x_2}\right)_w = -\left(\dfrac{s_1}{\rho_e \mu_e v_0}\right)^{\frac{1}{2}} \rho_w v_w \tag{7}$$

$$\eta = \eta_e : f_e' = \dfrac{v_{1e}}{v_0}, \quad g_e' = \dfrac{v_{2e}}{v_0} \qquad direct\ mode \tag{8}$$

$$f_e + f_e'(\delta_{1x_1}\dfrac{v_0}{\beta} - \int_0^{\eta_e} c\,d\eta) = f_w, \quad g_e + g_e'(\delta_{1x_2}\dfrac{v_0}{\beta} - \int_0^{\eta_e} c\,d\eta) = g_w$$

$$inverse\ mode. \tag{9}$$

The approximation of the derivatives tangential to the surface are based on implicit second-order accurate central difference operators and first-order upwind type operators. In addition, a matching between both discretisation methods can be applied. Normal derivatives are approximated by 4th-order Hermitian polynomials. Hence, various boundary conditions can be directly incorporated. The resulting system of coupled algebraic equations is nonlinear. Therefore an iteration process according to Newton-Raphson linearisation is applied. The quasi 4 x 4 block tridiagonal system of discrete boundary-layer equations is solved directly by using an LU-decomposition technique. Subsequently, the energy equation is solved providing the density distribution.

4.13.2 Space Marching Integration Scheme

The guiding principle to obtain a stable and accurate solution of the 3-D boundary-layer equation is to identify zones of influence and dependence at any gridpoint. A combination of Zig-Zag schemes and Box or Line scheme is applied for the inner domain and the outer domain of the boundary layer, respectively.

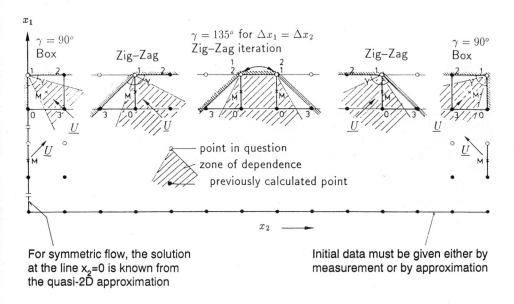

For symmetric flow, the solution at the line $x_2=0$ is known from the quasi-2D approximation

Initial data must be given either by measurement or by approximation

Fig. 2 Implicit second-order accurate integration scheme

The solution procedure starts with a similarity assumption for an arbitrary starting line x_1 = const. Further downstream, the integration follows the propagation of the flow. After detecting the location of diverging streamlines, the spanwise (x_2) integration of a new line x_1 = const. is initiated. In case the flow does not change its x_2 propagation direction, the integration starts with a similar solution at the appropriate spanwise boundary. A conflict of two converging spanwise integration processes caused by more than one diverging streamlines is solved by an iteration process between the two opposite Zig-Zag molecules. The distribution of gridpoints on the surface (x_1, x_2) is non-uniform. The spacing in spanwise direction is prescribed once. The step size Δx_1 in main stream direction is controlled according to the asymptotic behaviour of the velocity profiles at the outer boundary. In the normal direction a constant step size is assumed in the outer 15%. In the inner domain the normal distribution has a geometric progression.

4.13.3 Turbulence Models

The two original turbulence models considered are the algebraic model of Cebeci and Smith [CS] (Cebeci, 1984) and Abid's 3-D extension (Abid, 1988) of the two dimensional one-half equation model of Johnson and King [JK] (Johnson-King, 1984). Additionally, modifications of the velocity scale and blending function suggested by Johnson and Coakley [JC] (Johnson-Coakley, 1989) are combined with Abid's non-equilibrium model.

The Cebeci-Smith model differs from the description of sect. 3.1.4 in the formulation of the turbulent length scale. Here the normal derivatives of the resulting velocity profile

$$\left[\left(\frac{\partial v_1}{\partial x_3}\right)^2 + \left(\frac{\partial v_2}{\partial x_3}\right)^2 + 2\cos\left(\frac{\partial v_1}{\partial x_3}\right)\left(\frac{\partial v_2}{\partial x_3}\right)\right]^{\frac{1}{2}} \qquad (10)$$

is used rather than the magnitude of the vorticity as denoted in eq.(CS4). Referring to the outer viscosity the displacement thickness δ_i^* in eq.(CS7) is computed from the resulting velocity profile.

Abid's 3-D extension of the Johnson-King model differs from the description of sect. 3.2.2 in the formulation of the eddy viscosity. The normal derivative of the resulting velocity profile is used to determine the maximum turbulent shear stress in the definition of the inner viscosity (eq.(JK3)). The outer viscosity agrees with eq.(JK2) unless δ_i^* is computed from the resulting velocity profile, too. Due to the fact that we have to consider 3-D effects the simplified equation of the maximum turbulent shear stress (eq.(JK5)) contains partial derivatives with respect to x_1, x_2 in the convective part.

4.14 Description of Method Used by TU Denmark (J. Michelsen, J.N. Sørensen)

4.14.1 2D Navier-Stokes Method for Incompressible Flow

Numerical Method

The Navier-Stokes equations written in the primitive variable form are employed in the present study. As only low Mach numbers are involved, local Mach numbers ranging up to approximately 0.3 at the suction peak, the incompressible form is relied on. As a result, local under-prediction of the velocity, up to 3-4 percent, in the vicinity of the suction peak would be expected.

In contrast to most aerodynamics codes, the present code is based on a velocity-pressure coupling in the form of a predictor-corrector scheme. The momentum equations are employed as the predictor, while an integral form of the Poisson equation for pressure acts as the corrector.

The equations are discretized on cell-centered finite volumes, using central difference (CDF) expressions. The practice proposed by Rhie (1981) is employed to preclude pressure uncoupling, normally inflicting non-staggered schemes.

Solution of the implicit predictor-system demands non-negative coefficients, which do not in general result from CDF expressions. In the velocity-pressure coupled schemes, this problem is normally solved by use of low-order upwind differences. Preliminary results from the EUROVAL project showed, however, that the resulting amount of numerical diffusion had devastating effects on the solution.

Instead, the CONDIF approach (Runchal, 1986) was employed, with an extension similar to that proposed by Hedberg (1989). The negative coefficients are transferred to one of the directions, which already has a positive coefficient, in a way that does not alter residual calculation. The ratio of variable-gradients between neighbouring cells is involved in this practice. Whenever the variable in question does not exhibit a local extremum, this method is applied. At extrema, an upwind scheme is resorted to.

The CONDIF scheme is highly non-linear, and may during the progress of convergence produce wiggles, which calls for employment of the upwind scheme, which in turn removes the wiggles etc. In order to keep the numerical diffusion as low as possible, and for improved convergence, fourth-order dissipation is included. The dissipation coefficient is taken as 1/256. The dissipation term involves the local Reynolds number. Hence, the dissipation vanishes in the low-speed regions.

Mesh Adaptation

The present study considers only the Aerospatiale A-airfoil. The 64x384 C-mesh, delivered from NLR, and employed for one case by all partners in task 2.1, has been appropriately adapted to suit all 7 cases. The mesh was adapted to the wake behind the trailing edge. Furthermore, the mesh was stretched in the surface region to be employed at the higher Reynolds number. The first cell-height was made inversely proportional to the square-root of the Reynolds number, while still keeping the same distance to the far-field boundary, and approximately the same mesh expansion factor at the airfoil surface.

Turbulence Modeling

The Cebeci-Smith model is applied in the present study. The viscosity of the inner layer is given by (CS1-CS5) and the outer viscosity is given by (CS6-CS8). The boundary layer thickness, which enters the Klebanoff damping factor and limits the region of integration of momentum, and displacement thickness, is calculated in the Baldwin-Lomax fashion, proposed by Stock & Haase (1989).

The Clauser parameter in (CS6) takes the value 0.0168 for momentum-Reynolds numbers exceeding 5.000, while a correction based on Coles profiles takes place for the low momentum - Reynolds numbers,

$$K = 0.0168 \cdot [1+\Pi(\infty)]/[1+\Pi(z)],$$

with

$$\Pi(z) = 0.55 \cdot [1-\exp(-0.243z^{\frac{1}{2}}-0.298z)],$$

$$z = \frac{Re_\theta}{425} - 1.$$

The Clauser parameter takes the value 0.064 in the wake behind the trailing edge. As the inner layer disappears at the trailing edge, an exponential blending in the streamwise direction is employed here, based on the displacement thickness above the trailing edge.

As relatively low Reynolds numbers are involved in the present study, the transition may take place over a considerable distance. Hence, the transition intermittency factor by Cebeci & Smith (1974) is employed,

$$\gamma_{TR} = 1-\exp\left[-g\cdot\int_{x_{TR}}^{x} dx \cdot \int_{x_{TR}}^{x} \frac{dx}{U_e}\right],$$

with

$$G = \frac{U_e^3}{1200\nu^2 \, Re_{TR}^{1.34}}.$$

The intermittency factor is introduced into (CS2) and (CS6), it vanishes ahead of transition and approaches unity at the end of the transition region. In the present study, transition start is fixed in each case.

Boundary Conditions

No-slip conditions are employed on the airfoil surface. Dirichlet conditions for velocities are used for the farfield points, prescribing the freestream velocity, plus correction for circulation due to the presence of the airfoil.
Radiation conditions are employed at the outflow boundary.

4.14.2 3D Navier-Stokes Method for Direct Simulation of Vortex Breakdown

Numerical Method

For direct simulation of three-dimensional vortex breakdown the incompressible Navier-Stokes equations are formulated in vorticity-velocity variables and given in cylindrical coordinates.

Taking the curl of the Navier-Stokes momentum equations we get the transport equations for vorticity

$$\frac{\partial \omega}{\partial t} + \nabla \times (\omega \times V) = -\frac{1}{Re} \nabla \times (\nabla \times \omega) ,$$

where the vorticity is defined as

$$\nabla \times V = \omega ,$$

and Re denotes the Reynolds number. The system of governing equations is completed by introducing the equation of continuity

$$\nabla \cdot V = 0 .$$

In principle, the solution proceeds as follows: Having obtained the vorticity from the transport equations, assuming a known velocity field, an updated velocity field is found by solution of the vorticity-definition equations and continuity. These equations constitute a set of first order elliptic differential equations of the Cauchy-Riemann type which have the special property that they are overdetermined, i.e. the system contains more equations than unknowns.

The transport equations are discretized by central differences and solved by the ADI method developed by Douglas (1962). This method results in tridiagonal systems which are solved by standard techniques. The Cauchy-Riemann equations are discretized by second order accurate two-point differences. In order to avoid wiggles or spurious modes some sort of staggered arrangement is needed. Instead of full staggering a box scheme has been chosen. In this scheme the domain is subdiveded into small boxes with the variables defined at the vertices and the equations discretized about the center. As the Cauchy-Riemann equations are overdetermined special solution procedures must be sought. This is here done by finding the system's "least squares solution" which is defined as the vector that minimizes the L2-norm of the residual vector. It is well known that this vector is a solution of the associated normal system which is obtained by multiplying the original matrix equation by its transposed matrix. This results in a symmetric and positive definit matrix system which is solved by the conjugate gradient method. A detailed description of the method is given in Hansen et al. (1992).

Mesh

As the aim of the study is to analyse rotating flows by direct simulation a simple cylindrical mesh is employed, with 31 nodepoints in the radial direction, 51 in the tangential direction and 101 in the axial direction.

Boundary Conditions

A vortex is given in a cylindrical flow domain (r,Θ,z) with velocity components (u,v,w) and embedded in an axial shear flow. Initially, the vortex is defined by a tangential velocity distribution $V_o = V_o(r)$, an axial velocity field $W_o = W_o(r,\Theta)$, and zero radial velocity $U_o = 0$. At the lateral boundary the axial and tangential velocities are fixed, and the radial velocity component and the vorticity are determined from continuity and the definition of vorticity, respectively. At inflow the velocities are kept constant equal to their initial values and at outflow Neumann conditions are assumed for the axial velocity component and vorticity. The remaining variables are determined from the definition of vorticity and the equation of continuity.

As there is no a priori assumptions of axisymmetry, a special treatment has to be carried out at the center line. Owing to the polar singularity at $r = 0$ some of the operations of the transport equations become singular. This problem is circumvented by employing 3-point backward difference formulas at the first radial nodepoints, i.e. at $r = \Delta r$.

For the Cauchy-Riemann equations at the singularity, Cartesian components (v_{xo}, v_{yo}) are introduced by the transformation

$$u(r=0,\Theta) = v_{xo}\cos\Theta + v_{yo}\sin\Theta,$$

$$v(r=0,\Theta) = v_{yo}\cos\Theta - v_{xo}\sin\Theta,$$

and the solution is accomplished by substituting these expressions into the difference scheme at $r = 0$.

Acknowledgements

The synergetic atmosphere experienced in the EUROVAL project made it a great pleasure to participate and contribute.

4.15 Description of Method Used by UMIST
(G. Page and M.A. Leschziner)

4.15.1 Governing Equations

The flows considered are described by subsets of the usual general set of conservation equations governing continuity, momentum, energy and any other transported property associated with the turbulence model. These equations may be written, collectively, as:

$$\frac{\partial U}{\partial t} + \frac{\partial F}{\partial x} + \frac{\partial G}{\partial y} = \frac{\partial}{\partial x}[T_x] + \frac{\partial}{\partial y}[T_y] + S_U \tag{1}$$

where **U** is the vector of the dependent variables which, for inviscid or laminar flow, is:

$$U = [\rho, \rho U, \rho V, \rho e_0] \tag{2a}$$

while for turbulent flow, computed with the k-ε model or algebraic Reynolds-stress model, the vector expands to:

$$U = [\rho, \rho U, \rho V, \rho e_0, \rho k, \rho \epsilon]. \tag{2b}$$

The vectors **F** and **G** represent the inertial fluxes in x and y directions, respectively, and contain convection and pressure terms. The vectors T_x and T_y describe the diffusive fluxes of momentum, heat, turbulence energy and rate of turbulence-energy dissipation. In the case of the energy equation, the stress-work terms are also included in T_x and T_y. Finally, any terms which cannot be accommodated within the above flux terms are collectively represented by the 'source'-term vector S_U. Non-zero contributions to this vector only arise in the case of transport of turbulence quantities and are associated with generative and dissipative processes linking turbulence to the mean-strain fields.

4.15.2 Discretization

While the differential equations are expressed here in Cartesian coordinates, they are actually discretized and solved on a non-orthogonal curvilinear mesh such as sketched in Fig. 1, with the Cartesian decomposition of the velocity vector retained. All dependent variables are placed at the vertices of the finite volume or cell '1234'. It is over this cell that inertial terms are integrated to impose the conservation principle macroscopically. Focusing, say, on the term ∂F/∂x, the integration over the cell (1234) yields the value of the derivative at the centroid 'C'. Application of the divergence theorem gives:

$$\left.\frac{\partial F}{\partial x}\right|_C = \frac{1}{(\Delta A)_C} \int F dy \tag{3}$$

where the integration is to be carried out along the perimeter of the cell.

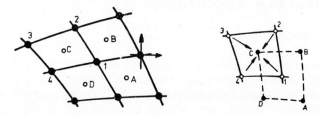

Fig. 1 Cell-vertex storage and cells used for integration

With the assumption of linear internodal variation of **F** along the cell faces (trapezoidal rule) the final result is:

$$\frac{\partial F}{\partial x}\bigg|_C = \frac{1}{2(\Delta A)_C}\left[(F_1-F_3)(y_2-y_4)+(F_2-F_4)(y_3-y_1)\right] \quad (4a)$$

where $(\Delta A)_C$ is the area of the cell (1234). Similarly:

$$\frac{\partial G}{\partial y}\bigg|_C = \frac{1}{2(\Delta A)_C}\left[(G_3-G_1)(x_2-x_4)+(G_2-G_4)(x_1-x_3)\right]. \quad (4b)$$

The inclusion of diffusive transport requires a choice of the locations at which the diffusive fluxes T_x and T_y are evaluated and the volume over which their derivatives should be integrated. An obvious option is to evaluate diffusive fluxes, typically containing property derivatives $\partial\phi/\partial x$ and $\partial\phi/\partial y$, at cell vertices, and to augment **F** and **G** appropriately so as to preserve cell conservation and maintain algorithmic simplicity. In this option, the derivatives of 'ϕ' at any grid node, say node '1', are evaluated from formulae analogous to equations (4) after integration over the cell (ABCD), while the required centroidal values are computed as arithmetic averages of surrounding nodal ones. It is easy to verify, however, that the diffusive contribution contains differences of 'ϕ' values at alternate nodes and cannot, therefore, smooth chequerboard oscillations. Indeed, it turns out that this practice requires artificial smoothing to cope with the steep property gradients in the boundary layer, with seriously detrimental effects on accuracy. Alternative locations for T_x and T_y are the centroids of the cell - (ABCD) in Fig. 1 - or the midpoints of the primary-cell faces. Here, the former option has been implemented, and the dashed line identifies the cell over which the balance of diffusive fluxes is considered. Thus, the fluxes at any centroid are obtained in terms of property values at surrounding nodes from equations (4). Subsequently, the viscous contribution at any node, say '1', is obtained by integration over the cell (ABCD). For example:

$$\frac{\partial T_x}{\partial x}\bigg|_1 = \frac{1}{2(\Delta A)_{ABCD}}\left[(T_x|_A-T_x|_C)(y_B-y_D)+(T_x|_B-T_y|_D)(y_C-y_A)\right] \quad (5a)$$

$$\frac{\partial T_y}{\partial y}\bigg|_1 = \frac{1}{2(\Delta A)_{ABCD}}\left[(T_y|_C-T_y|_A)(x_B-x_D)+(T_y|_B-T_y|_D)(x_A-x_C)\right]. \quad (5b)$$

With the above practice, physical diffusion acts to suppress oscillatory solutions in the shear layer, which then obviates the need for artificial damping. However, the penalty paid is loss of formal cell conservation, since inertial and diffusive fluxes are balanced over different finite volumes.

4.15.3 Time Marching

Following Hall (1986), we use the explicit Lax-Wendroff scheme to march the solution in time towards the steady state. The scheme relates the value of variable 'U' at any cell-vertex node at time level 'n' to the sought value of 'U' at time level 'n+1' through the first two terms of a Taylor series expansion in time. In the present approach, a third term is added to include the diffusive contribution, yielding, for node '1' in Fig. 1, the general formula:

$$(\Delta U)_1 = \delta t \left[\frac{\partial U}{\partial t}\right]_1 + \frac{\delta t^2}{2}\left[\frac{\partial^2 U}{\partial t^2}\right]_1 + (\Delta U^V)_1 \equiv (\Delta U')_1 + (\Delta U'')_1 + (\Delta U^V)_1 \quad (6)$$

where
$$(\Delta U)_1 \equiv (U)_1^{(n+1)} - (U)_1^{(n)}.$$

The first and second terms on the RHS of (6), referred to, respectively, as first- and second-order changes hereafter, are computed from a discretized equivalent of equation (1) with diffusive transport excluded. The last term represents the change due to diffusion.

The first-order change is initially evaluated at the centroid of the main cell as:

$$(\Delta U')_C = (\delta t)_C \left[-\left[\frac{\partial F}{\partial x}\right]_C - \left[\frac{\partial G}{\partial y}\right]_C + (S_U)_C\right] \quad (7)$$

where the spatial derivatives are obtained from equations (4). Subsequently, the first-order change at node 1 is obtained as an area-weighted average of surrounding changes:

$$(\Delta U')_1 = \frac{(\Delta U')_A(\Delta A)_A + (\Delta U')_B(\Delta A)_B + (\Delta U')_C(\Delta A)_C + (\Delta U')_D(\Delta A_D)}{(\Delta A)_A + (\Delta A)_B + (\Delta A)_C + (\Delta A)_D} \quad (8)$$

It is useful to remark that the evaluation of the source terms at centroids yields two benefits: First, stability is significantly enhanced. Without this practice or the introduction of significant artificial smoothing, the sources render the scheme unstable for a range of wave numbers, as has been verified by a von Neumann stability analysis. Second, the practice permits an accurate evaluation of the strain rates which appear in the production terms of the turbulence-energy equation.

The second order change is included for stability purposes only and has no effect on the accuracy of the converged. This change is evaluated at grid nodes by suitable manipulation and integration of (1) over the volume (ABCD). The final result is:

$$(\Delta U'')_1 = \left[\frac{\delta t^2}{4(\Delta A)}\right]_1 \left[\left[\frac{\partial F}{\partial t}\bigg|_C - \frac{\partial F}{\partial t}\bigg|_A\right](y_B-y_D) + \left[\frac{\partial F}{\partial t}\bigg|_D - \frac{\partial F}{\partial t}\bigg|_B\right](y_C-y_A) + \right.$$
$$\left.\left[\frac{\partial G}{\partial t}\bigg|_A - \frac{\partial G}{\partial t}\bigg|_C\right](x_B-x_D) + \left[\frac{\partial G}{\partial t}\bigg|_D - \frac{\partial G}{\partial t}\bigg|_B\right](x_A-x_C)\right]. \quad (9)$$

The terms containing time derivatives above are approximated as:

$$\frac{\partial F}{\partial t}\bigg|_C = \frac{F[U_C + (\Delta U')_C] - F[U_C]}{(\delta t)_C} \quad (10)$$

in which use is made of the first-order changes at surrounding centroids, as given by equation (8). The above approximation has been used in preference to the commonly adopted practice,

$$\left[\frac{\partial F}{\partial t}\right]_C = \left[\frac{\Delta F}{\Delta U}\right]_C (\Delta U')_C \quad (11)$$

which involves the Jacobian matrix ∂F/∂U of the flux vector. The two approaches are practically equivalent but equation (10) has been preferred as it avoids the calculation of the Jacobian. It will be observed from equation (9) that the second-order change contains contributions from inertial fluxes only. The consequence of the omission of diffusive terms has been investigated at an early stage of the research. With this option, diffusive fluxes are included in **F** and **G**, and their contribution to (9) can be easily accounted for. The effect on convergence history and the final solution has been found to be entirely negligible, which, on reflection, is expected since the additional terms involved are of order three and above. A similar attempt to include the Jacobians of the source terms was also found to yield no benefits to convergence characteristics. It is fortunate that the second-order change can be evaluated in terms of inertial fluxes only without any penalty to stability or convergence rate. The form of equation (9) not only avoids lengthy and expensive computations, but also simplifies the introduction of the multilevel method, as is explained in Dimitriadis & Leschziner (1989). Finally, the diffusive contribution at any node is given by:

$$(\Delta U^V)_1 = (\delta t)_1 \left[\frac{\partial T_x}{\partial x}\bigg|_1 + \frac{\partial T_y}{\partial y}\bigg|_1\right] \quad (12)$$

The spatial derivatives above are obtained from equations (4) which are consistent with storage arrangement in Fig. 1.

4.15.4 Time-Stepping Control

The time step is, in principle, restricted by the CFL criterion, modified to account for the presence of diffusion. For the subset $[\rho, \rho u, \rho v, \rho e_o]$, a spatially variable step is used, given at any cell by:

$$\delta t = \min\left\{\frac{\delta x}{|U|+a+2\nu/\delta x}, \frac{\delta y}{|V|+a+2\nu/\delta y}\right\} \quad (13)$$

where a is the speed of sound and ν is the effective kinematic viscosity. Strictly speaking, equation (13) is only valid for a Cartesian mesh, but it is sufficient for the geometries examined here, in which the grid lines are closely aligned with the x- and y-axes. In practice, δt given by (13) is multiplied by a factor of 0.85 to enhance stability.

In the case of the subset [$\rho k, \rho \varepsilon$], significant acceleration of convergence can be gained if a modified version of (13) is used in which the speed of sound is omitted, reflecting the absence of a pressure term from the transport equations. Use of a variable time step gives rise to instability, however, probably due to very steep spatial variations of 'δt'. Therefore, a uniform 'δt' is adopted, equal to the minimum value over all the cells of the computational domain. This time step is not held invariant, however, but allowed to adjust, via constraint (13), to the continually changing grid distances effected by a multigrid convergence-acceleration strategy whose details are documented in Dimitriadis & Leschziner (1989).

4.15.5 Artificial Dissipation

The second-order change of the Lax-Wendroff procedure allows the space-centered scheme to operate in a stable fashion in regions of smooth property variations without artificial dissipation. The boundary layer is highly sheared, however, and involves steep property gradients. In this region, stability can be maintained, again without dissipation, by use of a fine mesh with the assistance of physical diffusion.

The shock region presents a greater obstacle to stability, however, and requires the introduction of artificial smoothing. One highly attractive route is via the TVD concept which introduces, by way of a feedback mechanism, just sufficient dissipation to avoid oscillations. Dimitriadis and Leschziner (1991) have recently combined the cell-vertex scheme with the Yee/Roe type TVD scheme and have demonstrated the characteristics of the composite procedure on the basis of calculations for inviscid, laminar and turbulent flow. The practice adopted here is simpler and involves the addition of a carefully controlled measure of second order dissipation to the r.h.s. of equation (5), along the lines of Hall's (1986) inviscid method:

$$\Delta U^s = \sigma_x \delta x^2 \frac{\partial^2 U}{\partial x^2} + \sigma_y \delta y^2 \frac{\partial^2 U}{\partial y^2} \quad (14)$$

where, in the original method, the coefficients 'σ_x' and 'σ_y' were equal and constant. The numerical viscosity introduced by the above term can be estimated with the aid of equation (13). Assuming that the time step is determined by the grid size 'δy', as should be the case in the region close to the wall, the cross-stream numerical viscosity is:

$$(\nu_n)_y = \sigma_y [\delta y(|V|+a)+2\nu]. \quad (15)$$

Equation (15) indicates that grid refinement reduces the effects of artificial diffusion. Numerical experiments show, however, that, for practical grid densities, a uniform value of the coefficients 'σ', large enough to allow stable shock-capturing capabilities, has noticeable consequences on accuracy in the boundary layer. A remedy is to adopt variable 'σ_x' and 'σ_y' so that smoothing is only activated in the vicinity of the shock. To this end, the pressure field is used as a 'sensor' of the shock region, given that the pressure varies slowly in the boundary layer. Therefore, making the coefficients 'σ_x' and 'σ_y' proportional to pressure differences between neighbouring nodes provides, in principle, the desirable selective property. Additional desirable features of the activating mechanism are a smooth transition between 'active' and 'non-active' regions, and a bound on the maximum value of 'σ_x' and 'σ_y' to avoid shifting the shock position by excessive smoothing. After considerable experimentation with both inviscid shocks and boundary layers, the following formulation evolved:

$$\sigma_x = \min\{\sigma_{max}, \sigma_0 \Delta P_x\}; \quad \sigma_y = \min\{\sigma_{max}, \sigma_0 \Delta P_y\} \quad (16)$$

where
$$\Delta P_x = \frac{1}{4P_0}\Big[|P_{i,j} - P_{i-1,j}| + |P_{i+1,j} - P_{i,j}| + |P_{i-1,j} - P_{i-2,j}| + |P_{i+2,j} - P_{i+1,j}|\Big]$$

$$\Delta P_y = \frac{1}{4P_0}\Big[|P_{i,j} - P_{i,j-1}| + |P_{i,j+1} - P_{i,j}| + |P_{i,j-1} - P_{i,j-2}| + |P_{i,j+2} - P_{i,j+1}|\Big]$$

In equations (16), P_0 is the stagnation pressure at inlet, while σ_{max} and σ_0 are constants taking the values 0.0045 and 0.1, respectively. In the case of oblique shocks, equation (16) can be extended to include similar terms along the cell diagonals. The above formulation provides sufficient smoothing in the shock region, whilst keeping the artificial dissipation down to insignificant levels in the shear layer. It can be stated, therefore, that the present numerical scheme provides a framework for a reliable assessment of turbulence models.

4.15.6 Multigrid Acceleration

The time-marching process commences with a complete calculation for velocity, density etc (via their temporal changes) on the finest mesh, and proceeds to a user-chosen number of progressively coarser meshes. Each mesh is derived from its predecessor by omitting every other node in any spatial direction. The resulting doubling of the mesh size permits a doubling of the time steps, whilst maintaining the CFL criterion.

After restricting the fine-grid variable changes (which are equivalent to the residuals in an iterative sequence) to the next (coarser) mesh, the Lax-Wendroff scheme is applied to that mesh, and this results in coarse-grid changes which are linked to those restricted. This linkage prevents the generation of (artificial) property changes on coarser grids in the absence of fine-grid changes, which would corrupt the solution process. The above steps are applied to the entire sequence of coarsening meshes, and the total changes thus computed are subsequently prolongated to the finest mesh (via the progression of intermediate meshes) using bi-linear interpolation. Use of higher order interpolation has not been found to yield any advantages.

It is finally interesting to note that the multi-level method is, in terms of iteration number (or time steps), virtually insensitive to grid density. Specifically, doubling the number of nodes in each direction allows one further grid level to be employed so that the number of iterations performed on any one level remains roughly the same.

4.15.7 Turbulence Models

Three turbulence-transport models have been incorporated into the above numerical framework and all feature in the present validation study. These models are:

- the k-ε model of Chien (Section 3.4.3);
- the k-ε model of Launder and Sharma (Section 3.4.2);
- the algebraic Reynolds-stress model of Gibson and Launder, combined with Rodi's approximation for transport (Section 3.5.2).

All three models have been used in precisely the form documented in Chapter 3 and need not, therefore, be elaborated on further in the summary.

4.16 Description of Method Used by VU Brussel (M. Mulas, Ch. Hirsch)

4.16.1 Introduction

This section describes EURANUS, a general software system for the simulation of the 2D and 3D Reynolds averaged Navier-Stokes equations, under development for the European Space Agency, Hirsch et al. (1991). The code is written in a general framework with a novel system for the internal data management that allows one to choose a variety of different schemes and time integration options. The management of the data is done in two large one-dimensional work vectors, one for real and one for integer data, which appear to the programmer as a common tree structure in which all items are addressable through a system of pointers.

The heart of the solver is represented by the multigrid algorithm which is used to drive the solution to steady-state. Within this context different explicit and implicit time integration schemes are considered as "smoothers". In the present EUROVAL project it was decided to make use of high resolution TVD schemes rather than classical central plus artificial dissipation schemes, in order to present an approach which no other participant used.

4.16.2 Cell–Centered Finite Volume Formulation

The system of the Navier-Stokes equations in conservative, integral form is given by:

$$\frac{\partial}{\partial t} \int_{\Delta V} Q \, dV + \int_{\Delta S} H \, dS = 0 \qquad (1)$$

with Q representing the conservative variable array, and where $H = E n_x + F n_y$ represents the total flux through the surface. It is convenient to divide the total flux H in an inviscid and in a viscous contribution, H_i and H_v respectively, which are given by:

$$H_i = \begin{pmatrix} \rho V_n \\ \rho V_n u + p n_x \\ \rho V_n v + p n_y \\ \rho V_n h \end{pmatrix}$$

$$H_v = \begin{pmatrix} 0 \\ (\tau_{xx} n_x + \tau_{xy} n_y) \\ (\tau_{xy} n_x + \tau_{yy} n_y) \\ u(\tau_{xx} n_x + \tau_{xy} n_y) + v(\tau_{xy} n_x + \tau_{yy} n_y) + k \, \partial T/\partial n \end{pmatrix} . \qquad (2)$$

In a cell-centered finite volume formulation the conservative variables Q represent volume averaged values conventionally assigned to the cell centers. In the framework of a separate space and time discretization the governing equations (1) are semi-discretized as a system of ordinary differential equations in time

$$\frac{dQ}{dt} = \frac{1}{\Delta V} \left[\sum_{1}^{4} (H_v^* - H_i^*)_k \Delta S_k \right] \qquad (3)$$

where H_v^* and H_i^* represent the numerical fluxes.

4.16.3 High Resolution TVD Schemes

Two families of methods can be defined for the discretization of the inviscid fluxes, namely the classical central schemes with additional artificial dissipation and the modern high resolution schemes with hardly any adjustable parameters. Although these two approaches have largely different backgrounds, a uniform framework can be set up, whereby in all cases the numerical flux can be written as a central flux plus additional "dissipation" terms

$$H_{i+1/2}^* = \frac{[H(Q_{i+1}) + H(Q_i)]}{2} + D_{i+1/2} . \qquad (4)$$

Extension of Roe's first order upwind scheme, Hirsch (1990), to non linear system of conservation laws is represented by

$$D_{i+1/2} = -\frac{1}{2}|C_{i+1/2}|(Q_{i+1} - Q_i) \qquad (5)$$

where **C** represents the Jacobian matrix of the inviscid flux H_i

$$\mathbf{C} = \left(\frac{\partial H_i}{\partial Q}\right) = \left(\frac{\partial E_i}{\partial Q}\right) n_x + \left(\frac{\partial F_i}{\partial Q}\right) n_y = \mathbf{A}n_x + \mathbf{B}n_x \qquad (6)$$

and

$$E_i = \begin{bmatrix} \rho u \\ \rho u^2 + p \\ \rho uv \\ \rho uh \end{bmatrix} , \quad F_i = \begin{bmatrix} \rho v \\ \rho vu \\ \rho v^2 + p \\ \rho vh \end{bmatrix}$$

$$\mathbf{C} = \begin{bmatrix} 0 & n_x & n_y & 0 \\ q^2\beta n_x - uV_n & V_n + (1-\beta)un_x & un_y - \beta vn_x & \beta n_x \\ q^2\beta n_y - vV_n & vn_x - \beta un_y & V_n + (1-\beta)vn_y & \beta n_y \\ (-\gamma E + 2q^2\beta)V_n & (\gamma E - q^2\beta)n_x - \beta uV_n & (\gamma E - q^2\beta)n_y - \beta vV_n & \gamma V_n \end{bmatrix} \qquad (7)$$

with $q^2 = (u^2 + v^2)$ and $\beta = (\gamma - 1)$.

The matrix **C** can be diagonalized according to **L C R** = Λ , where the elements of the diagonal matrix Λ represent the eigenvalues λ of the matrix **C**. The first order upwind scheme, given by eq. (5), can be expressed by

$$D_{i+1/2} = -\frac{1}{2} \mathbf{R}_m |\Lambda_m| \mathbf{L}_m (Q_{i+1} - Q_i) . \tag{8}$$

This is the so called "local characteristic approach" where the additional "dissipative" term $D_{i+1/2}$, has been locally linearized and decomposed in four waves associated to the four characteristic variables:

$$\Delta W = (W_{i+1} - W_i) = \mathbf{L}_{i+1/2} (Q_{i+1} - Q_i) = \begin{bmatrix} \Delta\rho - 1/c^2 \Delta p \\ n_y \Delta u - n_x \Delta v \\ \Delta V_n + 1/\rho c \Delta p \\ -\Delta V_n + 1/\rho c \Delta p \end{bmatrix} \tag{9}$$

For each wave a Riemann problem is solved with prescribed constant initial values on both sides of the interface (W_{i+1} and W_i). The matrices in eq.(8) are considered locally constant and are to be evaluated at an intermediate state Q_m defined by Roe (1981). Second order accuracy is achieved by using flux extrapolation on the two sides of the interface together with various flux limiting mechanisms to avoid spurious oscillations of the solutions near discontinuities and to insure the TVD properties, Yee et al. (1983). The elements $|\lambda|$ of the diagonal matrix $|\Lambda|$ are substituted by

$$\alpha = |\lambda| - \lambda^+ \Psi(R^+) + \lambda^- \Psi(R^-) . \tag{10}$$

Eqs. (8) and (10) define an upwind TVD second order accurate scheme where $R^\pm$ define ratios of consecutives fluxes

$$R^+_{i+1/2} = \frac{\lambda^+_{i-1/2}(W_i - W_{i-1})}{\lambda^+_{i+1/2}(W_{i+1} - W_i)} \quad \text{and} \quad R^-_{i+1/2} = \frac{\lambda^-_{i+3/2}(W_{i+2} - W_{i+1})}{\lambda^-_{i+1/2}(W_{i+1} - W_i)} . \tag{11}$$

$\lambda^\pm$ are defined as $\lambda^+ = (\lambda + |\lambda|)/2$ and $\lambda^- = (\lambda - |\lambda|)/2$, and where $\Psi(R)$ represents one of the following limiters:

- minmod $\qquad\qquad\Psi(R) = \max [0, \min (1,R)]$
- van Albada $\qquad\;\;\Psi(R) = (R+R^2) / (1+R^2)$
- superbee $\qquad\quad\;\Psi(R) = \max [0, \min (2R,1), \min (R,2)]$

(with no limiter equivalent to: $\Psi(R) = R$).

A second type of high resolution TVD schemes is given by the symmetric TVD scheme developed by Yee (1987). Although it derives from a generalization of the Lax-Wendroff scheme made by Roe (1984), it can be recast to fit into the general framework given by eq.(8). In this case the elements $|\lambda|$ of $|\Lambda|$ are substituted by

$$\alpha = |\lambda|[1 - \Phi(r+, r-)] . \tag{12}$$

In eq.(12) Φ represents the limiting mechanism, which depends upon three consecutive gradients of the characteristic variables

$$r^+_{i+1/2} = \frac{W_{i+2} - W_{i+1}}{W_{i+1} - W_i} \quad \text{and} \quad r^-_{i+1/2} = \frac{W_i - W_{i-1}}{W_{i+1} - W_i} \quad . \tag{13}$$

The same limiters used above are reformulated as

- minmod $\qquad \Phi(r^-,r^+) = \max[0, \min(1,r^-)] + \max[0, \min(1,r^+)] - 1$
- van Albada $\qquad \Phi(r^-,r^+) = [r^- + (r^-)^2]/[1+(r^-)^2] + [r^+ + (r^+)^2]/[1+(r^+)^2] - 1$
- superbee $\qquad \Phi(r^-,r^+) = \max[0, \min(2r^-,1), \min(r^-,2)] +$
$\qquad\qquad\qquad\qquad \max[0, \min(2r^+,1), \min(r^+,2)] - 1$.

Unlike classical artificial dissipation schemes, in high resolution TVD schemes the additional numerical dissipation is not adjustable. In general, symmetric TVD schemes are simpler than upwind TVD schemes and computationally less expensive as well, but they introduce a larger amount of numerical dissipation. A general discussion on upwind and symmetric TVD schemes as well as a complete bibliography can be found in Yee (1989).

4.16.4 Time Integration

The semi-discretized formulation of the governing equations is integrated in time using a 5-stage Runge-Kutta method with coefficients given by: 0.059, 0.14, 0.273, 0.5 and 1.0. Local time-step

$$\Delta t_{i,j} = \frac{CFL \; \Delta V_{i,j}}{(\lambda_{max} \Delta S)_i + (\lambda_{max} \Delta S)_j} \tag{14}$$

is used, with a typical CFL number of 2.

A 4-level V-cycle multigrid algorithm is used to accelerate convergence to steady state. During the fine-to-coarse part of the cycle, on each level, starting from the finest one, respectively 2, 2, 3 and 10 5-stage RK iterations are performed. Coarse grid corrections are implicitly smoothed and interpolated back to the finest level. The implicit smoothing algorithm is the one proposed by Vatsa and Wedan (1989). In all coarse levels the inviscid numerical flux is only first order accurate and no physical viscous flux contribution is added.

4.16.5 Observations

In order to simplify the boundary treatment (dependent variables in dummy cells are fixed in order to have the correct boundary flux), the inviscid numerical flux given by eq.(4), is obtained with the averaged conservative variables

$$H\left(\frac{Q_i + Q_{i+1}}{2}\right) \tag{15}$$

rather than with the averaged flux

$$\frac{H(Q_i) + H(Q_{i+1})}{2} . \tag{16}$$

It has been found that this treatment can possibly cause convergence problems in flows with moderate to strong shock waves. Maximum residuals cannot be driven to machine accuracy level.

The implementation of the Baldwin-Lomax turbulence model follows strictly the original formulation given in section 3.1.2. However numerical experiments have been carried on concerning the value of the wake constant C_{WK} defined in eq.(BL7), and the calculation of the wall shear stress in eq.(BL5). Most of the authors using the BL model in the present EUROVAL context, make use of a value $C_{WK} = 1$ and of the maximum shear stress in each profile to avoid problems in separated flow regions.

The Onera bump case C test case described in section 5.2, has been used to test the BL model performance in presence of a remarkable separated flow region. No difference has been found when using τ_w rather than τ_{max}. Concerning the wake constant C_{WK}, it was found that use of the value 1 makes the wake formulation in bottom line of eq.(BL7) always bigger than the one on top line of eq.(BL7). Hence the C_{WK} is never utilized and as a result, the turbulent diffusivity is overestimated.

5 Application–Oriented Discussion of Results

AIRFOIL FLOW RAE 2822 CASES 9 AND 10

5.1

Colour figure - provided by UMIST - shows:

3D representation of two-dimensional transonic airfoil flow; shown are selected mesh lines, pressure-contours and velocity vectors.

RAE 2822 Airfoil - Test Case 9

Mach number = 0.734
Reynolds number = 6.5×10^6
Angle of attack = $2.54°$
Transition at x/c = 0.03
Turbulence model AS

5.1 Airfoil Flow RAE–2822 Cases 9 and 10

Author: **E.Elsholz,** Deutsche Airbus
Plot coordinator: **M.Tzitzidis,** Analysis S.R.

5.1.1 Introduction

Validation of Navier-Stokes codes applied to transonic airfoil flow in general means to resolve various numerical effects which normally are combined to produce reliable results. Effects like these not only arise from the code applied and its specific parameters, but the turbulence model chosen and the way it is implemented may also play a major role. Besides these, it is obvious that the numerical grid may affect considerably the solution which, in general, depends on the type of mesh, its numerical density with respect to the boundary layer resolution, the mesh extension and the treatment of the outer edge of the grid (far-field boundary conditions). Last but not least, when comparing to experimental results, question arises on the type and quantification of appropriate wind tunnel corrections that may be necessary to improve the experimental insight with respect to its free-flight equivalent.

In order to gain insight on how these effects influence the overall solution, the experimental work of Cook, McDonald, Firmin (1979) on transonic RAE-2822 airfoil flows was chosen as a basis against which to check several numerical effects. From the whole experimental material available, CASE 9 (no/small separation region) and CASE 10 (shock induced separation) have been chosen to be the mandatory test cases that most of the partners' work should be based on. The experimental data of these test cases include lift, drag and momentum coefficients as well as pressure and skin-friction, displacement and momentum thickness distributions along x/c. In addition, there are measured velocity profiles available for a number of stations.

5.1.2 Overview on Investigations

During the validation project, 10 partners in total have been engaged in the RAE-2822 transonic airfoil validation tasks. These partners are
- ASR (S.Perrakis, M.Tzitzidis)
- BAe CAL (J.Benton)
- CAPTEC (P.Duffy, E.Cox)
- CASA (A.Abbas)
- DA (E.Elsholz)
- DLR (D.Schwamborn)
- DO (W.Haase)
- NLR (F.Brandsma)
- UMIST (G.Page, M.Leschziner)
- VUB (M.Mulas, Ch.Hirsch)

This group of partners may be roughly divided into two sub-groups working either on effects of numerical parameters or on turbulence modelling effects. In Table 1, the individual investigations of the parners involved are summarized with respect to the numerical effects investigated and the various turbulence models under consideration.

Aiming to establish a common basis for further interpretation of results and tendencies, each party had been asked to deliver 'mandatory test case results' computed for a commonly agreed set of flow parameters (one test case at least, depending on the individual workload accepted). Furthermore, these results had been agreed to be computed within a common (i.e. also mandatory) grid delivered by BAe for this purpose which will, however, be discussed later in this section. If there were modifications to this mandatory grid or completely different meshes used (as CASA, for instance, generally uses triangular unstructured grids) an open symbol applies in the 'mandatory cases' column of Table 1. The choice of turbulence models used for these mandatory computations had been up to each partner.

As can be seen from the table, several partners performed additional investigations on distinct numerical effects such as mesh density/extension and far-field boundary condition, wind tunnel corrections including camber-correction and, last but not least, with different Mach-numbers and incidences.

On the other hand, there were 10 different turbulence models and variants thereof under consideration within the airfoil test cases, ranging from algebraic models up to Reynolds stress modelling. The models tested are also denoted in Table 1, the

Table 1 Overview on Investigations on Airfoil RAE 2822

partner	mand. cases 9	mand. cases 10	num. parameters mesh dens.	ext.	far-field b.c.	camb. corr.	phys. parameters Mach	alpha	turb. models
ASR	☐	☐	■				■	■	CS
BAe	■	■	■	■	■	■	■	■	BL, C, SJL-W
CAPTEC	☐	☐	■						CS
CASA	☐	unstruct'd grid					■	■	BL
DA	■	■	■		■		■	■	BL, JK
DLR	■	■							BL, GR, JK, JC
DO	■	■	■						BL, CS, JK, JK-E
NLR	■	■	■	■	■	■	■	■	BL, GB
UMIST	■	■							C, RS
VUB	■				■		■	■	BL

abbreviations used here, refer to the model descriptions in Section 3. In the following, these models are identified by:

- BL   - Baldwin-Lomax
- CS   - Cebeci-Smith
- C   - Chien
- GR   - Granville correction to Baldwin-Lomax
- GB   - Goldberg backflow model combined with Baldwin-Lomax
- JK   - Johnson-King
- JK-E   - Johnson-King equilibrium
- JC   - Johnson-Coakley
- SJL-W   - Standard high-Re Jones-Launder with 1-equ. sublayer (Wolfshtein)
- RS   - Gibson/Launder-Rodi Reynolds stress

Overviewing descriptions of the individual Navier-Stokes codes in use may be found in Section 4.

During the 26 months period of the validation project, some partners performed additional investigations concerning other than RAE-2822 test cases 9,10 as well as specific investigations on the individual method parameters employed such as the dissipation model etc. These investigations, however, will not be discussed in the context of this chapter; they are part of intermediate reports which are available on special request.

5.1.3 Mandatory Mesh and Flow Parameter Set, Effect of Camber-correction

In order to define a mandatory basis for computation, appropriate mesh and flow parameter settings had to be derived. So at an early stage of the validation project, BAe investigated a number of their meshes with emphasis given to

- moderate number of mesh points with respect to the different level computer power available, ranging from -386/Unix systems up to Cray-YMP supercomputers at the very beginning of the project.
- achievement of satisfying boundary layer resolution by introducing the minimum normal step size at the airfoil surface to be of order one in y^+.
- allowing free-stream, i.e. no vortex-corrected far-field boundary conditions to succeed to a satisfactory degree by establishing the outer mesh boundary far away from the airfoil (10 chords at least).

A mesh that is generated with respect to these restrictions was felt to be a good compromise between the two contradicting goals

i. to obtain reliable results and - at the same time -
ii. to make the grid acceptable to all partners.

BAe checked different meshes with respect to the results obtained for case 9 under several conditions. Finally, the so-called GRID 1 had been agreed to become the mandatory mesh in which each party should compute their case 9 and 10 results. A close-up view of this mesh is plotted in Fig. 1. It is a C-type grid of 10 chords overall extension. As can be seen from the figure, no special flow adaptation holds, i.e. no staggered streamwise stepsizes are introduced along the measured shock region which would treat the various computed shock locations in different manners. Along the leading and trailing edge of the airfoil, however, the stepsizes are condensed to resolve the strong gradients in the stagnation point area and to minimize pressure oscillations at the rear part of the airfoil. When reaching the downstream boundary, the wake cut line tentatively is aligned to a certain amount of incidence. This mesh is defined by 256x64 cells within the 10 chords area, allowing for 208 cells in total to be located along the airfoil's surface. The mesh has been generated along the measured, not designed, airfoil contour after applying a camber-correction as described below on both, upper and lower sides of the airfoil.

The wind tunnel correction procedure advised in Cook, McDonald, Firmin (1979) for M>0.7 has been changed to a slightly different method, which is based on a suggestion of P.D.Smith, DRA/RAE (1990) and mainly follows the procedure for lower Mach numbers.

First, when correcting the angle of incidence by

$$\Delta \alpha = c/h \{\delta_0 * C_L + \delta_1 * c/(\beta h) [C_L/4 + C_M] \} \qquad (1)$$

where the symbols are defined in Cook, McDonald, Firmin (1979), the experimental values of C_L, C_M of the specific test case should be used, but the constants δ_0, δ_1 now are recommended to be $\delta_0 = -0.065$ and $\delta_1 = 0.175$, respectively. Next, the free-stream Mach number now is advised to be increased by

$$\Delta M_\infty = 0.004 \qquad (2)$$

in general for the cases of main interest, i.e. cases 6,9,10. - Finally, a positive camber-correction applies to the measured airfoil contour, defined by

$$\Delta (z/c) = K * x/c (1-x/c) \qquad (3)$$

with K suggested to be constant $K=0.006$.

As a result of this revised wind tunnel correction procedure, the mandatory set of computational parameters had been set to

- Case 9: $M = .734$ $\alpha = 2.54°$ $Re = 6.5 \cdot 10^6$
- Case 10: $M = .754$ $\alpha = 2.57°$ $Re = 6.2 \cdot 10^6$.

Two partners (BAe, NLR) explicitly investigated test case 9 on own meshes of different density and extension with and without camber-correction, see Table 2 and Fig. 2. Both investigators employ the same algebraic turbulence model (Baldwin-Lomax).

Table 2 Influence of Camber-correction
Case 9:   M = 0.734 α = 2.54°   Re = 6.5·10⁶ x_{tr} = 0.03

partner	camber correct.	mesh	along airfoil	y^+_{wall}	mesh extens.	turb. model	C_L	C_D
BAe	no	272x 96	208	O(1)	30 c	BL	.816	.0169
	yes						.838	.0186
NLR	no	320x 80	256	O(1)	15 c	BL	.783	.0171
	yes						.808	.0186
Experiment:							.803	.0168

5.1.4 Effect of Turbulence Modelling: Mandatory Test Cases

In total, there have been 10 different turbulence models investigated that may be sub-divided into the following model classes, see also Section 3:
- algebraic models: Cebeci-Smith (CS), Baldwin-Lomax (BL) and extensions of the latter, such as the Granville-extension (GR) and the Goldberg backflow model (GB) combined with BL
- the 'half-equation' model of Johnson-King (JK) in different implementations (equilibrium/non-equilibrium) as well as the more recent variant by Johnson-Coakley (JC)
- transport models: the standard Jones-Launder model with an additional low-Re sublayer equation (Wolfshtein) incorporated (SJL-W) as well as Chien's low Reynolds variant (C)
- algebraic Reynolds stress model of Gibson/Launder-Rodi (RS) which is the most complex model under consideration within the airfoil test cases.

In order to extract the specific turbulence modelling performances, the mandatory grid and flow parameter sets that had been defined above strictly apply on the collection of results of both test cases, 9 and 10, given in Tables 3 and 4 and Figs. 3 to 6.

Some remarks on the non-standard grids used in this task may be added here: ASR coarsened the mandatory grid by dropping every second j-line (mesh lines parallel to the airfoil) whereas CAPTEC employed private meshes of 18 chords extension with different mesh spacing. Comparing here for turbulence model effects, CAPTECs grid "B" has been chosen. CASA uses unstructured triangular grid generation in general. They employed a grid of 16923 nodal points whereof 337 are situated on the airfoils surface. To apply an algebraic turbulence model more easily, they superimposed a regular sub-grid of 25 layers normal extension in the near wall region that corresponds to the mandatory grid distribution.

Tables 3 and 4 represent a complete collection of the test case results of all partners involved, grouped with respect to the applied turbulence models. The tables denote the parties invvolved, their computed lift and drag coefficients and, by indicating the

Table 3 Influence of Turbulence Modelling on Case 9

Mandatory mesh		GRID1: 256 x 64 cells (208 on surface) / 10 c			
Mandatory parameters		$M = 0.734$ $\alpha = 2.54°$ $Re = 6.5 \cdot 10^6$ $x_{tr} = 0.03$			

partner	turb-mod.	C_L	C_D	fig.	remarks
BAe	BL	.819	.0192	3	$C_{WK} = 1.0$
CASA		.756	.0167	"	$C_{WK} = 0.25$ unstruct.grid /13c
DA		.817	.0185	"	$C_{WK} = 1.0$
DLR		.790	.0192	"	$C_{WK} = 1.0$
DO		.837	.0179	"	$C_{WK} = 0.25$
NLR		.819	.0189	"	$C_{WK} = 1.0$
VUB		.816	.0181	"	$C_{WK} = 0.25$
DLR	GR	.754	.0179	"	$C_{WK} = 1.0$
ASR	CS	.647	.0194	5	coarsened grid 256x 32
CAPTEC		.740	.0175	"	priv.grid 256x 64 / 18 c
DO		.801	.0165	"	
DA	JK	.786	.0172	7	
DLR		.766	.0182	"	
DO		.764	.0157	"	
DO	JK-E	.764	.0156	"	
DLR	JC	.800	.0195	"	
BAe	SJL-W	.796	.0187	9	
BAe	C	.788	.0188	"	
UMIST	C	.786	.0171	"	
UMIST	RS	.781	.0168	"	
Experiment:		**.803**	**.0168**		

actual figure number, the reader may be guided to the appropriate cross-plots where the flow quantities are presented comparing to other partners' results. These cross-plots include the computed pressure, skin-friction, displacement and momentum thickness distributions, the latter of which, however, are restricted to the upper side of the airfoil. From the available experimental velocity profiles, six stations in the range of $x/c = .498$ to 1.0 on the upper surface have been chosen for comparing the computed profiles of partners with the measurements. Originally, these experiments had been

Table 4 Influence of Turbulence Modelling on Case 10

Mandatory mesh GRID1: 256 x 64 cells (208 on surface) / 10 c
Mandatory parameters M = 0.754 α= 2.57° Re= 6.2*10^6 x_{tr} = 0.03

partner	turb-mod.	C_L	C_D	fig.	remarks
BAe	BL	.813	.0299	4	C_{wk} =1.0
DA		.819	.0298	"	C_{wk} =1.0
DLR		.783	.0289	"	C_{wk} =1.0
NLR		.810	.0296	"	C_{wk} =1.0
DO		.831	.0297	"	C_{wk} =1.0
DLR	GR	.724	.0262	"	C_{wk} =1.0
NLR	GB	.729	.0260	"	C_{wk} =1.0
ASR	CS	.561	.0253	6	coarsened grid 256x 32
CAPTEC		.724	.0256	"	priv.grid 256x 64 / 18 c
DO		.762	.0259	"	
DA	JK	.772	.0272	8	
DLR		.761	.0275	"	
DO		.713	.0247	"	
DO	JK-E	.742	.0249	"	
DLR	JC	.796	.0298	"	
BAe	SJL-W	.767	.0280	10	
BAe	C	.742	.0277	"	
UMIST	C	.764	.0268	"	
UMIST	RS	.750	.0258	"	
Experiment:		**.743**	**.0242**		

designed as true boundary layer test cases and therefore they are non-dimensionalized by boundary layer edge values. Unfortunately, as the actual edge velocities are left unreferenced, they had to be recomputed from the experimental pressure distribution using isentropic assumptions in order to resolve for the skin-friction coefficient and velocity quantities that are normally non-dimensionalized in Navier-Stokes computations by free-stream values rather than by boundary layer edge values.

Depending on the actual turbulence model in use, Mach contour-plots can provide an insight on the resulting degree of interaction between the viscous shear layer and the inviscid near-field. Actually, the overall contour-plots turn out to look rather uniform for all models tested herein, so they are not included in this section.

5.1.4.1 Algebraic Turbulence Models

The mandatory outcomings of all partners involved are presented within four sets of cross-plots. Figs. 3 and 4 contain the results from BL, GR and GB models for both test cases 9 and 10, respectively, and in Figs. 5 and 6 the CS results are given.

Baldwin-Lomax model (BL)

This model denotes the most common turbulence model in use within Navier-Stokes computations and had been used in the validation context by 8 partners. It should be noted that the majority of the partners changed the model constant $C_{WK}=.25$ (see Sect.3) to $C_{WK}=1.0$ in their implementations. Those partners sticking to the original constant setting, are CASA, Dornier (case 9 only) and VUB. The reason for this change may be found in the fact that the model with $C_{WK}=1.0$ experiences much safer convergence in case of flow detachment. Unfortunately, none of the partners explicitly investigated on this feature within the airfoil test cases.

In general, the outcoming figures of the different implementations are of remarkable uniformity for both, case 9 and 10 results apart from a few curves differing from the majority. Especially, one cannot depict systematic differences due to the different coefficients C_{WK} used here.

Case 9: From Fig. 3, it becomes obvious that the results of most partners tend to locate the shock somewhat downstream of the experimental position. One effect may be the camber-correction applied as had been shown above. Small deviations from this feature are observed in DLR's results that are computed to a slightly lower pressure level upstream of the shock and then smoothly approaches the common computational shock location. CASA, however, computes a much better shock location in their private unstructured grid that in fact looks like a missing camber-correction in this grid. The differences in the computed shock location also become obvious from the skin-friction distributions. Despite of this, the results obtained are reasonably close to the experiments in general, although, certain scatter in lift and drag becomes obvious which often is overpredicted, see Table 3. Some partners tend to underpredict the displacement thickness to a certain degree. The velocity profiles obtained, are also in good agreement with the experimental profiles, Fig. 3. The only larger scatter is found at $x/c=0.574$ which is close to the shock which simply reflects the differences in the computed shock location.

Case 10: Here, larger discrepancies between the computed and measured shock locations arise from the BL model in general, Fig. 4, as the case 10 flow exhibits shock induced separation that induces non-equilibrium turbulence development that the BL model cannot account for. As had been found for case 9, the partners' results are remarkably close to each other again. Similar effects as in case 9 are observed: DLR computes slightly lower pressure in front of the shock and obtains a smooth approach into the shock which is also indicated by their skin-friction distribution. NLR also gets a slightly smoother shock entry than DA and Dornier. From Table 4, a common tendency to overpredict the lift coefficients as a result of the more downstream shock positions, becomes obvious. Again, there are some underpredictions of the displacement thickness upstream of the shock. The velocity profiles collected in Fig. 4 show some deviation from the experiments in the region of adverse pressure gradient ($x/c=.574$ to $.750$) as well as a scatter of results particularly within the shock region, indicating again the difficulties of the model to handle this test case.

Last but not least it becomes obvious from the velocity profiles that the normal extension of the separation region is predicted very small in general. This may give rise to a systematic underprediction of displacement and momentum thicknesses downstream of the shock. Hence, this is argued to be the main reason for the obvious deficiencies of this model.

Granville extension (GR) to the BL model

This model has been used by DLR for the airfoil test cases as well as for the channel bump flow, Sect. 5.2. DLR reports 'best overall performance' of this modification comparing to their BL, JK and JC implementations. Indeed, there are a number of obvious improvements compared to the parent BL model.

Case 9 and 10: As may be seen from Table 3 and 4, lift and drag are computed to a lower level, resulting in better agreement with the experiments at least for case 10. In both test cases, this model yields a more forward shock position than its BL parent (Fig. 3 and 4), but showing again the 'smooth' shock entry as had been observed before in DLR's BL implementation. In case 10, the resulting pressure at the trailing edge is computed to an improved degree of accuracy. Due to the improved shock position, the displacement thickness is obtained on a higher level (case 9: slightly higher level) and the profiles are computed with better accuracy in that area. For case 9, the GR model meets the experimental velocity profile at $x/c=.574$ and for case 10, an impressive accuracy is found also ($x/c=.650$). The latter, however, results in a moderate height of the separation bubble only. Downstream of the shock, the profiles in general seem to be slightly better than the majority using the BL model. From the case 10 computations, the overall performance of this model seems to be similar to the GB model which will be discussed next.

Goldberg backflow model (GB) in combination with BL

Case 10: This model combination has been used by NLR in their case 10 computations. From Table 4 it is observed that this variant results in lower lift and drag coefficients than their original BL implementation, and indeed, this model approaches significantly better to the experiment. In Fig. 4, the shock location has been improved as well as the velocity profiles in the shock region. This performance obviously is comparable to the GR model and almost comparable to the JK model class which will be discussed below. The main difference from the BL model seems to be a significant increase in the resulting height of the reverse flow bubble and increased accuracy of the trailing edge pressure level, again similar to the GR model performance.

Cebeci-Smith model (CS)

This model has been employed by 3 partners. Amongst these, however, Dornier's case 9 and 10 results are the only ones computed using the mandatory grid. Hence, the outcomings are not that uniform as had been observed for the BL model.

Case 9: Good agreement with the experimental lift and drag coefficients is obtained by DO. The pressure distribution as well as skin-friction and boundary layer thicknesses (displacement and momentum thickness) reasonably fit to the experiments, see Fig. 5, but the predicted shock location is slightly downwashed again. ASR and CAPTEC who computed in non-standard grids, both obtain a shock location slightly upstream of the experimental position coinciding with a relative underprediction of displacement and momentum thicknesses and considerably lower lift coefficients. The skin-friction distributions indicate small separation bubbles at the shock foot. Surprising, the velocity profiles of the three partners are close to each other and to the experiments as well, in general. The scatter at x/c=.574 again accounts for the different shock locations computed.

Case 10: As had been seen within the BL model solutions already, here again the shock location tends to be computed some distance downstream of the experimental position, Fig. 6 (two partners, CAPTEC and Dornier at least), resulting in overpredicted lift coefficients, although further downstream, the resulting pressure distributions deviate from the measurements in a way that compensates the overpredicted lift to some degree. In fact, these overpredictions seem to be considerably less than for the BL model. In this test case, ASR obtained a nice shock position but further downstream, there are the same systematic deviations in the pressure distribution as shown by the results of the other partners, hence yielding much lower lift coefficients along with overprediction of drag. Displacement and momentum thicknesses are underpredicted by the partners to a different degree although the separated region extends until the trailing edge. The profile collection shows some scatter in the partners' results. It identifies the height of the separation bubble varying but it remains moderately small for the CS model also.

5.1.4.2 Half-equation Turbulence Models

All mandatory results of this model class are cross-plotted in Fig. 7 and 8 but they will be discussed with respect to their different variants below.

Johnson-King model (JK)

Three partners tested this type of model but there are some basic differences in their implementations: Dornier used the original version while on the other hand, DA and DLR used different variants that are based on BL type formulations of the two eddy-viscosity regimes, i.e. utilizing vorticity and the well known F- function procedure, implemented in the non-equilibrium framework that identifies this class of models.

case 9: The results of the three partners are close to each other and there are slight but common improvements of the results when comparing to BL and CS models: the JK variants compute the shock location in clearly better agreement to the experiment, Fig. 7. However, as the case 9 flow is at least close to equilibrium, the non-equilibrium models here do not perform significantly better than an equilibrium model can do, except in the vicinity of the shock region. Due to the improved shock location, the computed profiles in that area (x/c= 0.574) now fit the experiments much better, and there is only small scatter between the plotted results of the partners which, however, increases at the trailing edge. Displacement and momentum thicknesses have been computed also to moderate scatter by the three model variants and certain differences in lift and drag computation are obvious. Concerning the skin-friction distribution, all model variants show the well known deficiencies when computing for equilibrium flow regions, resulting here in an overpredicted skin-friction downstream of the shock. The cross-plots again show a smoother shock entry computed by DLR (JK as well as JC model).

case 10: In the highly non-equilibrium flow of this test case, the models here mainly result in an improved shock location in comparison to the BL solutions, as well as improved profiles within the shock region, Fig. 8. Dornier's implementation meets the shock position remarkably and therefore results in the largest increase of displacement thickness downstream of the shock. Note, that the original JK formulation used by Dornier also results in very low skin-friction along the whole adverse pressure gradient region up to the trailing edge which is, indeed, close to separation. According to this, Dornier obtained an improved pressure level approaching the trailing edge.

In contrast to the algebraic models, all the model variants tested here, produce an increased thickness of the separation bubbles, see for instance, station $x/c=.650$ in Fig. 8, now clearly showing - although to different degrees - interactive response of the pressure distribution on the separated flow region. However, the experimental pressure distribution does not exhibit this locally restricted response as indicated by some of the computations.

Johnson-King model JK/equilibrium (JK-E)

Case 9 + 10: Dornier employed this variant. By dropping the non-equilibrium factor of the original model, this variant becomes a true algebraic equilibrium model differing from the CS model by incorporating the matching function, however. Although the shock positions computed for both, case 9 and 10, are comparable to those of the non-equilibrium JK model variants, this model shows the typical algebraic behaviour at the shock foot, i.e. no pressure response, thus leading to a lower pressure right at the shock foot, see Figs. 7 and 8. Indeed, this variant turns out to perform very similar to Dornier's CS implementation for both test cases.

Johnson-Coakley model (JC)

DLR uses this model that had been originally designed to overcome the shortness of JK concerning the skin-friction.

Case 9 + 10: Comparing the results to those of the JK variant of DLR, Fig. 7 and 8, a slightly increased downwash of the shock location is obtained but the model, indeed, improves the skin-friction to a certain degree in the non-equilibrium flow case 10 at least. This improvement, however, does not yet improve the computed pressure in the trailing edge region when comparing again to the JK model implementation of DLR. Due to the increased downwash of the shock position, lift and drag are computed considerably higher than from DLR's JK model implementation. So the overall performance of this model in the airfoil test cases is not really encouraging which, on the other hand, is in contrast to the results obtained for the channel bump flow test cases, see Sect. 5.2.

5.1.4.3 Transport and Reynolds Stress Turbulence Models (C), (SJL-W), (RS)

In total, there are four implementations of these different model classes: BAe employs the C and SJL-W model variants (with the Wolfshtein sublayer equation over the innermost 17 layers). UMIST uses the C and RS models. Here, these models will be discussed in close context to each other and therefore, the results are presented altogether in Figs. 9 and 10, respectively.

UMIST settled their transition position at $(x/c)_{tr}=0.05$ (erroneously instead of 0.03), so the skin-friction distributions in these figures show some initial deviations for both, case 9 and 10 results. The shock position has been computed to an impressive degree of coincidence by all four model implementations - but downstream of the shock, there is a large scatter in the skin-friction distributions. Moreover, the curves representing the two Chien model implementations are widely spread in this area. This observation holds for both test cases. BAe suggests erroneous behaviour in the near wall region of their model.

Despite of this, there are small differences in the pressure distributions computed by the two partners along the forward shock region. Certainly arising from the different transition positions, these differences are clearly observed between the solutions of different partners but not between the different models employed. This effect again holds for both test cases. The SJL-W model of BAe is the only one that results in a smoother pressure distribution at the shock foot (case 9 and 10 again), with c_f indicating nearly separated flow for case 9 and a small separation region for case 10. It is surprising to see the common tendency of these models not to predict a separation region of case 10 flow in general.

The displacement thicknesses differ more between the partners than between the models again, with BAe predicting slightly larger displacement than UMIST does. The same holds for the momentum thicknesses. These tendencies, however, are more obvious in case 9 than they are in case 10 results. Looking at the velocity profiles of case 9, Fig. 9, there is only small scatter between the computed and experimental profiles except at $x/c=0.574$ due to the slightly different downstream shock locations. In case 10, however, some scatter arises between the solutions due to larger differences in the computed shock positions, Fig. 10. The resulting pressure distributions in case 10, here in general, look similar to those typically obtained by algebraic models.

The computed lift and drag coefficients are close to the experimental values in general and there is only small scatter observed, Table 3 and 4.

Due to the fact that the results of these more complex models are very similar to each other it might be argued that the main reasons for the obviously inadequate deficiencies are to be found in the streamwise stepsize distribution of the mandatory mesh that do not, in general, allow a satisfactory resolution of the lambda structure at the shock foot and hence, common advantages expected from these models are suppressed here. When computing for the bump flow test cases (see Sect. 5.2), where much higher streamwise density of the grid had been introduced within the shock region, partially different assessment of the performance of the various models is obtained. In those test cases, the RS model seems to be superiour over the transport models and the algebraic formulations.

Unfortunately, due to the organisational taskwise engagement of partners within the airfoil test case applications discussed herein, there are no specific investigations available on this particular question by using flow adapted grids of increased streamwise density in the shock region with one of the more complex transport or stress models employed.

5.1.5 Effects of Mesh: Density, Extension and Far-field Condition

The results of all partners dealing with these mesh effects are summarized in Tables 5 and 6 with the special effect under investigation marked.

5.1.5.1 Mesh Density Effects

Increasing global density (streamwise and normal simultaneously)

ASR and Dornier investigated on this aspect, both based on the CS model but in different ways with respect to the overall mesh density changes. Dornier reports on a mesh dependence study performed by their multigrid code within the mandatory grid. This study shows a straightforward convergence of both, lift and drag coefficients

Table 5 Mesh Density and Extension
Case 9: M = 0.734 α = 2.54° Re = 6.5·10⁶ x_{tr} = 0.03

partner	mesh	along airfoil	y^+_{wall}	mesh extens.	turb.-model	C_L	C_D	effect
ASR	144x 48	94	O(2)	18 c	CS	.712	.0194	global dens.
	288x 64	192	O(1)	18 c		.695	.0169	
BAe	256x 64	208	O(1)	10 c	BL	.819	.0192	extension
	272x 72	"	"	30 c		.820	.0194	
	272x 96	"	"	30 c		.838	.0186	normal dens.
CAPTEC	192x 64	128	O(1)	18 c	CS	.756	.0179	circumf.dens.
	256x 64	192	O(1)	"		.740	.0175	
	256x128	"	O(.5)	"		.731	.0161	normal dens.
DO	64x 16	52	O(4)	10 c	CS	.665	.0320	global dens.
	128x 32	104	O(2)	"		.774	.0203	
	256x 64	208	O(1)	"		.805	.0170	global dens.
Experiment:						.803	.0168	

Table 6 Mesh Density and Extension
Case 10: M = 0.754 α = 2.57° Re = 6.2·10⁶ x_{tr} = 0.03

partner	mesh	along airfoil	y^+_{wall}	mesh extens.	turb.-model	C_L	C_D	effect
BAe	256x 64	208	O(1)	10 c	BL	.813	.0299	extension
	272x 96	"	O(1)	30 c		.827	.0298	
CAPTEC	192x 64	128	O(1)	18 c	CS	.735	.0264	circumf.dens.
	256x 64	192	O(1)	"		.724	.0256	
	256x128	"	O(.5)	"		.733	.0256	normal dens.
DA	320x 64	208	O(1)	10 c	JK	.774	.0263	normal dens.
	320x 64	"	O(.5)	5 c		.778	.0261	
Experiment:						.743	.0242	

towards the experimental values along with the mesh refinements, Table 5.

ASR computed results in two private 18-chord grids of different densities. They get decreasing drag at higher density as Dornier does and decreasing lift also while Dornier's study shows an increase of lift with the mesh-refinements. It may be argued that this contradiction might be due to the different ways the partners refined their meshes.

Increasing streamwise density

CAPTEC is the only partner isolating this particular aspect by investigating for both, case 9 and 10 with the CS model applied. Their results show decreasing lift and drag coefficients in the refined grid variant. As may have been expected, they obtain a slight steepening of gradients in the shock region but without affecting the shock location itself.

Increasing normal density

There are a few investigations for both, case 9 and 10.

Case 9: BAe obtaines increasing lift (BL model) but CAPTEC's lift coefficient (CS model) decreases in the finer mesh. Both partners obtain a slight decrease in drag. Fig. 11 shows the resulting pressure and skin-friction of BAe in their grids of different normal density distribution but maintaining the same extension of 30 chords. Their grid of 272 by 72 cells (GRID4) in fact is the mandatory grid within the inner 10 chords area but with outer cells added to extend the far-field boundary to 30 chords. GRID5, containing 272 by 96 cells is a really different mesh of 30 chords extension with increased normal density in the inviscid near-field region (see the close-up views in Fig. 11). Within the latter, slight improvements of the computed pressure level in the forward shock region are obtained (case 9, BL model) and small changes in skin-friction aft of the shock as well, see Fig. 11.

Case 10: CAPTEC and DA obtained the same tendencies in their private meshes: increasing c_L and nearly unaffected c_D, but they used different turbulence models (CS and JK) and different refinement techniques. While CAPTEC doubled their normal cell numbers, DA modified the far-field location without a change in the cell numbers, coming down to 5 chords mesh extension.

Table 7 Far-field Condition and Mesh Extension
Case 6 : M = 0.729 α = 2.31°   Re = 6.5·10^6 x_{tr} = 0.03

partner	mesh	along airfoil	y^+_{wall}	mesh extens.	turb.-model	C_L	C_D	remarks
NLR	240x 72	192	O(1)	10 c	BL	.759	.0129	with
	248x 76	192	O(1)	20 c		.758	.0128	vortex
	256x 80	192	O(1)	40 c		.757	.0127	corr.
				infinity		.757	.0127	
	240x 72	192	O(1)	10 c	BL	.706	.0148	no
	248x 76	192	O(1)	20 c		.731	.0138	vortex
	256x 80	192	O(1)	40 c		.743	.0133	corr.
				infinity		.755	.0127	
Experiment:						.743	.0127	

5.1.5.2 Mesh Extension Effects

BAe (case 9,10): They found increasing lift (case 10 more than case 9) but nearly unchanged drag with increasing extension. Due to the far-field vortex-correction in effect, the total mesh extension seems to be a minor effect.

NLR performed a complete investigation based on a case 6 parameter set (BL model applied, see Table 7). When increasing the mesh extension up to 40 chords, they get a nice convergence behaviour in decreasing lift and drag coefficients in case of vortex-corrected far-field conditions. On the other hand, when the vortex-correction is dropped, they start from a lower lift and higher drag level within 10 chords extension and consequently, converge by increasing lift and decreasing drag coefficients along with the extending computational domain. Extrapolating these series to infinite mesh extensions, NLR approaches the experimental drag coefficient impressively unless the lift coefficient remains slightly overpredicted.

Table 8   Influence of Far-field Condition
Case 9:   M = 0.734 α = 2.54°   Re = $6.5 \cdot 10^6$ x_{tr} = 0.03

partner	vortex correct.	mesh	y^+_{wall}	mesh extens.	turb.-model	C_L	C_D	remarks
BAe	no	256x 64	O(1)	10 c	BL	.768	.0204	mand. grid
	yes					.819	.0192	
	incompr.					.845	.0186	
	no	272x 96	O(1)	30 c	BL	.819	.0188	
	yes					.838	.0186	
VUB	no	256x 64	O(1)	10 c	BL	.770	.0198	mand. grid
	yes					.816	.0181	
Experiment:						.803	.0168	

5.1.5.3 Effects of Far-field Boundary Condition

case 6: From NLR's investigation (BL model) as was discussed above: Computing without vortex-correction, lower lift and higher drag are obtained comparing to the results including the vortex-correction procedure. This effect decays when increasing the mesh extension. However, NLR's extrapolation to 'infinite mesh extension' still yields a small effect, see Table 7 again.

case 9: Here, BAe and VUB investigated (BL model). Switching off the vortex-correction decreases lift but increases drag. This effect also decays with increasing mesh extension (BAe, 30c grid), see Table 8 and Fig. 12. The figure shows BAe's results with/without vortex-correction for the mandatory 10 chord grid as well as for their 30 chord. The significant influence from 10 chords extension is obvious but again, there is only a small effect in the 30-chord grid.

Table 9 Influence of Mach Number

Case 9: M = 0.734 α = 2.54° Re = 6.5·10⁶ x_{tr} = 0.03

partner	Mach	alpha	mesh	extens.	turb.-model	C_L	C_D	remarks
ASR	.730	3.19	192x 64	18 c	CS	.827	.0227	
	.735					.815	.0235	
	.740					.820	.0238	
	.745					.824	.0227	
BAe	.724	2.54	256x 64	10 c	BL	.816	.0144	mand. grid
	.734					.821	.0187	
	.744					.822	.0240	
	.754					.813	.0295	
CASA	.724	2.54	16923 pts	13 c	BL	.741	.0141	unstructured
	.734					.756	.0167	
	.744					.761	.0215	
DA	.724	2.54	256x 64	10 c	JK	.769	.0133	mand. grid
	.734					.786	.0172	
	.744					.787	.0220	
	.754					.769	.0268	
	.764					.735	.0309	
NLR	.726	2.54	320x 80	15 c	BL	.800	.0153	
	.730					.804	.0168	
	.734					.808	.0177	
VUB	.730	2.54	256x 64	10 c	BL	.810	.0163	mand. grid
	.734					.816	.0181	
Experiment:						.803	.0168	

5.1.6 Effect of Mach Number

A number of partners investigated on the influence of changing Mach numbers on the basis of the case 9 parameter set (standard setting: M=.734). These investigations are mainly done for the BL model except ASR used the CS model and DA the JK model, respectively, see Table 9. Some partners checked for the basic tendencies only by slightly changing the Mach number (NLR, VUB) and some partners extended the range of their investigations up to approximately M=.745 (ASR, CASA). In addition, there are two partners (BAe, DA) who included the Mach number regime of M≈.757, which is the nominal Mach number of test case 10.

From most of the results collected herein, a common main tendency is observed: In the lower Mach number regime, lift and drag are growing with increasing Mach number up to approximately M= .744. In this regime, all partners find the shock position moving downstream with increasing Mach number and the pressure level upstream of the shock being changed in the same way.

Table 10 Influence of Mach Number
Case 10: M = 0.754   α = 2.57°   Re = 6.2·10⁶ x_{tr} = 0.03

partner	Mach	alpha	mesh	extens.	turb.-model	C_L	C_D	remarks
BAe	.734	2.57	256x 64	10 c	BL	.826	.0191	mand.grid
	.744					.825	.0243	
	.754					.813	.0299	
	.764					.802	.0356	
DA	.724	2.57	256x 64	10 c	JK	.774	.0135	mand.grid
	.734					.790	.0175	
	.744					.790	.0224	
	.754					.772	.0272	
	.764					.738	.0314	
Experiment:						.743	.0242	

When further increasing the Mach number beyond M= .744, Fig. 13, the tendencies become different: now the lift comes down but drag is still growing. The results of CASA, computed in their unstructured grid up to M=.744 (BL model also), again show a more forward shock position than the BAe results leading to lower lift coefficients.

Along with the Mach number rising beyond M.744, a shock induced separation region of increasing extension has been found by both partners (BAe, DA). Note that the Mach number now reaches the value of case 10 (M=0.754). The JK model series of DA converges to fixed shock and separation point locations, while the BL model of BAe shows a more continuous downwash of the shock postion and separation point as well. BAe and DA computed the same variation based on the case 10 parameter set (Table 10, Fig. 14), each of them resulting in nearly the same figures as had been found for case 9 parameters but slightly increased lift and drag. From this it may be concluded that neither the small changes of incidence nor the 5% change of the Reynolds number really affect the flow. Hence, case 9 and 10 are identified to be just different members of the more general class of flow around the RAE-2822 airfoil - but the different turbulence models give different answers on the development of the flow in parameter space. Finally, in Fig. 15, lift and drag coefficients of all partners that have investigated on Mach number effects are presented, regardless whether they are computed for case 9 or 10 parameter set. As can be seen, the lift coefficients resulting from the different BL implementations of the partners are close together in general. ASR's results (CS model) show some irregularities the reason of which has not become clear. The lift predictions by the JK model (DA) are somewhat less than those of the BL model, mainly due to the different shock positions. CASA also predicts lower lift due to their forward shock position. Note the case 9/10 comparisons by BAe and DA that have the same tendencies although they are on different c_L levels: both show the maximum lift at approximately M=0.74 with only small differences between case 9 and 10. As far as drag is concerned, the predictions of all partners are close together in this Mach number variation.

Table 11 Influence of Numerical Incidence
Case 9:   M = 0.734   α = 2.54°   Re = 6.5·10⁶ x_{tr} = 0.03

partner	alpha	mesh	extens.	turb.-model	C_L	C_D	remarks
ASR	2.54	192x 64	18 c	CS	.707	.0176	
	2.79				.802	.0187	
	2.96				.762	.0164	M = .730
BAe	2.54	272x 96	30 c	BL	.838	.0186	
	2.79				.875	.0215	
CASA	2.54	16923 pts	13 c	BL	.756	.0167	
	2.79				.798	.0198	
DA	2.54	256x 64	10 c	JK	.786	.0172	mand.grid
	2.74				.820	.0193	
	2.94				.851	.0218	
	3.14				.880	.0247	
NLR	2.41	256x 64	10 c	BL	.797	.0177	mand. grid
	2.54				.819	.0189	
VUB	2.54	256x 64	10 c	BL	.816	.0181	mand. grid
	2.74				.845	.0199	
Experiment:					.803	.0168	

5.1.7 Effect of Numerical Incidence

There are also various investigations on this effect on both, case 9 and 10. Some partners tested systematically up to the experimental value of 3.14 degrees while others again just tested for small deviations from the mandatory values to check for the effects. Most of the partners employed the BL model again. The only exceptions from this are ASR (CS model employed), NLR (using the BL-BG model combination for case 10) and DA (JK model). Further details may be found in Tables 11 and 12.

The main results here are as may have been expected: lift and drag are growing with increasing angle of attack. All equilibrium models result in a fixed shock position, unaffected by the variation of incidence in both test cases, 9 and 10. The only deviation from this is observed in DA's non-equilibrium (JK) investigation: they also find the shock to be unaffected in the attached flow regime (case 9, Fig.16) but a slightly forward moving shock in the separated flow regime of case 10 when increasing the angle of attack. Along with the latter, considerable increase of the separated region can be observed in Fig. 17. Interesting to see the tendencies of this JK investigation: the increasing streamwise extension of the separation region brings the trailing edge

Table 12   Influence of Numerical Incidence
Case 10:   M = 0.754   α = 2.57°   Re = 6.2·10⁶ x_{tr} = 0.03

partner	alpha	mesh	extens.	turb.–model	C_L	C_D	remarks
ASR	2.57	192x 64	18 c	CS	.704	.0221	
	2.81				.769	.0274	
	2.96				.776	.0273	
	3.19				.730	.0270	M = .750
	3.25				.743	.0271	M = .750
BAe	2.57	272x 96	30 c	BL	.827	.0298	
	2.82				.857	.0336	
	3.07				.888	.0374	
DA	2.57	256x 64	10 c	JK	.772	.0272	mand. grid
	2.76				.794	.0295	
	2.95				.812	.0317	
	3.14				.828	.0342	
NLR	2.12	256x 64	10 c	BL	.738	.0237	mand. grid
	2.57				.810	.0296	
	2.57	256x 64	10 c	GB	.729	.0260	mand. grid
	2.63				.738	.0267	
Experiment:					.743	.0242	

pressure closer to the experiments. This, in fact is the same observation as had been discussed in Sect. 5.1.4.2.

Finally, in Fig. 18 and 19 the partners' lift and drag coefficients versus angle of attack are presented for case 9 and 10 computations. Again, similar tendencies are obvious as had been found from the Mach number variations. The BL model predicts higher lift than the JK model does in general, and CASA again slightly underpredict due to their forward shock position. The CS implementation of ASR again exhibits some erraneous behaviour. It is remarkable that nearly all partners found the same gradient in lift increase. The drag coefficients of all partners are close together for case 9 showing also very similar gradients of drag increase, but in case 10, considerable scatter in the drag predictions becomes obvious. Here, the GB-BL model of NLR results in lower lift and drag predictions than NLR's original BL implementation. Unfortunately, this investigation had been restricted to small changes of incidence only.

5.1.8 Summary / Conclusions

Multiple investigations have been performed on the main influences that may affect Navier-Stokes solutions such as turbulence modelling, grid effects and wind tunnel correction parameters. It has been shown that the quality of the results obtained indeed may strongly depend on these influences.

In general, the functionality of the Navier-Stokes codes and turbulence models used in this chapter becomes evident. However, as there remains some significant scatter in the results presented - possibly arising already from the various methods - general conclusions with respect to a rise of quality of future results are difficult.

Nevertheless, concerning the mesh influences and the performance of turbulence models, some marginal tendencies became visible:

The mandatory grid that has been used in most of the tests, still seems to affect the solutions to a certain degree in general. The main disadvantage of this grid may be seen in the absence of adequately condensed streamwise stepsizes within the shock region, preventing by this a more satisfactory resolution of the lambda structure at the shock foot. Hence, some partners achieved better agreement between their computations and the experiments just by using different grids of higher density.

The turbulence models employed, clearly have the most significant impact on the solutions. There obviously are drawbacks in all models and their performances sometimes turned out to be different in different flow situations. Modifications of the most common Baldwin-Lomax model, such as the Granville extension or the Goldberg backflow model, have been shown to perform to a significantly improved level of accuracy in the separated flow regime. At the present state the 'half equation' non-equilibrium models of the Johnson-King class turn out to be very promising, although the well known deficiencies that cause an unsatisfactory skin-friction representation still prevent a 'break through' of this model type. The Johnson-Coakley variant, originally designed to overcome this drawback, had been tested also, but, for the test cases investigated here, this model did not really succeed. However, a more definite statement still suffers from the lack of a larger number of carefully computed series of results with really just a single parameter varying. Such series like those presented on incidence and Mach number variations by some partners, employing different turbulence models are shown to give more insight in the overall performance of a model than a single test case, i.e. a single parameter set can do.

Due to the non-condensed mandatory grid that most of the computations have been performed within, the more complex turbulence models such as transport and stress models, here, did not succeed to a really superiour degree over the algebraic models, which is obviously caused by the impossibility to resolve for the shock lambda structure.

Last but not least, the results computed here, clearly depend on the way the wind tunnel corrections are introduced. This crucial question, however, is not really a numerical effect acting on the solutions, and so there is only little that can be done as long as the applications aim for free flight conditions rather than to account for the wind tunnel environment - which, however, is possible (see for instance Sect. 5.8) but is connected with considerable increase of computational effort.

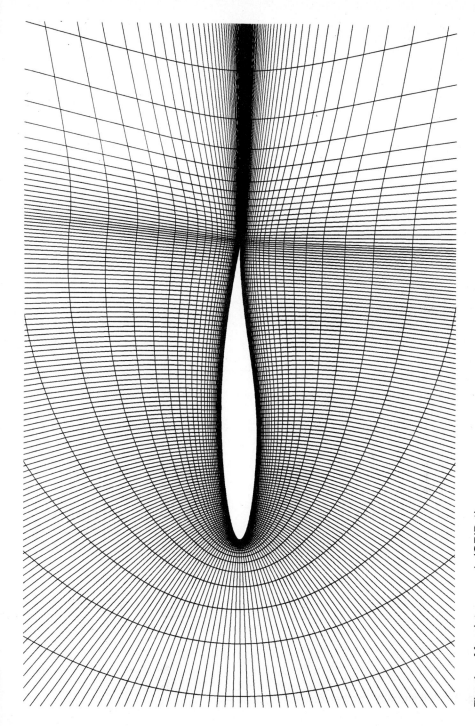

Fig. 1 Mandatory mesh (GRID 1)

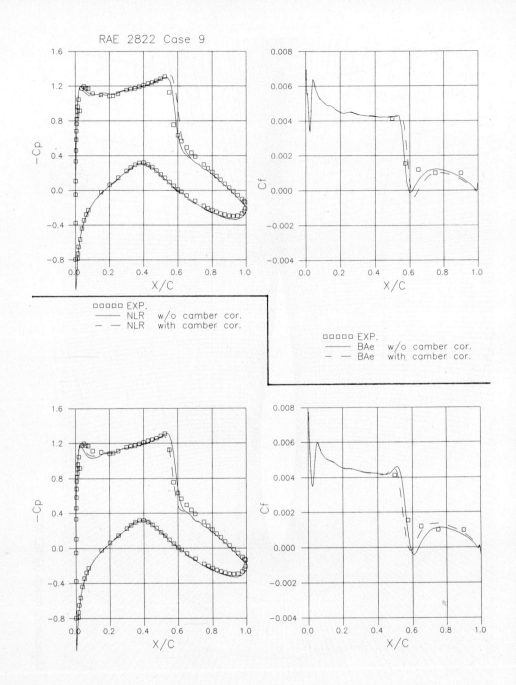

Fig. 2 Influence of camber-correction, case 9.

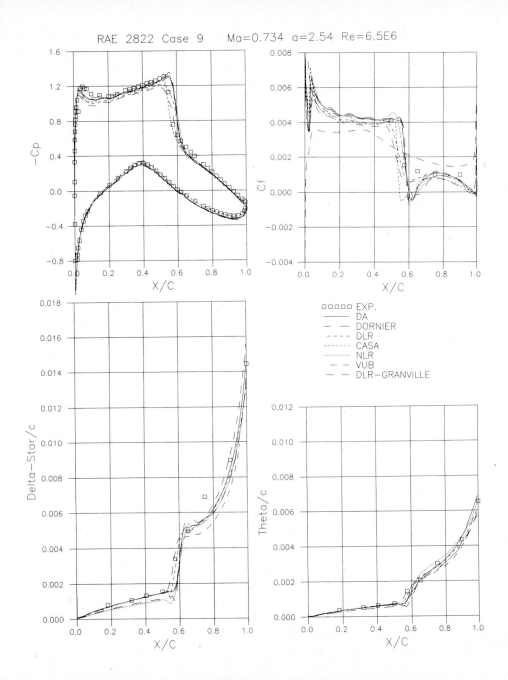

Fig. 3 Influence of turbulence models: Baldwin-Lomax and variants, case 9. Pressure distributions and upper side boundary layer parameters

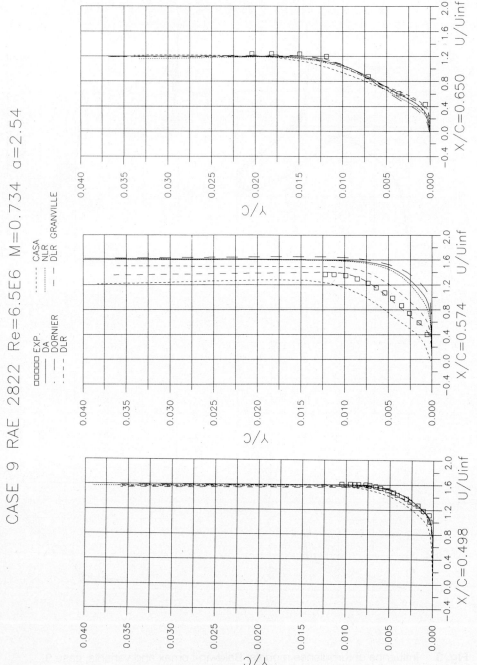

Fig. 3 Influence of turbulence models: Baldwin-Lomax and variants, case 9.
(continued) Velocity profiles on upper side of airfoil

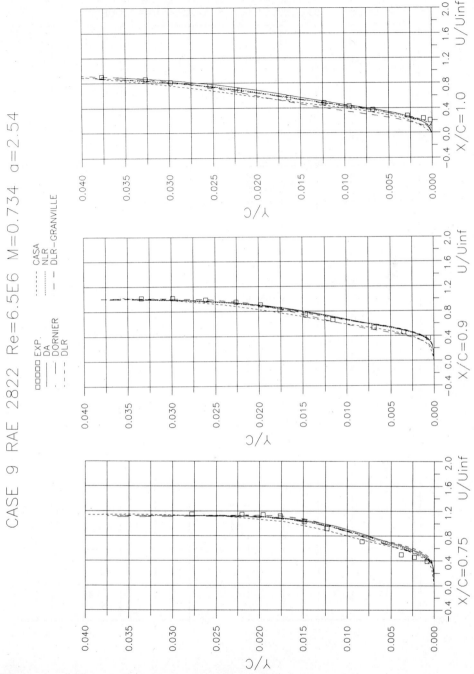

Fig. 3 Influence of turbulence models: Baldwin-Lomax and variants, case 9.
(continued) Velocity profiles on upper side of airfoil

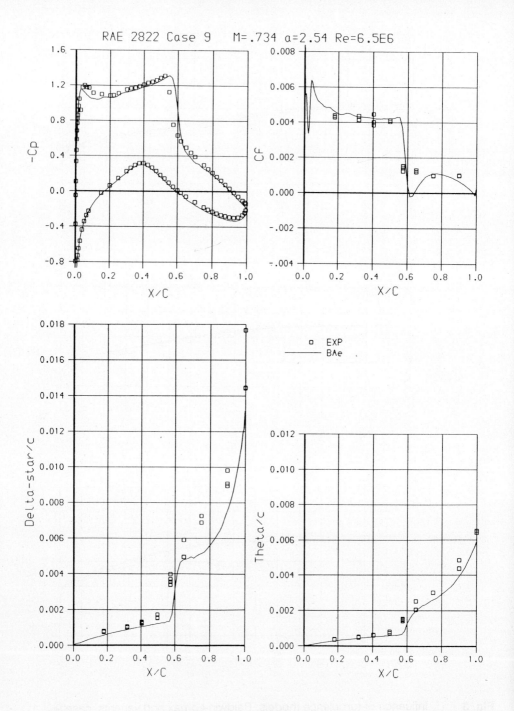

Fig.3 Influence of turbulence models: Baldwin-Lomax and variants, case 9.
(continued) Pressure distributions and upper side boundary layer parameters.

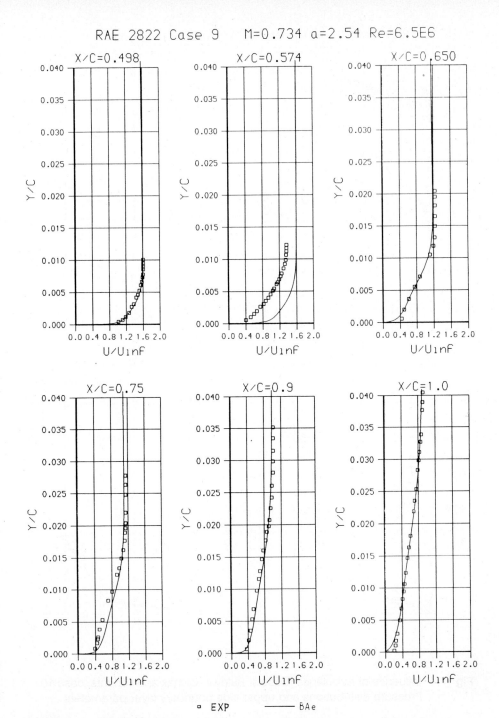

Fig. 3 (concluded) Influence of turbulence models: Baldwin-Lomax and variants, case 9. Velocity profiles on upper side of airfoil.

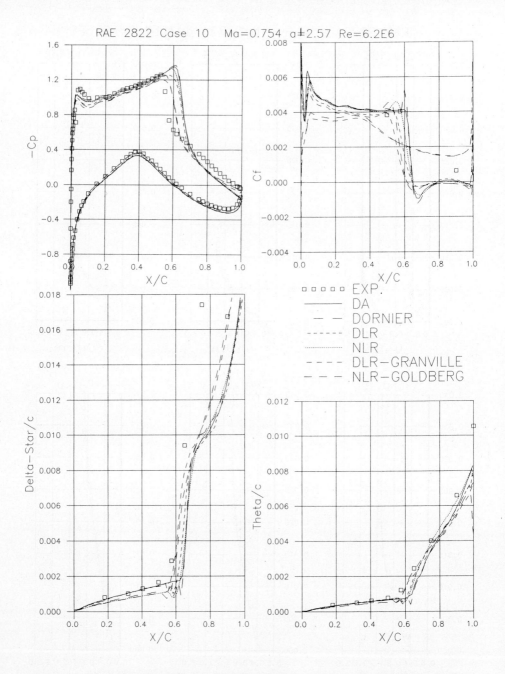

Fig. 4 Influence of turbulence models: Baldwin-Lomax and variants, case 10. Pressure distributions and upper side boundary layer parameters

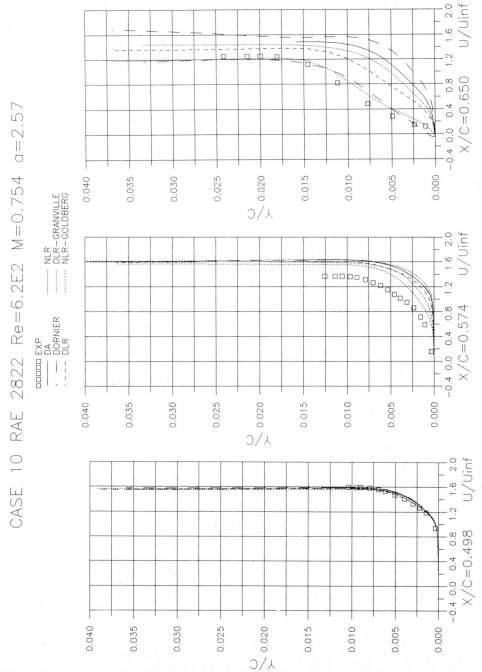

Fig. 4 Influence of turbulence models: Baldwin-Lomax and variants, case 10.
(continued) Velocity profiles on upper side of airfoil

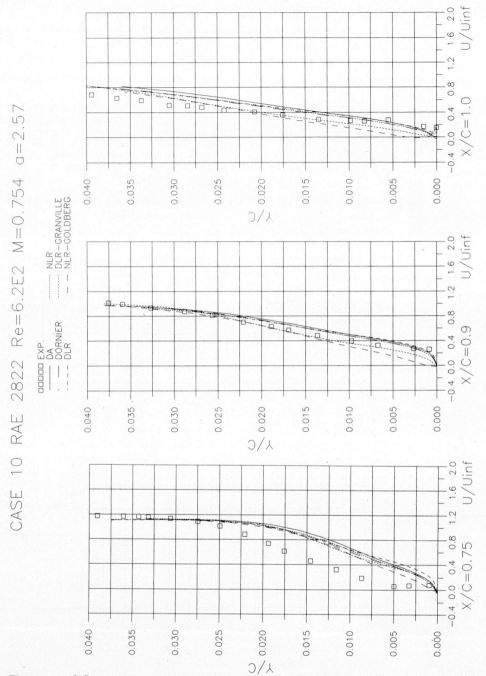

Fig. 4 Influence of turbulence models: Baldwin-Lomax and variants, case 10.
(continued) Velocity profiles on upper side of airfoil

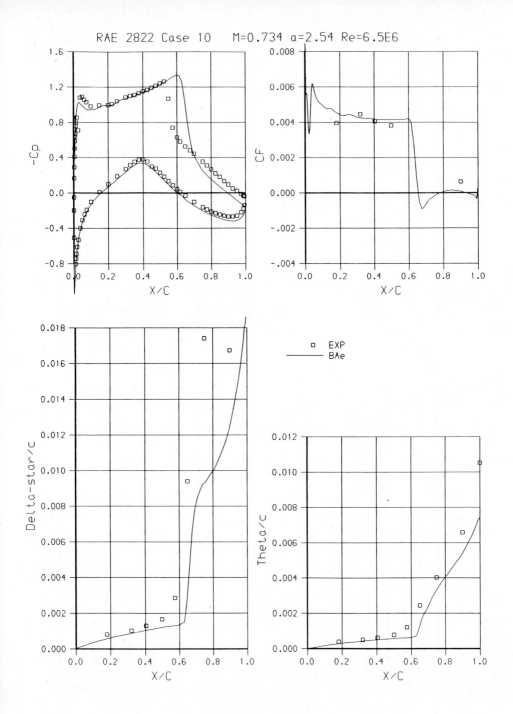

Fig. 4 Influence of turbulence models: Baldwin-Lomax and variants, case 10.
(continued) Pressure distributions and upper side boundary layer parameters.

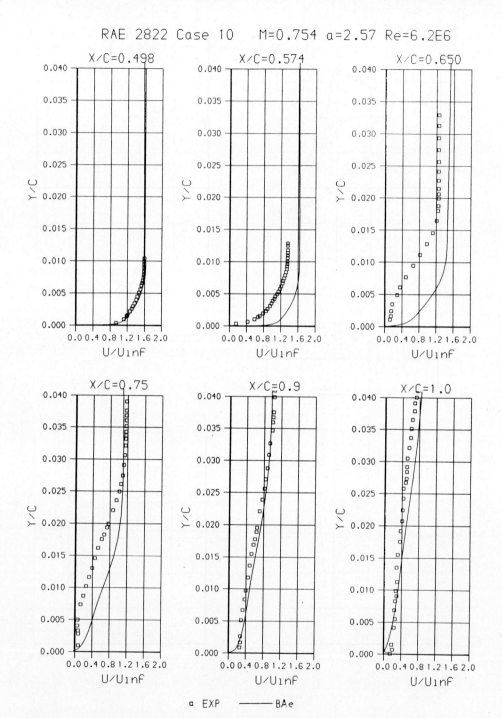

Fig. 4 Influence of turbulence models: Baldwin-Lomax and variants, case 10.
(concluded) Velocity profiles on upper side of airfoil.

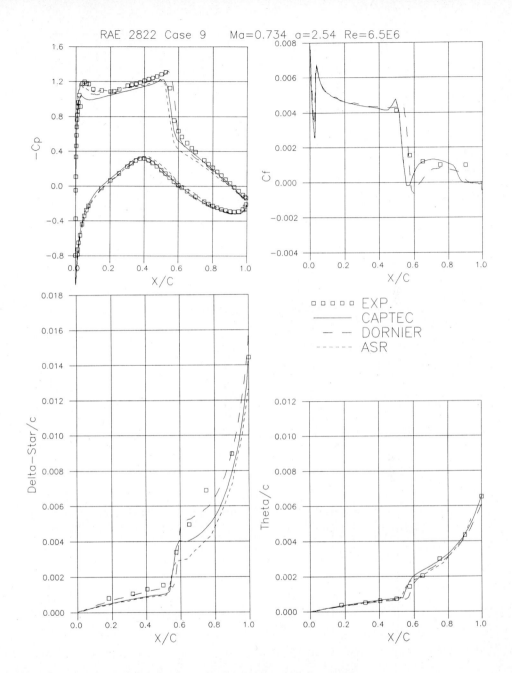

Fig. 5 Influence of turbulence models: Cebeci-Smith, case 9. Pressure distributions and upper side boundary layer parameters

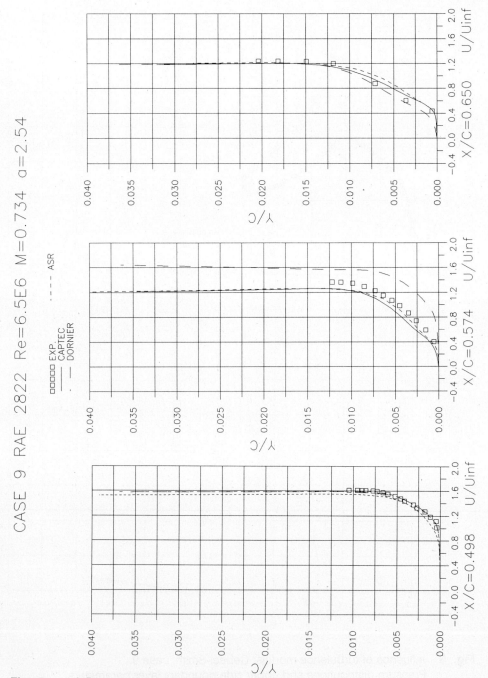

Fig. 5 Influence of turbulence models: Cebeci-Smith, case 9.
(continued) Velocity profiles on upper side of airfoil

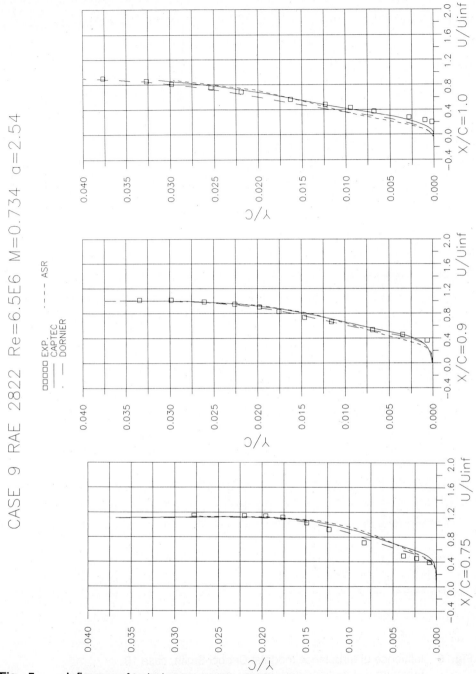

Fig. 5 Influence of turbulence models: Cebeci-Smith, case 9.
(concluded) Velocity profiles on upper side of airfoil

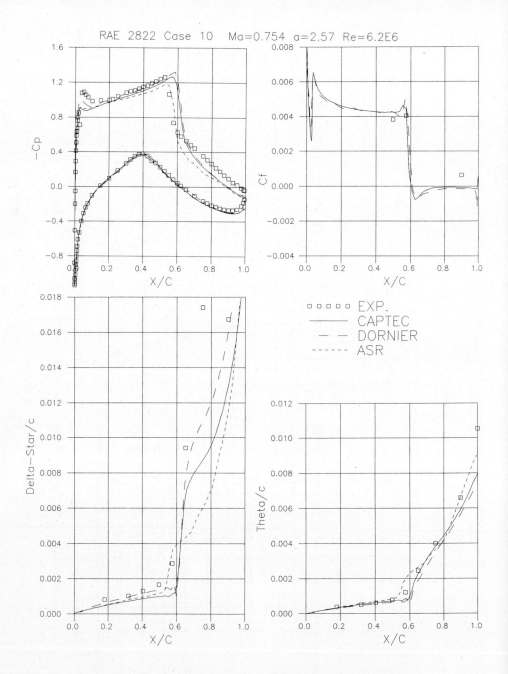

Fig. 6 Influence of turbulence models: Cebeci-Smith, case 10.
Pressure distributions and upper side boundary layer parameters

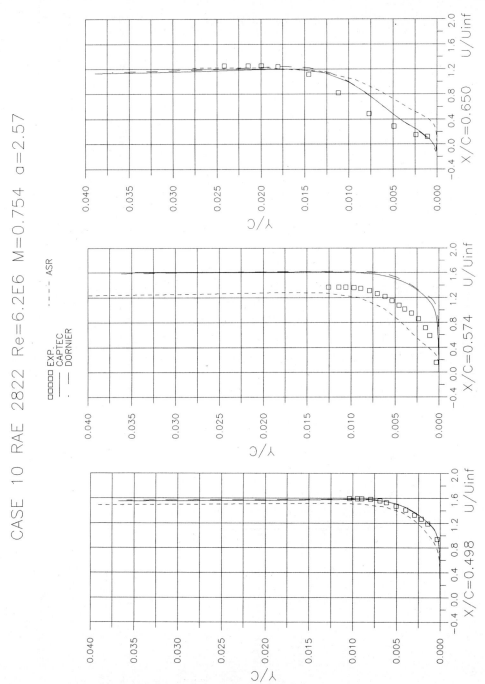

Fig. 6 Influence of turbulence models: Cebeci-Smith, case 10.
(continued) Velocity profiles on upper side of airfoil

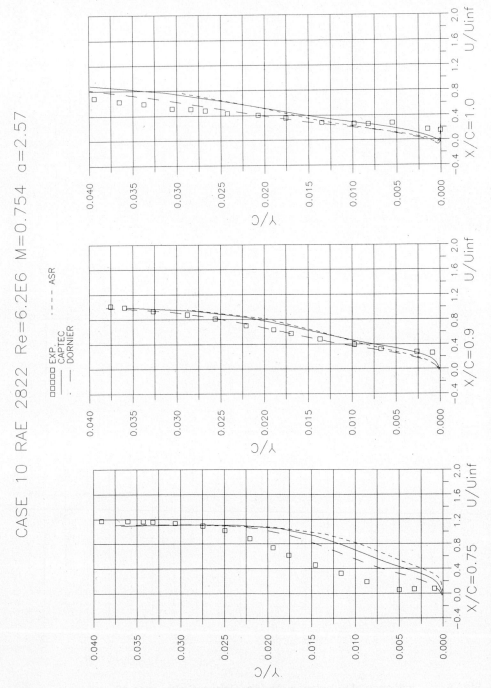

Fig. 6 Influence of turbulence models: Cebeci-Smith, case 10.
(concluded) Velocity profiles on upper side of airfoil

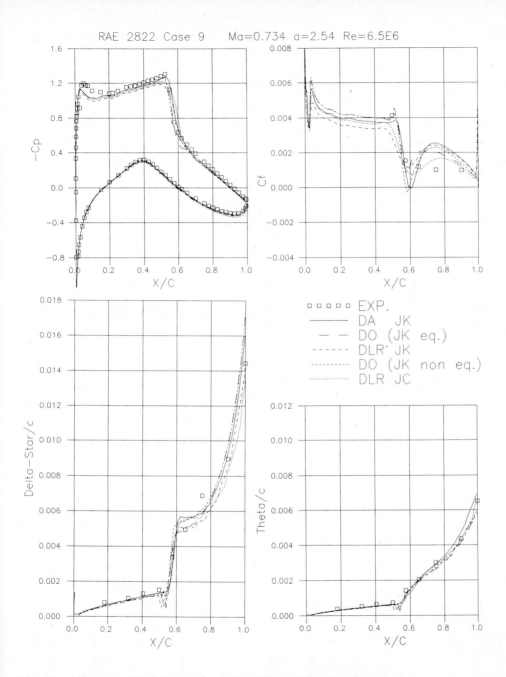

Fig. 7 Influence of turbulence models: half-equation models, case 9. Pressure distributions and upper side boundary layer parameters

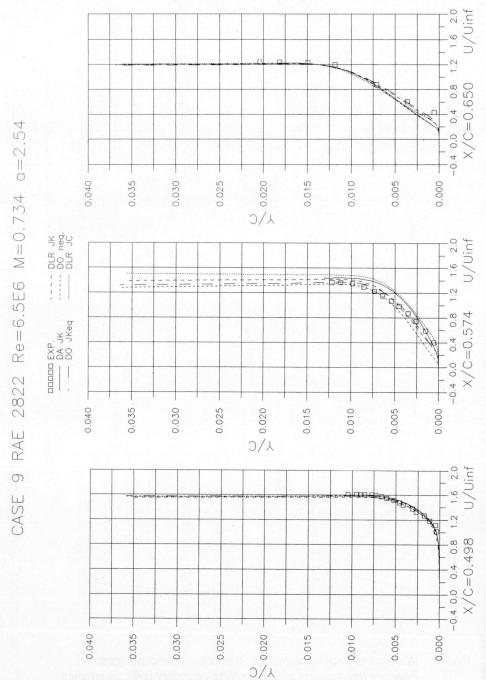

Fig. 7 Influence of turbulence models: half-equation models, case 9.
(continued) Velocity profiles on upper side of airfoil

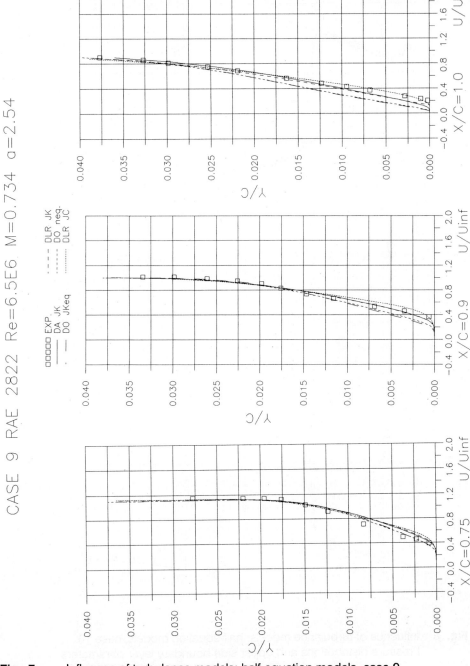

Fig. 7 Influence of turbulence models: half-equation models, case 9.
(concluded) Velocity profiles on upper side of airfoil

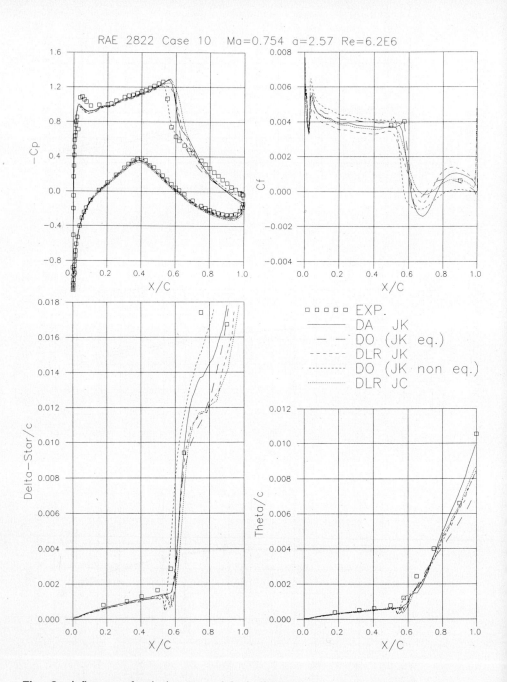

Fig. 8 Influence of turbulence models: half-equation models, case 10. Pressure distributions and upper side boundary layer parameters

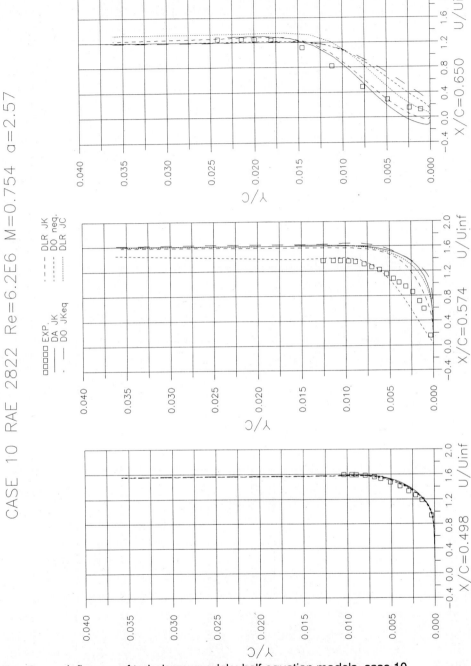

Fig. 8 (continued) Influence of turbulence models: half-equation models, case 10. Velocity profiles on upper side of airfoil

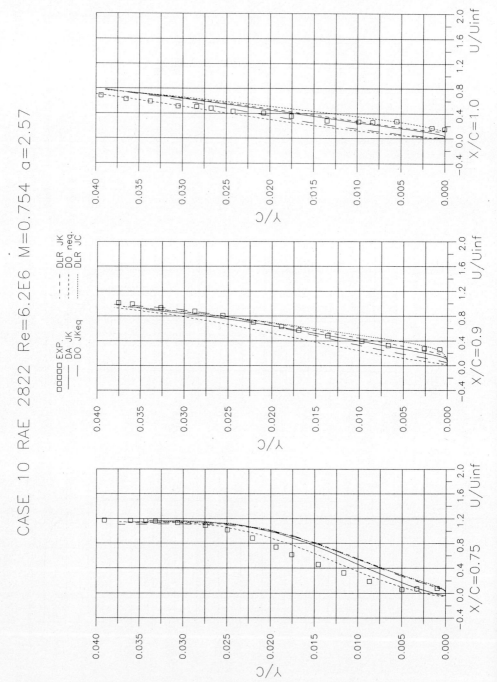

Fig. 8 Influence of turbulence models: half-equation models, case 10.
(concluded) Velocity profiles on upper side of airfoil

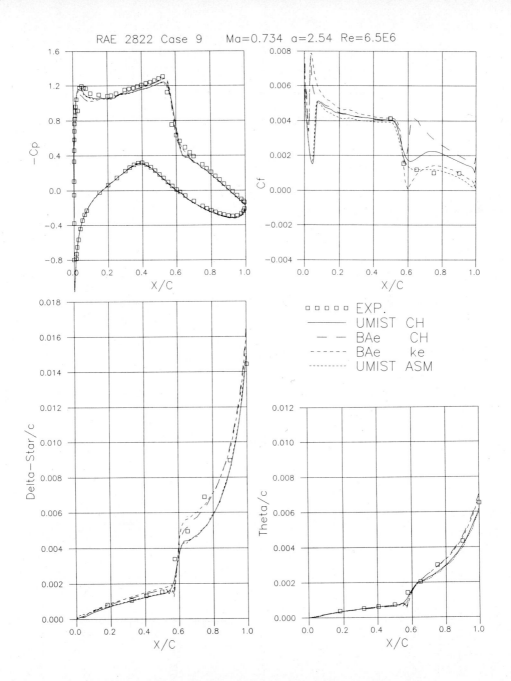

Fig. 9 Influence of turbulence models: transport and stress models, case 9. Pressure distributions and upper side boundary layer parameters

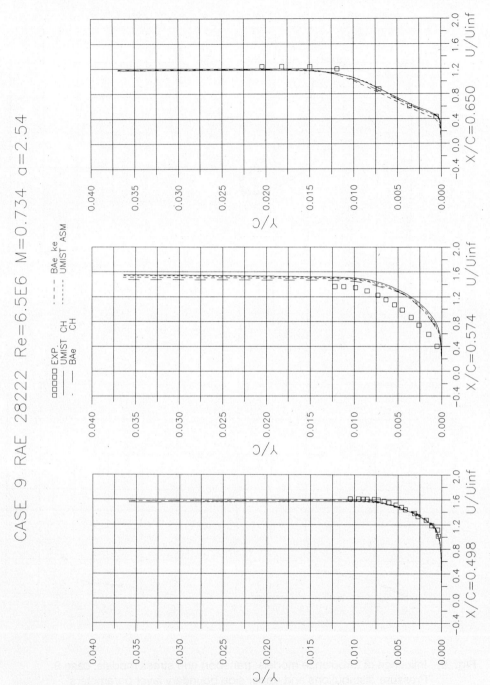

Fig. 9 Influence of turbulence models: transport and stress models, case 9.
(continued) Velocity profiles on upper side of airfoil

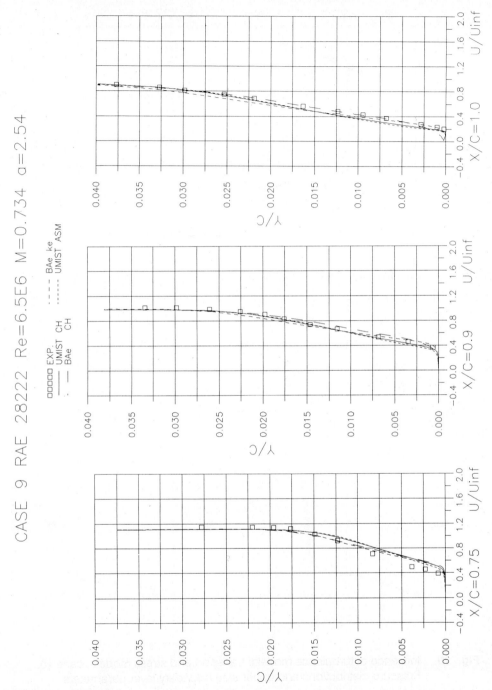

Fig. 9 Influence of turbulence models: transport and stress models, case 9.
(concluded) Velocity profiles on upper side of airfoil

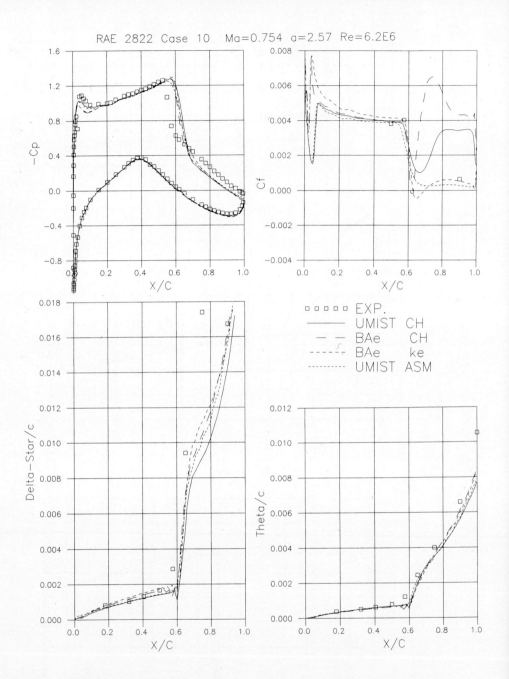

Fig. 10 Influence of turbulence models: transport and stress models, case 10. Pressure distributions and upper side boundary layer parameters

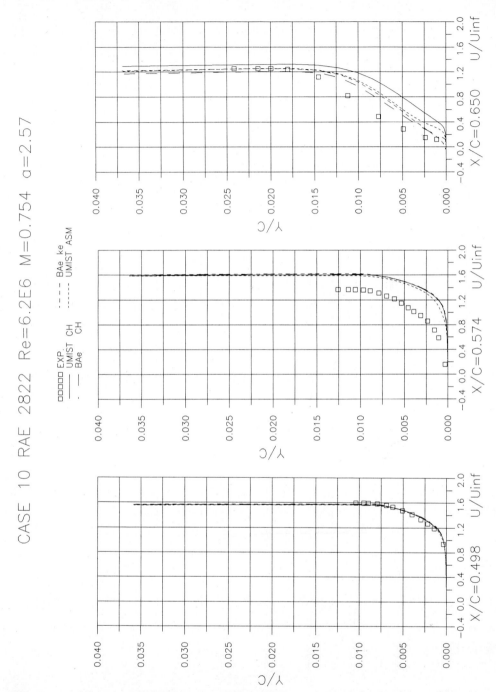

Fig. 10 Influence of turbulence models: transport and stress models, case 10.
(continued) Velocity profiles on upper side of airfoil

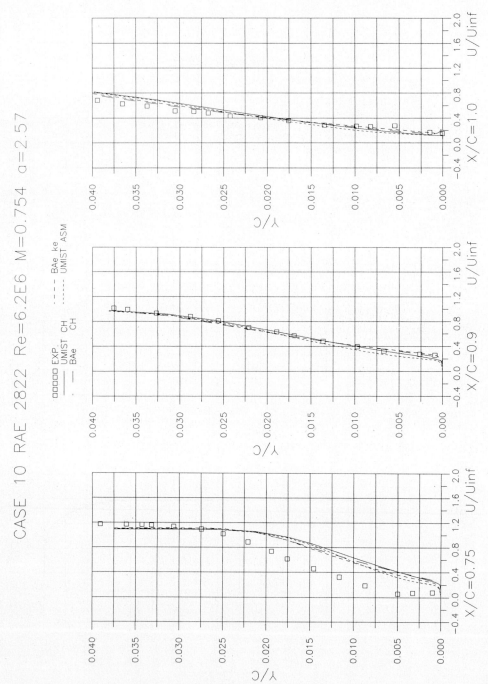

Fig. 10 Influence of turbulence models: transport and stress models, case 10.
(concluded) Velocity profiles on upper side of airfoil

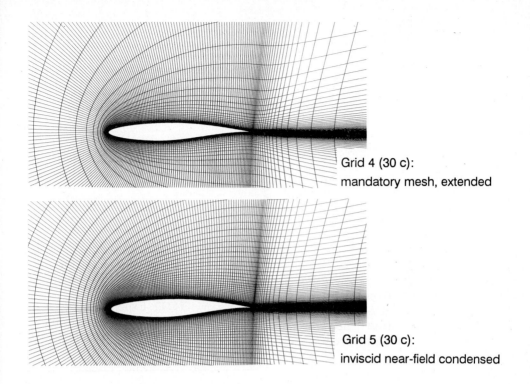

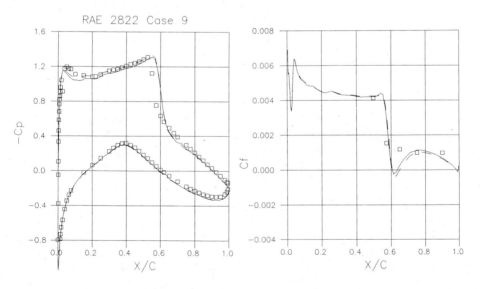

Fig. 11 Effect of normal grid density

□□□□□ EXP.
– – BAe grid 5
——— BAe grid 4

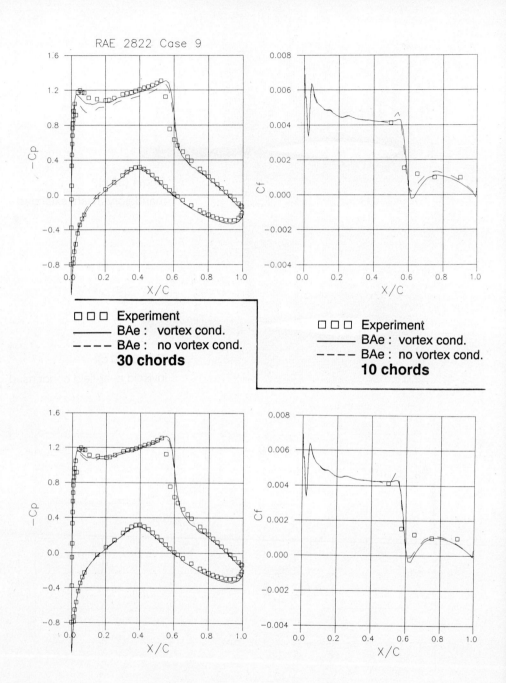

Fig. 12 Effect of far-field condition

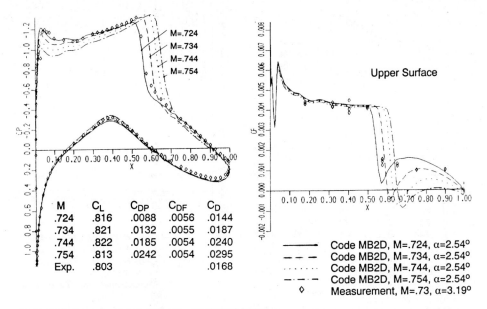
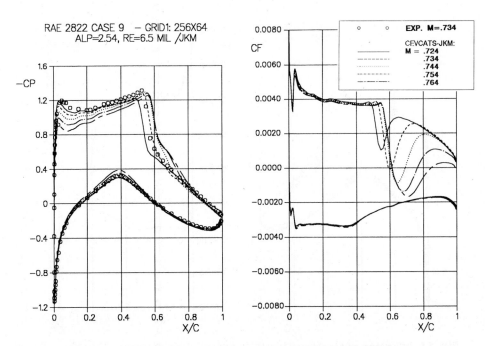

Fig. 13 Influence of Mach number based on case 9 parameters
BAe results for BL model (top) and DA results for JK model (bottom)

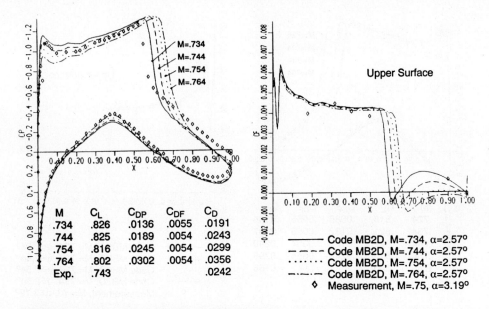

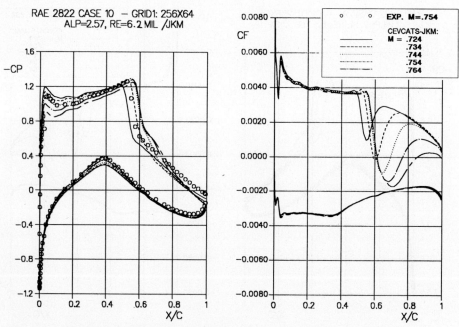

Fig. 14 Influence of Mach number based on case 10 parameters
BAe results for BL model (top) and DA results for JK model (bottom)

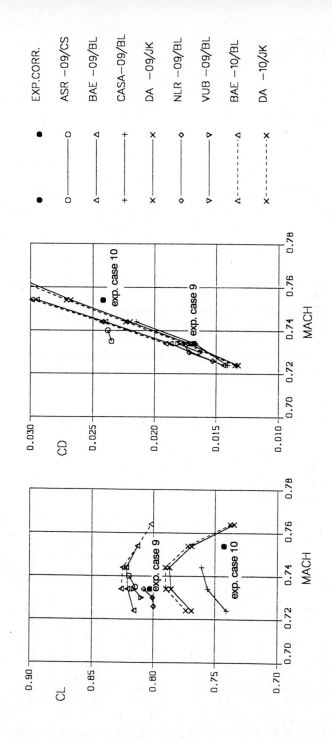

Fig. 15 Lift and drag due to Mach number variation

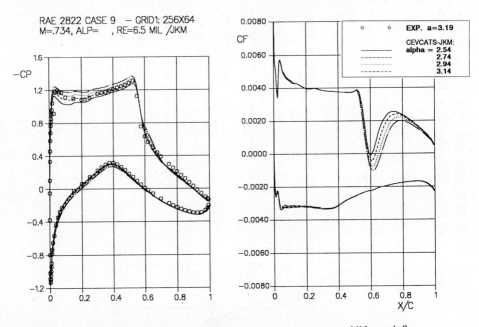

Fig. 16 Effect of incidence based on case 9 parameters (JK model) DA results

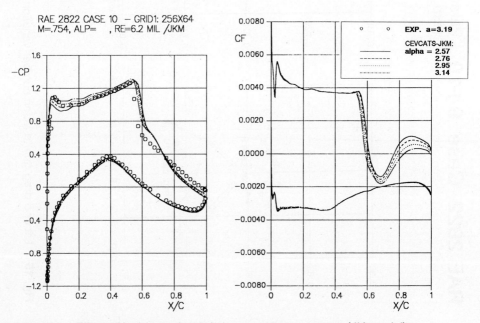

Fig. 17 Effect of incidence based on case 10 parameters (JK model) DA results

RAE 2822, Case 9   Parameters (exp./corr.):   ALFA=2.54 M=.734 Re=6.5 mio

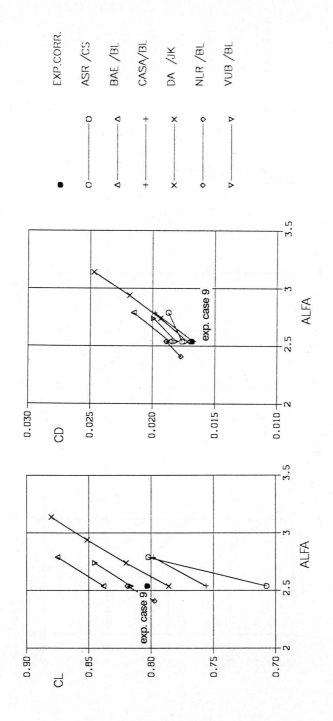

Fig. 18 Lift and drag due to variation of incidence, based on case 9 parameters

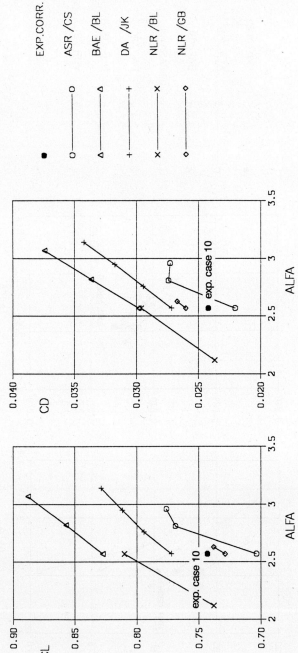

Fig. 19 Lift and drag due to variation of incidence, based on case 10 parameters

ONERA BUMPS
A and C

5.2

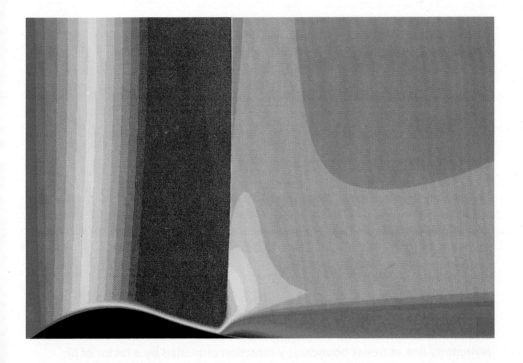

Colour figure - provided by Dornier - shows:

**Mach number contours for 2D channel flow
(symmetry line at upper boundary; y-direction stretched by a factor of 5):**

ONERA bump - Test Case A: Mach number   = 0.68
Isentropic Mach number target (experimental
value) on lower wall   = 1.047 at x = 0.158m
Total pressure, p_t = 0.96x10^5 N/m^2
Total temperature, T_t = 300 K
Pressure ratio, p_t/p_{exit} = 1.522
Turbulence model: CS

5.2 ONERA Bumps A and C
Author: **M.A. Leschziner**, UMIST
Plot Coordinator: **M. Mulas**, VUB

5.2.1. Introduction

The response of a turbulent boundary layer developing on an airfoil to a shock can have profound consequences in relation to the airfoil's overall performance characteristics and is, therefore, of considerable interest in the context of aeronautical design. This response - the main feature being rapid thickening and possibly separation - is essentially dictated by the structure of boundary layer ahead of the shock and the contribution of turbulence transport to the balance of transport processes in the boundary layer as it traverses the shock. It follows, therefore, that the model of turbulence used to predict the turbulence structure is an important element in any solution algorithm for airfoil flows. This element is the focus of the present investigation of bump flows.

While the performance of a turbulence model is, from a practical point of view, primarily of interest in the context of airfoil and wing applications, a study involving a detailed performance comparison of alternative models is best undertaken in the simplest possible, most tightly controlled flow environment which possesses all fundamentally essential features of the practical configuration. This philosophy underpins the present study of shock/boundary-layer interaction over channel bumps.

The recognition that optimising airfoil and wing performance is, ultimately, the main practical objective is reflected by the pursuit of a parallel study focusing on the RAE 2822 airfoil, which is documented in Section 5.1.

The geometries considered here are shown in Fig. 1a and identified as "Case A" and "Case C". In both cases, a bump accelerates a subsonic flow to supersonic conditions and generates a moderately strong shock in the divergent portion. Case A is symmetric, while in Case C the flow is asymmetric, being confined between the bump and an upper flat plate. Experiments performed by Delery and his colleagues (1980, 1981) at ONERA for both cases show that Case A is marginally attached (incipiently separated), while that of Case C is separated and contains a thin, elongated recirculation zone on the downstream bump face. Detailed data, much of it derived from LDA measurements, are available in both cases for surface pressure, displacement thickness, momentum thickness, velocity, turbulence energy and Reynolds stresses. It is this exceptional level of resolution which has motivated the choice of the present flows for assessing turbulence-model performance. A difficulty with both cases is the presence of close spanwise confinement which appears to generate non-negligible span-wise property variations. This poses, unfortunately, a degree of uncertainty in relation to comparisons between the data and two-dimensional solutions.

For the present comparison exercise to be meaningful, contributed calculations with different solution algorithms and models have adopted identical numerical meshes and virtually identical boundary conditions. Although this framework cannot preclude variations due to issues unrelated to turbulence-model characteristics, it does, nevertheless, offer the best prospect of arriving at a consensus on which modelling approach is superior to others; this is the principal objective pursued herein.

A matrix juxtaposing contributors, cases examined, solution strategies and turbulence models is given in the following section. Briefly, solutions for either Case

A alone or for both Case A and Case C have been contributed by:

- BAe CAL (J. Benton)
- CERFACS (L. Davidson)
- Dornier (W. Haase)
- DLR (D. Schwamborn)
- UMIST (G. Page and M.A. Leschziner)
- VUB (M. Mulas and Ch. Hirsch)

A total of ten turbulence-model variants feature in comparisons to be presented. A "variant" may either be a distinct model or a particular combination of a base model with a correction. The majority of contributors examined two or three models. The simplest closure used was the algebraic Baldwin-Lomax model, while the most elaborate involved the solution of equations for all active Reynolds-stress components. All solutions were generated with identical mandatory meshes and with closely similar boundary conditions. Any variations in the latter area are held (in some cases demonstrated) to be uninfluential and highly unlikely to obscure the assessment of the turbulence models used. Results presented and discussed below were plotted to a common format and identical scaling rules.

The remainder of this report is divided into 8 sub-sections relating, respectively, to a summary of contributions, turbulence models, numerical practices, grids, boundary conditions, resource requirements, comparison of computations with experimental data, and conclusions.

5.2.2 Overview of Contributors and Contributions

Table 1 gives an overview of contributors, cases computed, turbulence models investigated and numerical practices adopted. Models are identified by codes which draw upon designations adopted in Chapter 3. Because modelling practices often involved combinations of base models with corrections or the application of two models in different portions of one and the same solution domain, codes consist of combinations of designation or contain qualifiers. To avoid confusion, the codes appearing in Table 1 are interpreted below:

- BL   Baldwin/Lomax
- C   Chien
- CS   Cebeci/Smith
- GR   Granville correction to Baldwin/Lomax[+]
- JL   Jones/Launder or Launder/Sharma
- JC   Johnson/Coakley
- JK-E   Johnson/King-equilibrium
- JK   Johnson/King (standard non-equilibrium)
- RS   Algebraic stress model
- SJL   Standard high-Re Jones/Launder
- W   Wolfshtein;

+ In figures to follow, the letter G has been used instead of GR.

Table 1 Contributors, cases computed and practices adopted

Contributors	Cases	Turbulence Models	Numerical Methods	Grids
CERFACS	A, C	C; SJL-W; RS-W	Time-marching; Cell-centred FV; 4-stage Runge-Kutta; 4th-order dissipation; 1st-order Hybrid scheme for k-e	193x65; 193x97 + 7 cross-flow lines ahead of bump.
DLR	A, C	BL; BL-GR; JK; JC	Time-marching; Cell-centred FV; central differencing; 4-stage Runge-Kutta; 2nd/4th-order dissipation; Multigrid (FAS); Implicit residual smoothing (with MG).	193x65; 193x97; own
Dornier	A	BL; CS; JK-E; JK	Time-marching; Cell-centred FV; 3-stage Runge-Kutta; 2nd/4th-order dissipation; Implicit residual averaging; W-cycle Multigrid.	193x65
UMIST	A, C	C; JL; RS-W;	Time-marching; Cell-vertex FV; Centred differencing; Lax-Wendroff; 2nd-order dissipation; Multigrid.	193x65; 193x97
VUB	A, C	BL	Time-marching; Cell-centred FV; Central differencing or Roe/Yee upwinding with TVD limiter; Implicit relaxation or 4-stage Runge-Kutta; Multigrid.	193x65; 193x97
BAe	A, C	BL; SJL-W; C	Time-marching; Cell-centred FV; Central differencing; 4-stage Runge Kutta; 2nd/4th-order dissipation; Residual Smoothing and W-cycle Multigrid.	193x65; 193x97

5.2.3 Turbulence Models

Table 1 shows that the overlap of turbulence models among contributors is modest, except for the Baldwin-Lomax model - the simplest and most popular practice in external aerodynamics applications. A more wide-ranging overlap would undoubtedly have been conducive to assessing the influence of numerical factors on predicted solutions, but the imposition of mandatory models, in addition to mandatory grids, was deemed excessively constraining within the present study. Encouragingly, most baseline closures - that is those which are not combined with specialised corrections - have each been applied by at least two contributors. This includes transport variants right up to the Reynolds-stress model which has been applied by CERFACS and UMIST. This does offer a degree of security in judging the performance of models against the background of inevitable numerical uncertainties.

Base-line models identified by "BL", "CS", "C", "JK", "JC", "JL", "RS" and "SJL" are documented in detail in Chapter 3 and require no further elaboration. Some comments are held to be instructive, however, in relation to specialised corrections or other modifications to standard practices.

The Granville correction (1987) used by DLR is described in Section 3.1.3 by relations (GR1) to (GR5). The correction consists, in essence, of modified constants C_{KLEB} and C_{CP} which feature in the outer-layer eddy-viscosity formulation of the BL model. The modification involves a dependence of the constants on Coles' wake factor which is itself empirically related to the Clauser pressure-gradient parameter. Hence, the formulation takes account of experimentally established similarity laws. A further variation introduced in the Granville approach for low Reynolds numbers is that the Clauser parameter 'k' relating to the outer-layer eddy viscosity in the BL model is made dependent on the displacement-thickness Reynolds number.

A non-standard feature of DLR's JK and JC models is the use of the Baldwin-Lomax approach to determining the outer-layer viscosity by reference to the maximum "moment of vorticity" F_{max} at the position y_{max}. In contrast, the standard version, used by Dornier, is based on Cebeci and Smith's formulation which uses the outer velocity and boundary-layer displacement thickness. The variation thus involves a replacement of $\delta_i U_e$ by $C_{CP} y_{max} F_{max}$ in equations (JK2) and (JK7) of Section 3.2.2. A further procedural variation in DLR's implementation, not affecting the model's characteristics at steady state, is the addition of a time-derivative of g to (JK10). Finally, DLR use the mixing-length approach instead of the Clauser formulation in the velocity scale (JC3) of the inner-layer viscosity; further details are given in Section 4.8.

Dornier employs two variants of the JK model. One is the standard non-equilibrium ("half-equation") model. The other omits transport of shear stress and is thus algebraic in nature (denoted by JK-E in Section 5.2.2). In both variants, as well as in Dornier's CS model implementation, the displacement thickness was computed from y_{max} on the assumption that the velocity profile was identical to Coles' variation ($\delta = 1.936 y_{max}$). Dornier reports a particular computational implementation of the JK models which differed from Johnson & King's original practice, but this appears to have had no consequences to the statement derived from the models at steady state. Dornier also adopted a mildly non-standard variant of the BL model. First, a limiting process and second-order interpolation are used in determining the position and magnitude of the "moment of vorticity" F_{max} in relationship (BL10). (For details consult Dornier's report in Section 4.9). Second, transition is ignored through setting CMUTM=0 in the transition switch (BL13). Third, to avoid problems in separated zones, the wall shear stress in the definition of y^+ is replaced by the maximum value in the profile of the laminar shear stress at the axial position being considered. These variations are said to exert little impact on the fundamental characteristics of the BL model.

CERFACS and BAe have investigated a particular hybrid two-equation/one-equation eddy-viscosity model denoted by SJL-W in Table 1. This consists of the standard high-Re k-ε model of Jones and Launder, applied to the fully turbulent outer region, with Wolfshtein's one-equation k-l model used to cover the semi-viscous near-wall region. Both models are documented in Chapter 3 (Sections 3.4.5 and 3.3.2). This approach has its attractions as regards economy and numerical stability, but poses uncertainties with respect to interface location and continuity of variables. Both CERFACS and BAe adopted a particular near-wall grid line as the interface, along most of which y^+ was recorded as having a value of order 100.

UMIST and CERFACS have applied an identical combination of a high-Re Reynolds-stress model and Wolfshtein's one-equation model. Here, the use of a hybrid approach is hardly avoidable as there are no well-established low-Re Reynolds-stress closures available for separated conditions. Coupling between the two models presents its problems here too, for solutions are not entirely insensitive to the choice of interface along which y^+ varies considerably. As before, UMIST and CERFACS have both adopted a fixed cross-flow grid line as the interface on the basis of the criterion $y^+ = O(100)$. Tests conducted by UMIST have shown the solution to respond weakly to the position of the interface line, provided this line is not pushed too far into the outer flow.

Some final comments must be made in relation to the BL model - referred to earlier as a 'standard' formulation. Towards the end of the collaborative project, inconsistencies came to light in relation to the implementation of this model by different contributors. Barring one case, these inconsistencies amounted to departures from the standard values of the constants C_{WK} and C_{KLEB} which appear in equations (BL7) and (BL10), respectively (see Section 3.1.2). The exception is CERFACS which entirely ignored the conditional evaluation of F_{WAKE} in equation (BL7) and which simply used $F_{WAKE} = y_{max} F_{max}$ in conjunction with the standard C_{WK} value. As regards the values of the two constants above, it is noted that test calculations performed by DLR, BAe and VUB, in which C_{WK} was arbitrarily varied in the range 0.25-1, have shown the predicted solution to be, unfortunately, very sensitive to C_{WK}, while sensitivity to C_{KLEB} was demonstrated by DLR to be marginal. VUB investigated the sensitivity of the solution to a variation of this constant from 0.25 to 0.35, 0.5 and 1.0, while DLR and BAe only obtained solutions for the extreme values 0.25 and 1.0. Dornier confined its attention to the standard value. As will transpire later, the higher (non-standard!) values of C_{WK} depress the intensity of the interaction below that observed experimentally, which then leads to adverse conclusions on the model's capabilities. In contrast, the standard value yielded considerably stronger interaction, but also gave rise to unrealistic features in the velocity field. The general preference for the higher values stems from the fact that the standard value leads to instability in separated airfoil flows. It is finally interesting to report DLR's observation that setting $C_{WK}=1$. results in the conditional evaluation in (BL7) always returning $F_{WAKE} = y_{max} F_{max}$, which is the form used by CERFACS (but with the standard constant). It may be this dependence, rather than simply the value of C_{WK}, which primarily sensitises the solution to the constant.

5.2.4 Numerical Algorithms

In terms of numerical methodology, there is a remarkable level of conformity between the algorithms as far as fundamental principles are concerned. All but one variant - one of two used by VUB - combine explicit time-marching with the conservative finite volume approach applied over the same structured non-orthogonal cell system. The schemes of CERFACS, DLR, Dornier and BAe are cell-centred and advance the solution with three-step or four-step Runge-Kutta methods. UMIST's algorithm is the only one based on the less conventional cell-vertex scheme. This marches the solution in time by way of the Lax-Wendroff method.

In all schemes, spatial discretisation of mean-flow convective transport is second-order accurate, based either on a centred approximation or, in one of VUB's two variants, on a Roe/Yee-type Flux-difference-splitting/TVD upwind scheme. In the centred schemes, smoothing is achieved either by 2nd-order dissipation (UMIST), 4th-order dissipation (CERFACS) or a combination of both (DLR, BAe, Dornier, VUB). The exclusion of explicit 4th-order dissipation in UMIST's scheme will later be seen to manifest itself by unusually pronounced oscillatory features in the solution. All schemes adopt local time-stepping and multigrid acceleration (either V- or W-FAS cycles) to enhance convergence. Multigrid acceleration typically reduced CPU requirements by factors of 5-10 (see Section 5.2.7). Dornier's, BAe's and DLR's scheme also incorporates implicit residual averaging, and the first two partners report this practice to permit the maximum Courant number limit to be pushed up to approximately 3.5.

The procedures of CERFACS, BAe and UMIST involve the solution of differential transport equations for turbulence quantities from which either the eddy viscosity or the Reynolds stresses are determined. In all three schemes, the turbulence equations are solved as a sub-set separate from that describing density, mass fluxes and energy. The schemes of BAe and UMIST treat the turbulence-model equations consistently, in terms of order of accuracy, as any other equation for mean-flow quantities, but the latter scheme advances the turbulence-variables fields with a much larger time step than that used for the mean-flow equations. In contrast, the CERFACS scheme solves the turbulence-model equations by a semi-implicit ADI scheme with convection discretised by way of the first-order hybrid upwind/central-differencing scheme.

5.2.5 Grids

All contributors used the 'mandatory' non-uniform grids supplied by UMIST (193x65 for Case A and 193x97 for Case C, Fig. 1b). These grids arose from grid-independence tests performed by UMIST in the initial stages of the collaborative exercise. It should be pointed out that these tests did not simply involve successive doubling of grid-line numbers in each direction, for this tends to lead to a rather misleading convergence behaviour. Because the flow behaviour is particularly sensitive to the near-wall grid distribution, lateral grid refinement very close to the wall and stream-wise refinement in the interaction region are the key issues dictating grid independence. Towards the end of the collaborative exercise, Dornier re-examined grid-independence for Case A by using a succession of 3 grids, each twice as fine (linearly) as the previous one. Dornier used the BL model and focused attention on C_L and C_D. Although, as argued above, this test is somewhat crude, it has shown that the level 192x64 supports a solution which, in terms of mean-flow quantities, is within 1% of being grid-independent.

CERFACS was not able to start its computations at the upstream edge of the bump and extended its solution domain to x/L=-0.17 and -0.33 for Case A and Case C, respectively. This entailed the addition of 7 grid lines upstream of the agreed solution domain which was covered by the mandatory grid. DLR performed calculations with 3 grids for Case A, the mandatory one and two own grids containing 255x51 and 255x101 lines and extending upstream of the bump. The principal objective was to investigate the influence of inlet conditions on the predicted solutions; this was found to be insignificant, confirming earlier tests of UMIST.

5.2.6 Boundary Conditions

A 'standard' set of boundary conditions had been issued in the initial phase of the collaborative exercise, and all contributors broadly adhered to this prescription. The main elements of the standard set consisted of experimental stagnation conditions at the inlet, prescribed static pressure at outlet, no-slip velocity conditions and imposed zero wall-normal gradients for pressure and temperature at walls and zero-gradient conditions at the plane of symmetry of Case A. Notwithstanding the experimentally prescribed value of exit pressure, contributors were explicitly instructed to perturb this value in an effort to adjust the shock position so as to achieve agreement with a particular experimental value of isentropic wall Mach number in the shock region (Case A: M=1.047 at x/L=0.158 on bump wall; Case C: M=1.07 at x/L=0.2936 on upper wall). The aim was to obtain results which would allow the predicted structure of the interaction region to be compared without the obscuring influence of a shift in shock position. With the above prescription uniformly adopted, different contributors used slightly different practices to extract relevant quantities at inlet and outlet.

At inlet, all computors assumed the transverse velocity to be zero and evaluated the stream-wise velocity from total energy and local density via standard isentropic relationships. Most contributors using cell-centred approximations extrapolated either density or Mach number from the inner field to the inlet boundary. UMIST obtained the inlet density by solving the continuity equation at the inlet, simply because the cell-vertex scheme gives rise to nodes on all boundaries at which the conservation laws must be satisfied. Dornier experimented with two principal alternatives. In the first - also used by DLR - the stream-wise velocity U (rather than density) was extrapolated to the inlet (subject to U > 0), the lateral velocity V was kept at zero and the density was then computed from the appropriate isentropic relationship. In the other, both U and V were extrapolated to the inlet subject to the realizability constraint that the density computed on the basis of isentropy remained positive. The merit of both treatments is that they yield a continuous inlet velocity profile, rather than a discontinuous step at the wall. CERFACS, BAe and UMIST also needed to prescribe inlet conditions for the transported turbulence quantities. All adopted the same, simple approach of setting the turbulence energy and its rate of dissipation to insignificantly low values, subject to the condition that the turbulent viscosity be O(1)-O(10) times the laminar value.

At the exit plane, all contributors adopted an explicit prescription of the static pressure, subject to the shock-position criterion described in the first paragraph of this sub-section. Two practices were used in respect of other variables, although this is unlikely to have had more than a marginal effect on the predicted solutions. DLR, BAe, Dornier, VUB and UMIST determined the exit density and mass fluxes by linear or zero-order extrapolation, while CERFACS opted for one-dimensional Rieman invariants from which velocities and density were obtained.

5.2.7 Resource Requirements

Information on CPU resources was provided by all contributors. A comparison of requirements must be viewed with extreme caution, however, because of the strong dependence of computing times on a whole host of factors, among them flow type, complexity of turbulence model, convergence criterion, details of multigrid implementation, computer type, processor configuration and compiler efficiency. It is nevertheless instructive to convey a notion of resource requirements for refined NS solutions of the type performed in this exercise.

Interestingly, most computors providing information on resource requirements *without* multigrid acceleration report similar numbers of time steps, ranging from 8000-15000 to achieve a reduction of residual norms by 3-4 decades. CERFACS reports execution times of 2.1 hours on a 1-processor CRAY 2, 9 hours on a 8-processor ALLIANT FX/80 and 6.2 hours on a CONVEX 220 for a total of 15000 time steps. DLR reports figures of 0.2 s per iteration on a CRAY YMP and 13 s per iteration on a SUN SPARC 1+, which gives, for 10000 iterations, total execution times of 0.55 hours and 36 hours, respectively. VUB gives a figure of 1.8 s per iteration on a CRAY XMP (equivalent to about 0.9 s on the CRAY YMP) and about 8 s per iteration on a HP 730 workstation. At 10000 iterations, respective total CPU figures of 2.5 YMP hours and 22 HP hours arise. UMIST states a time of 0.16 s per iteration and a total execution time of about 2.6 hours on the AMDAHL VP1200 computer (at a speed of about 1/4 of the CRAY YMP), while on a 4-processor ALLIANT FX/2808 corresponding requirement were 1.3 s per iteration and 21 hours. In summary, while requirement varied considerably, a figure of 1-2 CPU hours on a CRAY-type machine and 5-10 times that figure on an advanced workstation give a fair idea of the order of magnitude of the computational task without multigrid acceleration.

Multigridding yields dramatic benefits in terms of CPU requirements. UMIST and DLR report speed-up factors of order 5, while VUB states an acceleration of up to 10. BAe and Dornier only provide resource figures with multigrid acceleration included, and typical CPU requirements are 5 minutes on a CRAY YMP in BAe's case and 2 hours on a CONVEX 220 in Dornier's case, the latter figure being roughly equivalent to 8 minutes on a CRAY YMP. This resource level is very close to DLR's CRAY YMP figure, while VUB's requirements, with 10-fold multigrid acceleration, are of the order 22 minutes. There is thus a reasonable degree of consensus that typical CPU needs of the most efficient multigrid algorithms on present-generation supercomputers are of order 5-10 minutes. However, it must be stressed again that requirements depend considerably on the turbulence model, smoothing practices and convergence criterion. Generally, k-ε-model computations require CPU times exceeding those with the Baldwin-Lomax model by factors of 2-3. Stress-model calculations are even more demanding, with the CPU factor being of order 5.

5.2.8 Results

5.2.8.1 Introductory Comments

Results are presented in this section for a number of flow parameters which, together, are held to convey the principal physical processes at play, as well as the predictive capabilities of the turbulence models employed. The availability of a range of results generated within a tightly coordinated framework for both Case A and Case C is exceptionally fortunate, for Case A is not far removed from equilibrium, while Case C is. Hence, the turbulence models have been investigated in two quite different physical environments which span a fair range of practical circumstances.

The comparison presented below is greatly facilitated by being uniformly processed and plotted, an activity undertaken by VUB. Participants provided the author with their own plots of Mach-number contours to strictly prescribed scales. These plots are valuable in that they provide an instant qualitative view of the flow, in particular of the strength of the interaction between the shock and boundary layer, which is reflected by the extent of the λ-structure of the shock close to the boundary layer. All other quantities - wall Mach number, symmetry-plane Mach number (in Case A), displacement thickness, momentum thickness, skin friction, velocity profiles and shear-stress profiles - have been cross-plotted by VUB from numerical data supplied by participants on magnetic media.

Despite the high degree of coordination exercised in the comparisons to follow, some uncertainty arises in relation to integral boundary-layer parameters (displacement and momentum thickness). These are evaluated by integrating the velocity profiles to the edge of the boundary layer, the problem here being that this edge is ill-defined, particularly in the presence of oscillations and strong local accelerations in the λ-shock region. Different contributors used different integration practices, although the majority have followed a procedure recommended by Stock and Haase (1989) in which the edge is held to be at $\delta = 1.936 y_{max}$, where y_{max} is the location of maximum 'moment-of-vorticity' value. This practice, supported by high-order integration and interpolation, yields smoother variations than alternative approaches.

The choice of integral quantities plotted was dictated entirely by the availability of corresponding experimental data. In the case of local data (profiles), more could have been plotted and discussed than contained below: measurements have been made of velocity profiles at more stream-wise positions, and data also exists for normal stresses and turbulence energy. As regards velocity profiles, it was felt that not a great deal would be gained by presenting more than 4 profiles for each case - one ahead of the shock, two in the interaction region and one well aft of the shock, in the recovery region. Profiles for normal stresses and energy can be informative, of course, but arise only from transport models of turbulence; indeed, meaningful normal-stress predictions arise only from Reynolds-stress models. In the interest of uniformity and relative conciseness, it was decided not to include such profiles. Readers interested in these quantities are advised to approach relevant contributors (BAe, CERFACS and UMIST) for further information or to request from the editors copies of Interim Reports produced in 1990 and 1991.

5.2.8.2 Delery Case A

It will be recalled that this flow is presumed to be just attached, as suggested by the experimental velocity distributions to be presented shortly. However, there is no conclusive evidence that separation is entirely suppressed, for there are no experimental data for skin friction. In fact, all but one calculated variations of skin friction indicate the existence of at least a small reverse-flow region close to the wall. Nevertheless, the flow is virtually attached, and algebraic (equilibrium) models are thus expected to yield at least a respectable correspondence with experimental data for this case.

An overall view of the flow is given in Fig. 2 in terms of computed Mach-number contours. A total of 18 plots are provided, and all reflect the interaction region at the foot of the shock by the (weak) λ-structure and the contours 'connecting' the severely thickened aft-shock boundary layer with the shock itself (The unusually prominent oscillatory features in UMIST's plots are due to the exclusion of 4th-order dissipation). Plots resulting from the BL model with non-standard constant C_{WK}=1. (BAe, DLR and VUB) are very similar and all suggest a relatively weak interaction. VUB and Dornier demonstrate a moderate increase in the level of interaction arising the use of C_{WK}-values of 0.25 and 0.35, as will emerge later from a comparison for displacement thickness. Results obtained by BAe and DLR for Case C, discussed in Section 5.2.8.3, support the trends observed here by VUB and Dornier as a result of decreasing the value of C_{WK}. Curiously, however, this increased interaction is hardly reflected in the Mach-contour field, as can be seen from Dornier's plot included in Fig. 2. Neither does the use of the standard value enhance the λ-shock structure in Case C, as will transpire later. This inconsistency gives rise to the suspicion that the enhanced interaction is effected via a wrongly modelled mechanism which does not correctly represent the physical state.

The level of interaction tends to increase as modelling progresses through the algebraic variants CS, BL-GR and JK-E to the transport models JK (i.e. the standard version), JC, SJL-W, JL and finally RS. Qualitatively, the sensitivity seems to be relatively modest, which is consistent with the expectation that all models should perform reasonably well in this near-equilibrium flow. It will be seen shortly, however, that important differences do arise between different models. In common with RS, the JK model gives a more pronounced interaction than do other models, but there are differences between the results obtained by Dornier and DLR, presumable as a result of the different outer-viscosity formulations adopted by the respective contributors. This will become clearer shortly.

Arguably, the most informative results relate to the isentropic Mach number and displacement thickness, shown in Figs. 4 and 5 respectively. Before discussing these variations, however, it is instructive to glance at Fig. 3 which shows distributions of the Mach number along the symmetry line of the channel. The distributions serve to verify that almost all contributors adhered to the requirement of adjusting the exit pressure so as to achieve correspondence between computed and experimentally recorded (inviscid) shock location. A notable exception is UMIST's calculation with the JL model. This model is known to be an exceptionally good low-Re variant, but is numerically very fragile due to a number of highly non-linear terms in the dissipation-rate equation. The implementation of the model in transonic conditions is particularly challenging, and UMIST has not pursued the adjustment of the exit pressure to the required extent due to the very slow convergence of the related calculation. All calculations are seen to have returned crisp shocks with mild to moderate levels of oscillations at the shock extremities. All calculations also result in a post-shock Mach number lying well below the experimental level (experimental data have, unfortunately, been omitted from Fig. 3), and this is an unambiguous reflection of three-dimensional features arising from the boundary layers thickening on the span-wise walls. This under-estimation of Mach number can be clearly recognised from Fig. 4. The local Mach-number maxima aft of the shock reflect the sudden and steep rise in boundary-layer displacement and its influence on inviscid stream, which is then followed by a (mild) decline in displacement as the boundary layers settles into a recovery process.

Reference to Fig. 4 shows a remarkable similarity between most predicted variations of isentropic bump-wall Mach number, the exceptions being JK, RS and JL. Strictly, the last model can be recognised to return a variation very similar to other k-ε variants, if account is taken of the unfortunate displacement of the inviscid shock predicted by that model. This displacement, it is recalled, was eliminated in all other cases by a careful adjustment of the exit pressure. The remaining two models return a considerably stronger level of interaction than the others, and appear to produce superior agreement with the experimental variation. That the better agreement is somewhat fortuitous, however, is implied by corresponding variations of displacement thickness shown in Fig. 5 (discontinuities in some plots are due purely to peculiar features of the integration practice used to obtain displacement and momentum thickness values - refer to Section 5.2.8.1). The BL solutions reported by BAe and DLR with C_{WK}=1. are seen to give an insufficient level of interaction, and there is fairly close correspondence between the two variations. Dornier and VUB used the standard value 0.25 and achieve closer agreement with the experimental data, with Dornier's displacement thickness being somewhat lower than VUB's. It is thus evident that the BL model, when used with C_{WK}=1.0, does not adequately capture the interaction. Using the standard value *appears* to improve matters considerably, but it will be seen shortly that this improvement is at the expense of aberrant features in the velocity field, which then shed light on the curious insensitivity of the Mach contours to the modification of the constant C_{WK}. The CS model and the 'corrected' BL-GR model yield respectable representations, as do the k-ε models SJL-W, JL and C. The fact that the latter class, although being more sophisticated and resting on much firmer fundamental ground, return results of similar quality to those arising from simpler algebraic closures may simply be a reflection of the greater empirical content of the algebraic variants and their careful tuning to the class of flows being considered. With reference to earlier comments on the variations of the wall Mach number, it is observed that the RS model and, even more so, the JK (non-equilibrium) model, predict an excessive sensitivity to the shock. Identical forms of the former model have been implemented by CERFACS and UMIST, and both give rise to similarly excessive maxima of displacement. As regards the JK model, it may be recalled from Section 5.2.3 that DLR's and Dornier's implementations are somewhat different. Both are seen to predict solutions which imply qualitatively similar model defects, however, with Dornier's (standard) implementation returning greater deviations from the experimental data. The apparent inconsistencies in the predictive quality of wall Mach number and displacement thickness by the JK and RS models are again due to three-dimensional features in the experiment. The wall Mach number is sensitive to processes across the entire flow section, while the displacement thickness merely reflects the structure of the boundary layer across the span-wise symmetry plane. In reality, the boundary layers on the span-wise walls tend to elevate the Mach number recorded on the bump wall, and hence a correct two-dimensional calculation would produce a variation lying below that experimentally recorded. The differences in the experimental and computed Mach-number level well downstream of the shock are, again, clear indicators of three-dimensional processes. Finally, the Coakley and, even more so, the equilibrium variants of the JK model - identified as JC and JK-E, respectively - are seen to yield a dramatic improvement relative to the parent JK model. The origin of this improvement is unclear, however, as the variations involved a number of modifications - in the case of JC, both to the inner-layer formulation and to the interface of the inner and outer layers. The sensitivity of the JK model to the outer-layer formulation observed earlier in relation to DLR's and Dornier's implementations suggests that the most influential fragment of the variation is the manner in which the outer and inner layers are blended. In any event, the JC model is clearly preferable to the parent JK variant. It will be seen in relation to Case C, however, that the modification 'robs' the model of its original advantageous predictive capabilities in the presence of separation.

Variations of momentum thickness on the bump wall are shown in Fig. 6. There is a considerable level of consistency between these results and those for displacement thickness: the BL model with $C_{WK}=1$. underestimates the level, the 'corrected' algebraic variants, the equilibrium JK implementation and the JC model return close agreement, while the JK model and the RS closure overestimate the momentum thickness. As regards the k-ε models, there are non-negligible discrepancies between the results of BAe CERFACS and UMIST. BAe's momentum-thickness variation is somewhat low, and this is particularly evident from the plot comparing solutions arising from the high-Re k-ε model (SJL-W); similar differences are also observed in relation to the displacement thickness shown in Fig. 5. UMIST's variations are, on the whole, fairly close to those of CERFACS, although care has to be exercised in comparing results derived from different k-ε variants. The origin of the differences is not clear. It is possible that they mainly reflect the consequences of different integration practices used to arrive at integral quantities. Momentum integration is particularly prone to error. This issue will be revisited when velocity profiles are examined.

Contrasting displacement-thickness and momentum-thickness variations leads to the conclusion that all models return broadly similar shape-factor values, i.e. similar tendencies towards separation and similar recovery behaviour.

Predicted variations of skin friction are shown in Fig. 7. As there are no corresponding experimental data available, little can be said, in absolute terms, about the performance of the turbulence models. However, the distributions are, nevertheless, useful as they reveal a number of interesting features. It is first recognised immediately that all models, except the k-ε variant JL implemented by UMIST, predict a small stream-wise region of negative skin friction. Hence, the flow is predicted, by most models, to be mildly separated, rather than on the verge of separation. The response of the JL model is curious and implies a rather low sensitivity of the boundary layer to the strong adverse pressure gradient associated with the shock. There is no other evidence to suggest, however, that this model has very different properties from those of other k-ε variants; for example, the JL-predicted displacement thickness is in accord with the values arising from other models). This suggests that the JL model returns an unusually elevated level of turbulent viscosity very close to the wall aft of the shock, which naturally favours a high value of skin friction in the recovery region further downstream. A second prominent feature is the aberrant variation returned by the Chien model (C). Clearly, this is a true manifestation of a serious model weakness, for similar curves have been predicted by BAe and CERFACS and UMIST, although BAe's peak level is considerably lower than that predicted by the other two contributors. Indeed, this defect has also been recorded in earlier studies, and reflects an erroneously steep rise in turbulent length scale ($k^{1.5}/\varepsilon$) in and above the semi-viscous sublayer when the boundary is subjected to an adverse pressure gradient. This rise then leads to a corresponding increase in eddy viscosity and hence shear stress. It is a regrettable feature of virtually all low-Re k-ε models (evidently including JL) that they return length-scale values in excess of the experimentally observed equilibrium variation 'κy' just outside the viscous sublayer. This defect is far more pronounced in the Chien model than in other k-ε variants, and the difference between the levels predicted by the JL and C variants is a measure of the seriousness of the defect in the latter model. Fig. 7 shows all remaining models to give broadly similar variations. Exceptions include VUB's BL implementation which suggests strong reverse flow - in contrast to the results of BAe, Dornier and DLR - and the performance of the JK model, examined by DLR and Dornier, which predicts an exceptionally long region of reverse flow followed by an exceptionally fast recovery further downstream, paralleling the behaviour returned by the Chien model. In the absence of experimental data, it is not possible to pass judgement on the realism of any one variation. However, it would seem, in view of gentle variations of the displacement and momentum thicknesses observed earlier, that the relatively slow recovery returned by the SJL-W and RS models, as well as the algebraic variants BL and BL-GR, is more realistic.

Velocity and shear-stress profiles at four stream-wise stations are shown in Fig. 8 and Fig. 9, respectively. The choice of streamwise locations for which profiles are presented is rooted in the wish to convey the salient flow features in four distinct regions: ahead of the shock, within the λ-structure, in the incipiently separated portion and aft of the shock where recovery takes place. The largest differences between profiles are seen to occur at the location x/L=0.8, This position is within the λ-shock structure, so that even slight differences in predicted shock position and structure tend to produce large changes in velocity; in fact, it is precisely this interaction which motivated the inclusion of profiles at this location. On the whole, it may be said that all models, except JK, BL and CERFACS' implementation of the k-ε models return a fair to good representation of the velocity field. A weakness of the BL model, consistently predicted by Dornier, BAe, DLR and VUB but not immediately apparent from Fig. 8, is a seemingly unrealistic separation pattern, whereby the velocity drops dramatically to its minimum value within an oddly thin region very close to the wall. This abnormal feature is particularly pronounced in VUB's and Dornier's solutions which show intense reverse motion very close to the wall (refer also to the associated skin-friction variation of VUB in Fig. 7). The JK model shows a particularly disappointing post-shock behaviour, yielding an unusually large separation region and displacement characteristics. It is instructive at this stage to refer back to Figs. 5 and 6 and to note that while the peak displacement thickness predicted by this model is far too high, the momentum thickness predicted in the same region tends to be too low. Hence the conclusion is that the predicted shape factor is much too high, consistent with an excessively separated flow. Further downstream, at x/L=1.25, the JK model is seen to return better (though not satisfactory) correspondence with the experimental data, suggesting a rapid recovery, as is also implied by the JK-predicted variations of skin friction shown in Fig. 7. It thus appears that the JK model has fairly serious weaknesses in conditions not far removed from equilibrium. The JC model returns, as anticipated from the displacement characteristics, a far better velocity field than the parent JK model. However, examination of the near-wall region reveals unfavourable features, especially an excessive tendency towards separation, which betray the kinship of this variant to its JK parent. Interestingly, the stress model RS tends to yield features which bear some resemblance to those arising from the JK model, although the latter is clearly inferior; here again, sensitivity to the shock is high and displacement is excessive. It is tempting to speculate that this commonality is linked to the fact that both models involve an equation directly describing the behaviour of the shear stress. The nature of this linkage is obscure, however, as the model formulations differ drastically in detail. One further noteworthy feature in Fig. 8 is that the aft-shock rate of recovery predicted by the RS model, particularly as implemented by UMIST, is too low; this contrasts with the behaviour returned by most other models, especially the k-ε variants. Close examination of the velocity profiles predicted by all two-equation models (to which, strictly, the present RS variant belongs, because of the algebraic approximation of stress transport) shows all to share this weakness. In the eddy-viscosity variants this weakness is not prominently displayed, however, because the shock-induced displacement predicted by these models is a little too low so that a delay in recovery remains hidden within the recovery range considered. The fact that the JK-predicted profiles at x/L=1.25 do show defects similar to those arising in the RS-profiles reflects a very rapid recovery following reattachment, as implied by the associated skin-friction variations in Fig. 7.

Finally, for this case, shear-stress profiles at the same locations as those pertaining to velocity are given in Fig. 9. The picture presenting itself here is rather disappointing and, seemingly, at odds with the much more favourable outcome in respect of velocity. Specifically in relation to this contrast, it is appropriate to point out, first, that experimental uncertainties are likely to be particularly high when measuring stresses, and second, that the behaviour of velocity in the shock region and immediately thereafter (although not in the recovery region) is not decisively sensitive to the local turbulence structure; it is rather more sensitive to the structure of the boundary layer approaching the shock. Hence, some fairly substantial differences in shear stress levels will not provoke correspondingly dramatic differences in velocity. Fig. 9 shows virtually all models to return excessive shear-stress levels in or immediately aft of the shock region and too low values in the recovery zone. No clear correspondence can be detected here between model complexity and predictive quality. It is also difficult to detect a clear relationship between stress levels and previously discussed features. There are, however, a few features arising from Fig. 9 which do contribute to understanding aspects considered above. First, it is noted that the largest sensitivity to the shock and the largest displacement arises from models which tend to return low levels of shear stress, particularly at the location x/L=0.8; such low levels are predicted especially by the JK and RS models. This is an evident linkage, for a low level of shear stress, especially near the wall, favours separation. The same models also tend to produce the largest lateral (upward) shift of the shear-stress maximum, and this is a direct result of the strong upward displacement of the boundary layer; clearly, the shear-stress maximum is indicative of the position of maximum strain in the displaced boundary layer. Conversely, those models returning high levels of shear stress inhibit displacement; this applies particularly to the BL and k-ε variants, although it will be recalled that the former yields a very thin (but fairly intense) separation zone, probably due to defects in the details of the inner-layer formulation. Finally, most models return far too low values of shear stress in the recovery region. This is a curious defect, for it suggests that the shear strain is seriously underestimated. Yet, Fig. 8 shows this not to be the case; indeed, the RS model predicts an excessive strain level, but this merely reduces the seriousness of the shear-stress defect. A low shear-stress level would necessarily result in an insufficient rate of recovery, which is a weakness identified in the case of the k-ε-type models and the RS variant. No such clear-cut defect can be identified in relation to the other models, however. The (contestable) assumption that conditions at x/L=1.25 are very close to equilibrium then leads to the conclusion that all models should give an erroneous representation of the turbulence structure in an ordinary boundary layer, which is known not to be the case. Hence, either shock-related history effects are still very important at x/L=1.25, obscuring the interaction between local stress and strain, or the measurements are erroneous.

5.2.8.3 Delery Case C

This case is considerably more challenging than the previous one, because the shock is sufficiently strong to provoke a significant area of recirculation. The strong interaction raises further the danger of three-dimensional features obscuring an objective assessment of turbulence-model performance, although the presence of the upper wall seems to counteract three-dimensionality.

The intensity of the interaction can again best be gleaned, qualitatively, from the Mach-number contours shown in Fig. 10 (As in Case A, UMIST's plots contain pronounced oscillatory features due to the exclusion of 4th-order dissipation). The expectation here is for the shock to display a much more pronounced λ-structure than in Case A, and this is, indeed, implied by the plots resulting from the JK, JC, BL-GR and RS models, all of which will later be demonstrated to give superior agreement with experimental data. In contrast, the BL-generated plots of BAe, DLR, VUB and CERFACS - the first three obtained with C_{WK}=1. or >0.5 - and, to a lesser extent, those resulting from the k-ε variants (JL, SJL-W, C) suggest a weaker interaction. The BL-fields, in particular, herald a poor predictive representation of this strongly separated flow. It must be added here that the use of low C_{WK} values (0.25 and 0.35) by VUB and BAe yields a much stronger interaction, but significantly <u>without</u> enhancing the λ-shock structure. As noted in Section 5.2.8.2, this inconsistency suggests that the enhanced interaction is produced by an incorrect mechanism, a fact which will become transparent upon a consideration of velocity distributions. CERFACS's RS-generated λ-structure is surprisingly weak, compared to that reported by UMIST, as both solutions were obtained with the same model. This difference may be due to the diffusive nature of CERFACS's first-order approximation of turbulence convection which may have suppressed the normal-shock leg of the λ-structure by introducing a significant level of second-order artificial diffusion.

Variations of isentropic Mach number along the lower bump wall and along the upper flat wall are shown in Figs. 11 and 12, respectively. Attention is drawn to the fact that VUB, in using the BL model, has investigated and reported the sensitivity of the solution to varying C_{WK} in equation (BL7), Section 3.1.2, between 0.25 and 0.5 (no changes arise for higher values), while BAe has provided solutions for 0.25 and 1. only. DLR used the value 1.0 (for which VUB obtained results very close to those arising for C_{WK}>0.5), while CERFACS entirely bypassed the C_{WK}-related fragment of the "wake function" F_{WAKE}, equation (BL7) in Section 3.1.2. DLR has also performed a calculation with C_{WK}=0.25 and observed trends similar to those reported by VUB and BAe; DLR's solutions have not, however, been included in the figures discussed herein. The Mach-number variations relating to the upper wall serve, as in Case A, to confirm that all contributors matched the shock reasonably well to the data point at M=1.07. These comparisons also provide a qualitative, global measure of three-dimensionality. As seen, the predicted post-shock Mach number asymptotes to a value approximately 7% lower than the experimental level, and this reflects the contribution of the side-wall boundary layers which accelerate the flow beyond the level predicted by any two-dimensional treatment.

Of much greater interest from a physical standpoint is the Mach-number variation along the lower wall, given in Fig. 12. This shows the BL model, as implemented by CERFACS, DLR, BAe and VUB (the second and third with C_{WK}=1. and the last with C_{WK}>0.5), to seriously misrepresent the actual interaction. The model displays only very weak signs of capturing the Mach-plateau which is associated with the separation zone, and this weakness carries over to all other flow parameters, leading rapidly to the conclusion that this model (with C_{WK} =1.0) is entirely inappropriate for separated flows. The results of BAe and VUB for the standard BL variant (with C_{WK}=0.25) are very similar and imply a far stronger interaction, although the mode and position of this interaction are evidently wrong. An elevation of C_{WK} to 0.35 weakens the interaction, in accord with expectations. The Granville variant BL-GR and the JC model give slightly better predictions, as do the k-ε variants (JL, SJL-W and C) examined by CERFACS, BAe and UMIST. As regards the k-ε models, the behaviour here is broadly consistent with that observed in Case A, and reinforces the conclusion that two-equation eddy-viscosity models yield an insufficient response of the boundary layer to shocks. Also consistent with Case A is the observation that the JK and RS models return the most sensitive response to the shock. These two models also attain a quantitative correspondence with the experimentally observed Mach plateau, but it will be recalled from the discussion in the previous subsection that this is not necessarily a true reflection of the model's predictive quality, in view of the importance of three-dimensional features.

Results for displacement thickness and momentum thickness are given in Figs. 13 and 14, respectively. The peak thickness values are of order three times higher than those obtained in Case A, reflecting the much stronger interaction and consequent separation process. Here again, the BL model with C_{WK} = 1.0 (or >0.5) is seriously deficient, reinforcing the validity of earlier comments. In contrast, VUB's and BAe's calculations with C_{WK}=0.25 give an excessive displacement over an extensive portion of the post-shock region, and this is associated, as will emerge later, with a misrepresentation of separation. The other algebraic variant, BL-GR, performs better, but will also be shown to give rise to erroneous velocity variations. The performance of the k-ε models is in line with expectations derived from the wall-Mach-number variations. This group of models give qualitative correspondence, but insufficient sensitivity and hence displacement. Encouragingly, all k-ε results for displacement thickness are similar, and this justifies confidence in the validity of the message conveyed by the distributions reported. Consistency in respect of momentum thickness is less good, but this parameter is more difficult to determine accurately from a given velocity field. Both thickness parameters predicted by the JK model come close to the experimental data, implying a correct level of the shape factor and confirming the widely held view that this model is well suited to separated flows which are very far from a state of equilibrium. It will be recalled, however, that this model fared badly in conditions much closer to equilibrium. The JC variant of the JK model displays a disappointing performance, faring not much better than the algebraic models. On reflection, this is not surprising, for the modifications introduced into the JK parent will be recalled to have led to a strong reduction in the predicted strength of interaction in Case A. It is precisely this reduction which deprives the so modified JK model of its ability to return the correct level of response to the shock. In common with the JK model, the stress closure RS gives a credible response to the shock, resulting in a broadly correct maximum level of displacement and momentum thickness. It is obvious, however, that the rate of recovery is, here again, too low - a consequence of an insufficient level of turbulent mixing downstream of the shock.

Prior to a consideration of velocity profiles, attention is turned briefly to the predicted skin-friction variations shown in Fig. 15. As in Case A, there are, unfortunately, no experimental data allowing a quantitative assessment. Apart from the very extensive reverse-flow region arising from the standard BL-model implementation (with C_{WK}=0.25), three qualitative features merit attention; these have also been highlighted in Case A. The first is the wholly unrealistic variation returned by the Chien model following separation. As in Case A, there is here too a qualitative agreement between variations returned by several contributors, which hightens confidence in the validity of the statement derived from the predictions. It will be recalled that this feature has been attributed to a serious over-estimation of the turbulent length scale very close to the wall. Clearly, the JL model also displays tendencies in this direction, although the defect is far less serious. The second facet worth highlighting is the rapid rise in skin friction predicted by the JK models downstream of the shock; this will be seen later to correspond to a rapid recovery of the velocity profile in the aft-shock region. Finally, it is noted that the RS model returns a post-shock recovery which is considerably slower than that predicted by the k-ε models; here again, this will be shown to be associated with consistent features in the velocity field, and is also compatible with the low rate of recovery of the displacement thickness seen in Fig. 13.

Velocity profiles at four representative axial locations are next shown in Fig. 16. The first location included is x/L=0.943. This is immediately behind the shock, and the erratic behaviour of some of the profiles reflects the fact that this location is still within the rapidly varying λ-shock region. The profiles predicted with the BL model by CERFACS, DLR, BAe and VUB, either with C_{WK}=1 or >0.5, reinforce earlier comments on serious model weaknesses. Separation is confined to a very thin region near the wall, and the displacement of the boundary layer is grossly under-estimated. This appears to reflect an excessive level of turbulence in the highly sheared region above the immediate near-wall layer, which inhibits separation and enhances downstream recovery, the latter resulting in low velocity gradients in the outer layer. Use of the standard BL-model constant C_{WK} = 0.25 by BAe and VUB gives rise to a much stronger interaction and results in seriously excessive separation; an increase in C_{WK} to 0.35 is shown by VUB to improve matters. The other algebraic variant, BL-GR, displays weaknesses which are similar to those returned by the BL model with C_{WK}=1., namely insufficient displacement and an aberrant velocity-reversal pattern close to the wall. Similar comments also apply to the JC model which is not materially superior to the algebraic formulations. The k-ε variants implemented by UMIST, CERFACS and BAe all result in similar characteristics: a qualitatively correct response to the shock, but insufficient levels of separation and displacement, followed by a seemingly reasonable level of correspondence in the recovery region. It is important to point out, as has been done in Case A, that all models actually return an insufficient rate of recovery, for otherwise the predicted velocity would significantly exceed the measured values in the boundary layer, considering the smaller defect in the predicted profiles in the separation zone. The JK (non-equilibrium) model and the stress closure RS again stand out as predicting a much more extensive separation and higher displacement than other models, although the lateral extent of the separation zone is still too low. In common with the other two-equation transport models, the RS closure produces an insufficient rate of recovery. Only here, this weakness manifests itself more prominently, because of the better resolution of the separated zone, and the predicted profiles in the recovery zone are at greater variance with the experimental data.

The fact that the recovery rate is too low is consistent with discrepancies in shear stress. Profiles at the same locations as those for velocity are given in Fig. 17, and all models show, albeit to a different extent, insufficiently high shear-stress levels beyond $x/L=0.943$. The BL model returns particularly low levels in the recovery region (for any C_{WK} value) due to the flatness of the associated velocity profiles. The higher level of shear predicted by the RS model in the outer region and far downstream is associated with the fact that velocity profiles in these regions are relatively steep due to a combination of higher displacement and insufficient recovery. A fair measure of agreement is observed between profiles predicted by different contributors for identical k-ε variants. Just aft of the shock, all k-ε variants - particularly C - return seriously excessive stress values, which obviously reduces the intensity of interaction and depresses the amount of displacement, separation and recirculation. More generally, those models which return the lowest levels of shear stress aft of the shock, particularly RS and Dornier's JK implementation, tend also to yield elevated values displacement and reverse flow, although the correlation is not complete.

5.2.9 Conclusions

This has been an almost unique collaborative study in so far as it involved tightly co-ordinated computations contributed by a number of research groups spread all over Europe for carefully chosen and uniformly prescribed test cases. The key objective of the study has been to arrive at a statement on the adequacy of a wide range of turbulence models for describing shock/boundary-layer interaction in marginally attached as well as strongly separated transonic flows over bumps in channels. A fair overlap of turbulence-modelling practices among the contributors has allowed consistency issues to be examined and numerical factors to be delineated. Whilst the study has been far from exhaustive, it has yielded a number of valuable lessons and indicators which are likely to have a considerable impact in a more practical engineering context.

Before summarising major conclusions, it is appropriate to strike a cautionary note with regard to the generality of conclusions presented below. The cases considered here have been nominally two-dimensional. It is a frequent and frustrating observation in CFD that turbulence modelling practices which prove themselves in a two-dimensional strain field often fail in three-dimensional conditions. The likelihood of such failure increases dramatically with the level of flow-class-specific empirical input into the formulation of a modelling practice. Conversely, the likelihood of universality increases if a model is based on general, fundamentally sound and well-established physical concepts. This by no means guarantees success. However, this approach provides a sound, rational framework within which to strive for improvement and breadth of applicability.

The main conclusions of the study may, thus, be summarised as follows:

Case A

In conditions not far from equilibrium, i.e. in the absence of separation, the majority of models yield satisfactory predictive realism. Algebraic models can do as well if not better than differential models, for they are carefully 'calibrated'. The Baldwin-Lomax model - the simplest variant examined - has given a reasonable representation, but has shown weaknesses even in the incipiently separated flow. It must be stressed that the performance of this model depends sensitively on the value of its "wake coefficient". Strictly, any departure from the standard value constitutes an inappropriate case-specific 're-calibration' which changes the nature of the model and frustrates a rational assessment of its performance. The 'corrected' algebraic model involving the Granville correction has performed very well. Similarly, the two-equation k-ε turbulence-transport models gave satisfactory results, although the sensitivity of the predicted flow to the shock is somewhat too low, indicating that turbulent mixing is excessive. The Johnson-King model has displayed unambiguous weaknesses which must be considered serious. This model is steadily gaining in popularity in external aerodynamics, and it is important to point out that it predicts, in near-equilibrium conditions, a seriously excessive sensitivity to the shock, premature separation and excessive displacement. The JC variant of the JK formulation has been demonstrated to be far superior to its JK parent, but the penalty is a serious loss of predictive realism in the presence of separation (see Case C below). The Reynolds-stress modelling strategy - the most complex considered herein - also returns an excessive sensitivity, although it is superior to the JK model. This is a rather disappointing observation, but it must be borne in mind that this modelling strategy is principally suited to curved shear regions in which the curvature-related secondary strain strongly affects the shear-stress level. There is no clear evidence that the Reynolds stress model is generally superior to two-equation eddy-viscosity models in incompressible flows subjected to strong adverse pressure gradient. A problem with all two-equation models - and this includes the Reynolds-stress model - is that the rate of recovery of the boundary-layer aft of the shock is too low. This may well be rooted in defects in the dissipation-rate equation which is common to all two-equation models examined herein.

Case C

In strongly separated conditions, the Baldwin-Lomax model, as implemented by the majority of contributors, is practically worthless. It gives a wholly insufficient level of interaction and, at the same time, a strong reverse flow region which is, however, confined to an unrealistically thin layer close to the wall. Use of the standard "wake coefficient" dramatically elevates the level of interaction, but also results in an abnormal elongation and a general mis-representation of the reverse-flow region. It is generally arguable that the need to introduce major case-specific 'adjustments' to influential constants seriously detracts from the value of a model. The Granville-corrected variant results in considerable improvements, but some inherent defects remain. These models cannot, therefore, be recommended for strongly non-equilibrium flows. The k-ε eddy-viscosity models give a fair representation of the separated flow, but the level of interaction is too weak; displacement is too low, separation is inhibited and the recirculation zone is too small. In this separated flow, the Johnson-King model has displayed its predictive strengths, yielding correct displacement and reverse-flow characteristics. The JC variant returned a disappointing performance, at a level similar to that of the Granville-corrected algebraic Baldwin-Lomax formulation. In common with the JK model, the Reynolds-stress closure has performed well, appropriately predicting a large recirculation zone and a very strong interaction. Here too, however, the model suffers from defects in the post-shock region, predicting too low a rate of flow recovery. This points, yet again, to defects in the dissipation-rate equation which is highly intuitive in character.

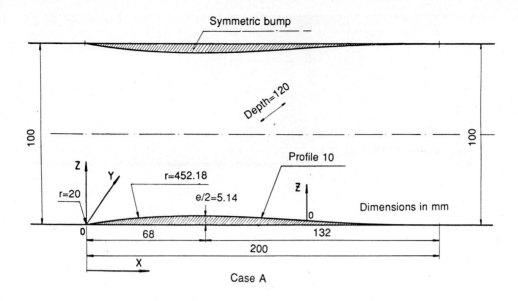

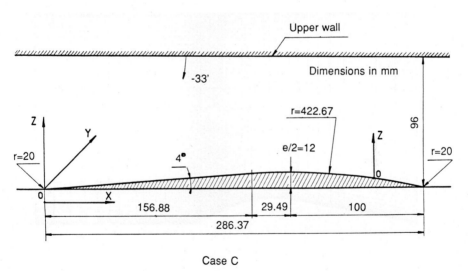

Fig. 1a Computed bump geometries

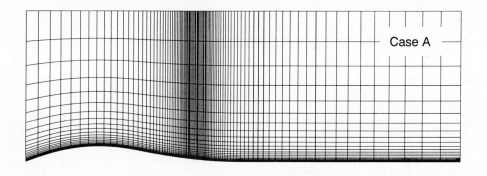

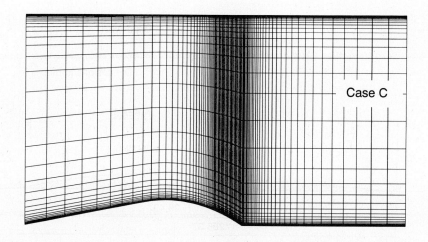

Fig. 1b Mandatory meshes for Case A and Case C as supplied by UMIST (Channel height enlarged; **every other grid line omitted in plot**)

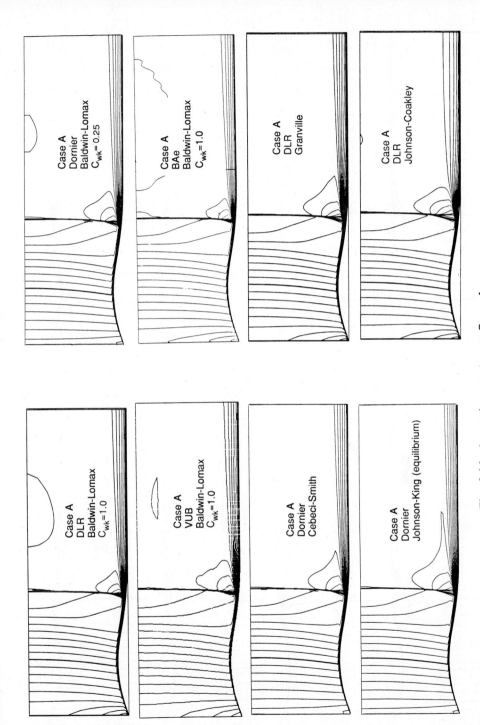

Fig. 2 Mach-number contours, Case A

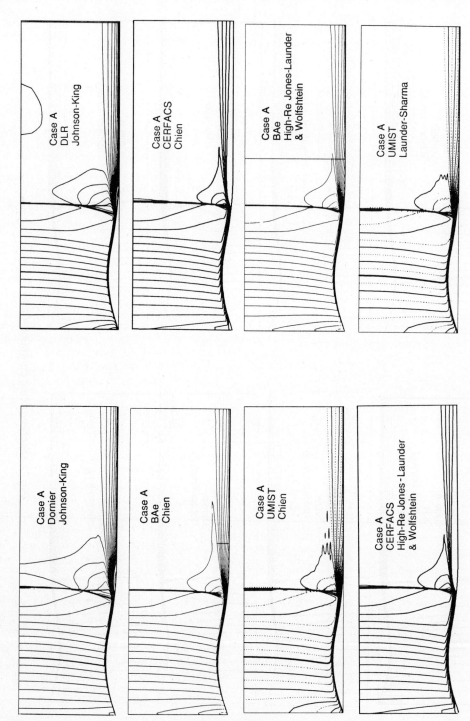

Fig. 2 (cont) Mach-number contours, Case A

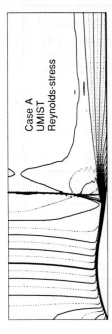

Fig. 2 (cont) Mach-number contours, Case A

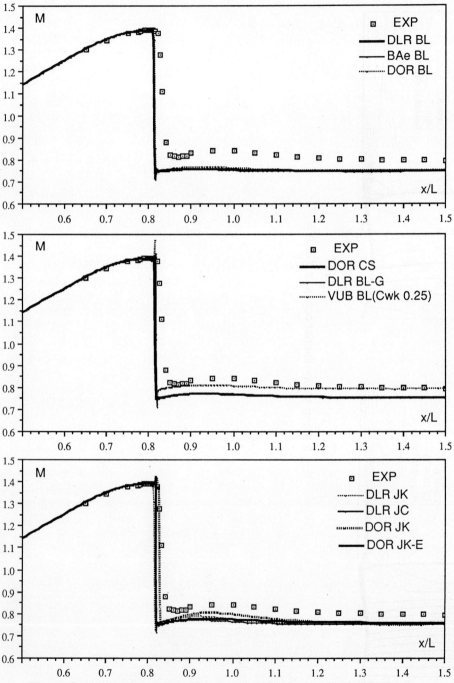

Fig.3 Case A. Centerline isentropic Mach number distribution

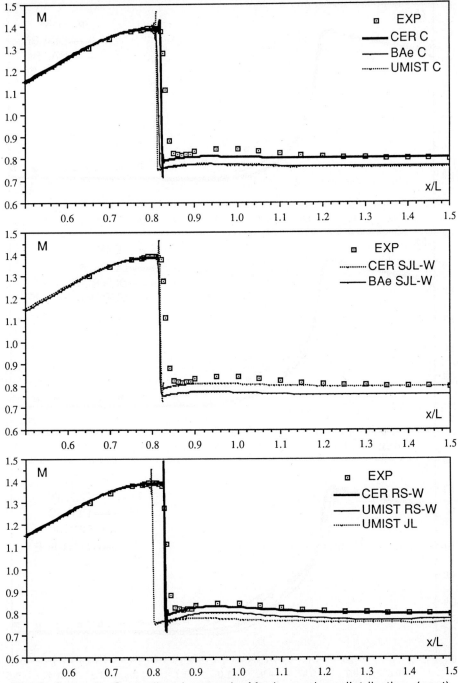

Fig.3 Case A. Centerline isentropic Mach number distribution (cont)

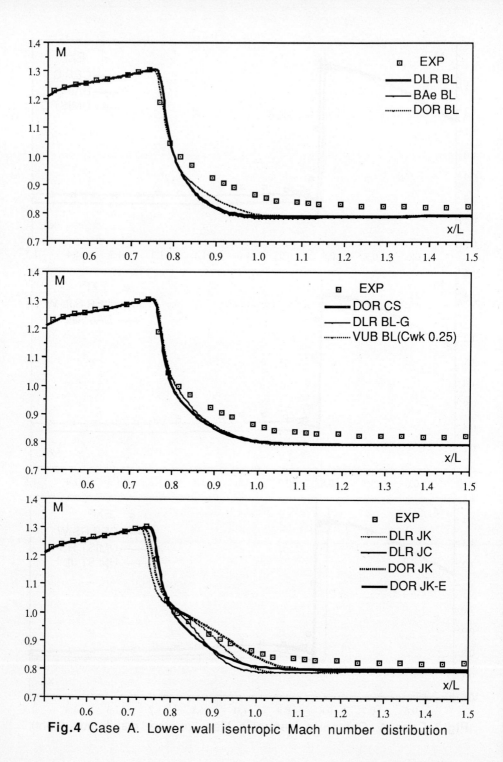

Fig.4 Case A. Lower wall isentropic Mach number distribution

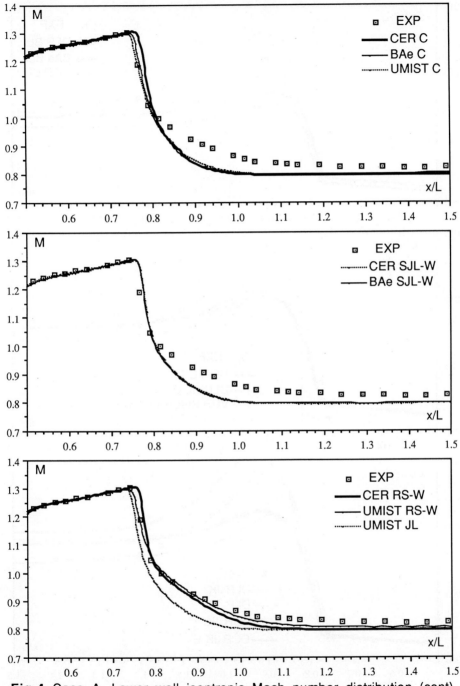

Fig.4 Case A. Lower wall isentropic Mach number distribution (cont)

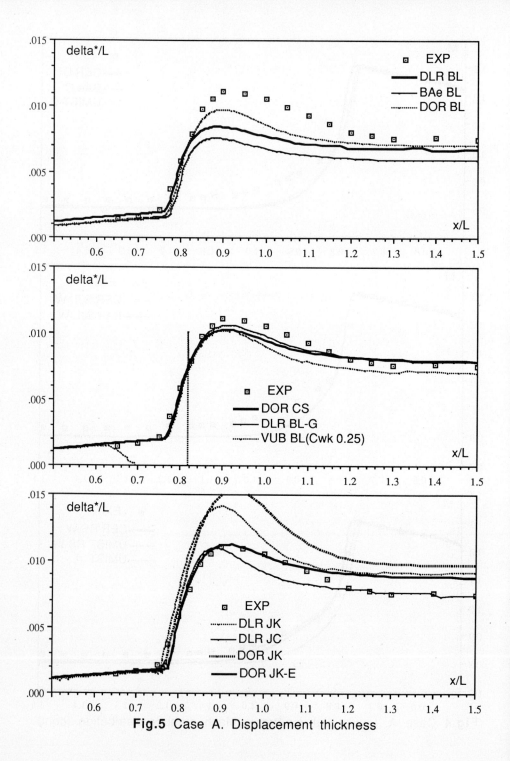

Fig.5 Case A. Displacement thickness

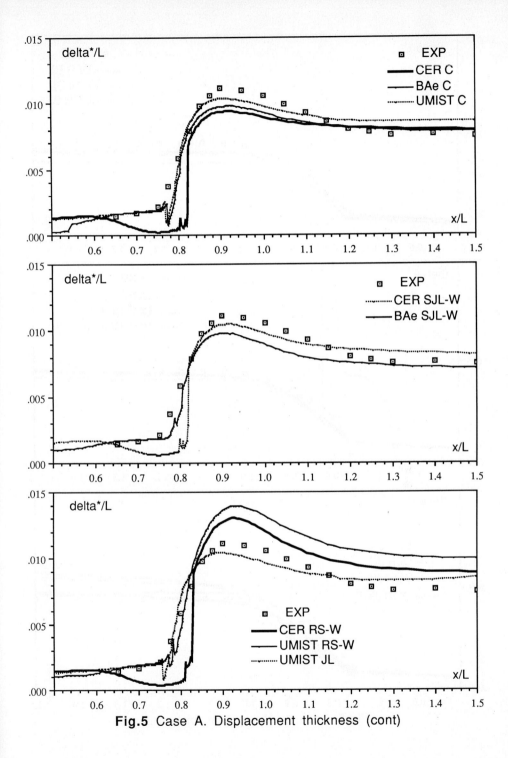

Fig.5 Case A. Displacement thickness (cont)

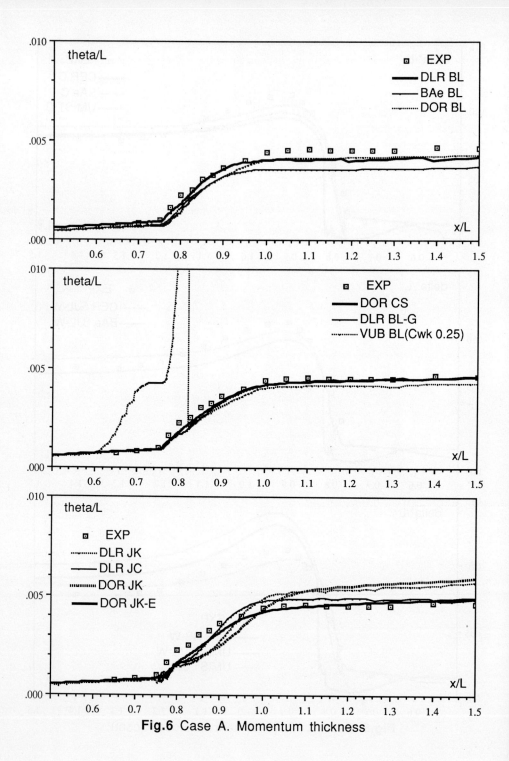

Fig.6 Case A. Momentum thickness

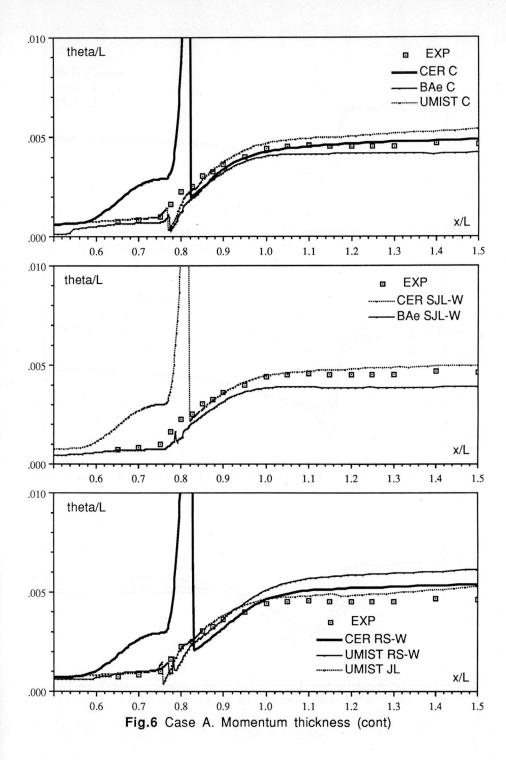

Fig.6 Case A. Momentum thickness (cont)

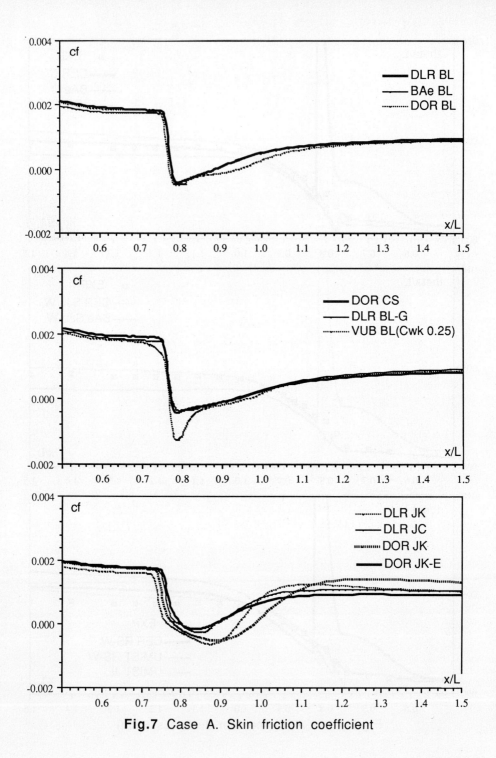

Fig.7 Case A. Skin friction coefficient

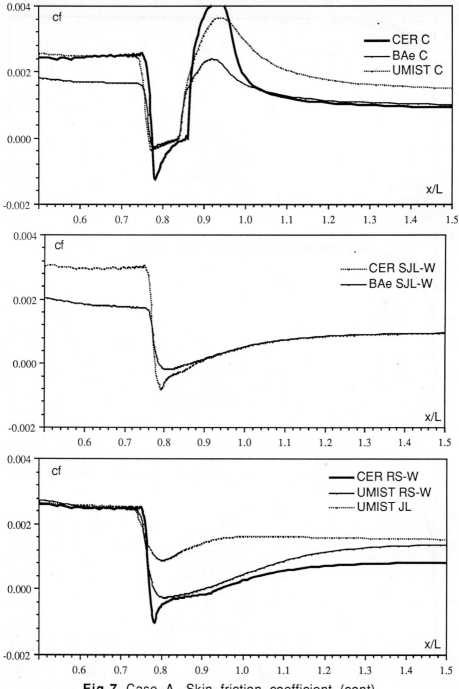

Fig.7 Case A. Skin friction coefficient (cont)

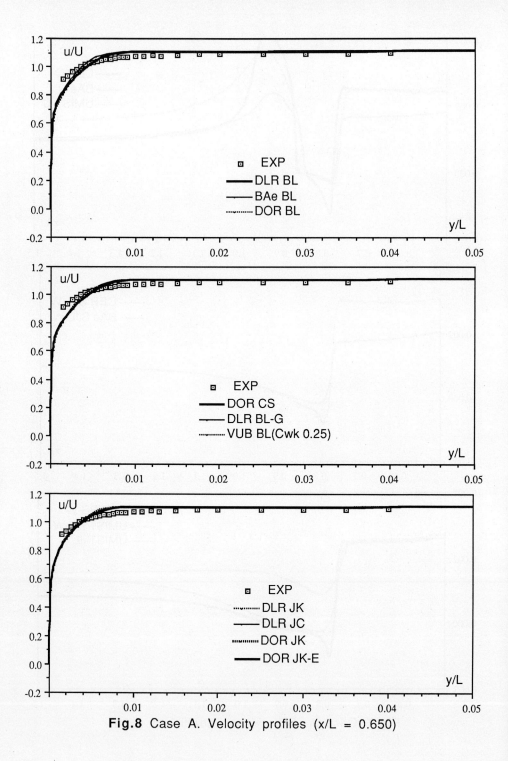

Fig.8 Case A. Velocity profiles (x/L = 0.650)

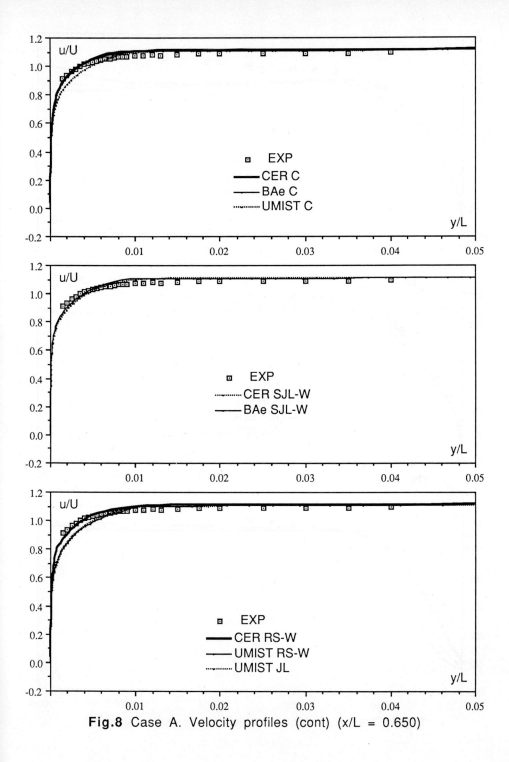

Fig.8 Case A. Velocity profiles (cont) (x/L = 0.650)

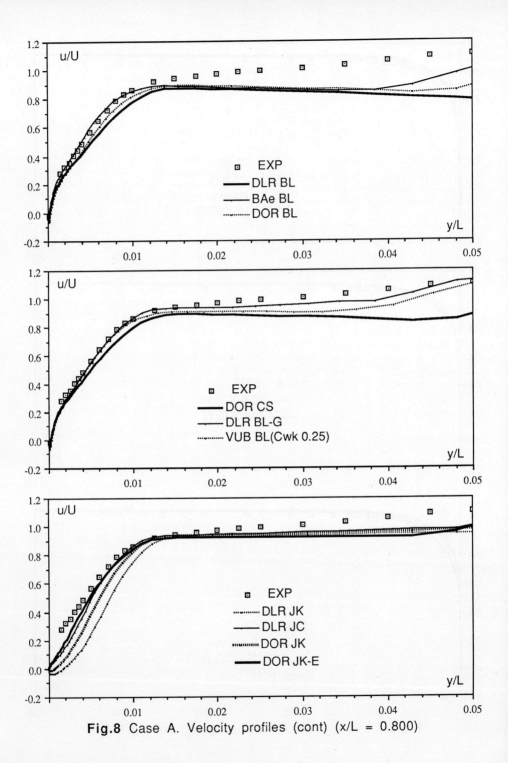

Fig.8 Case A. Velocity profiles (cont) (x/L = 0.800)

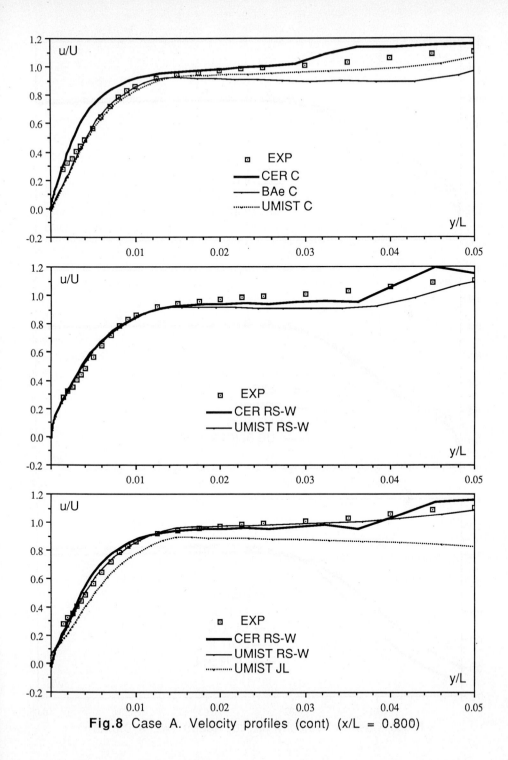

Fig.8 Case A. Velocity profiles (cont) (x/L = 0.800)

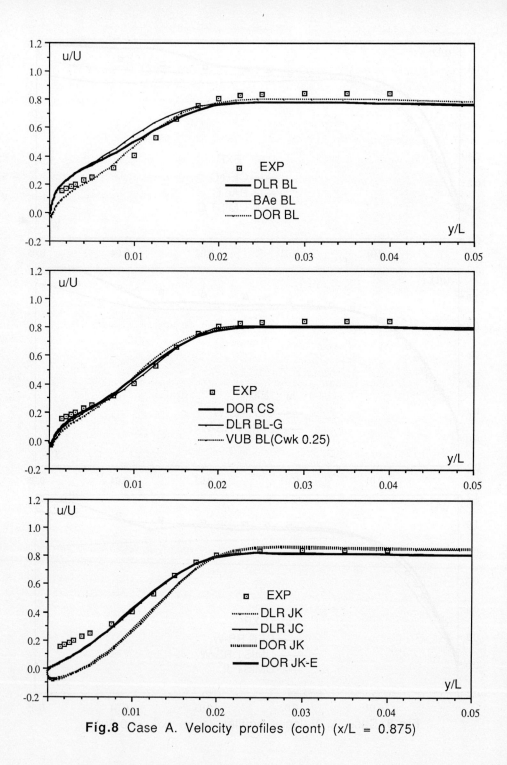

Fig.8 Case A. Velocity profiles (cont) (x/L = 0.875)

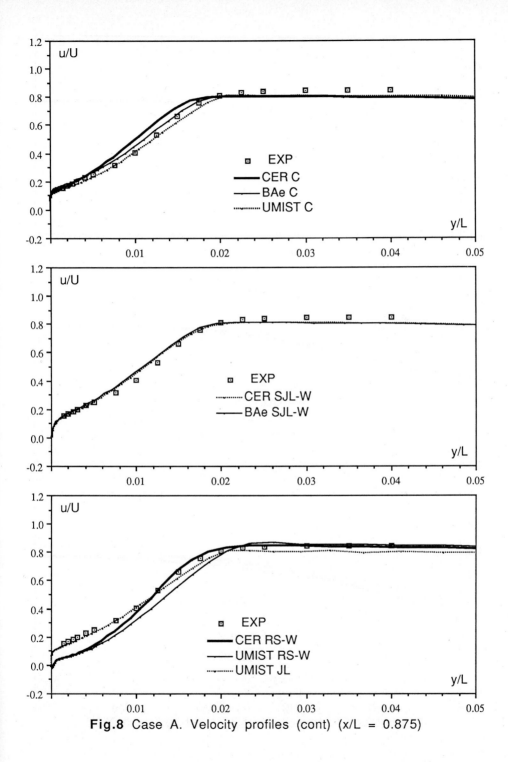

Fig.8 Case A. Velocity profiles (cont) (x/L = 0.875)

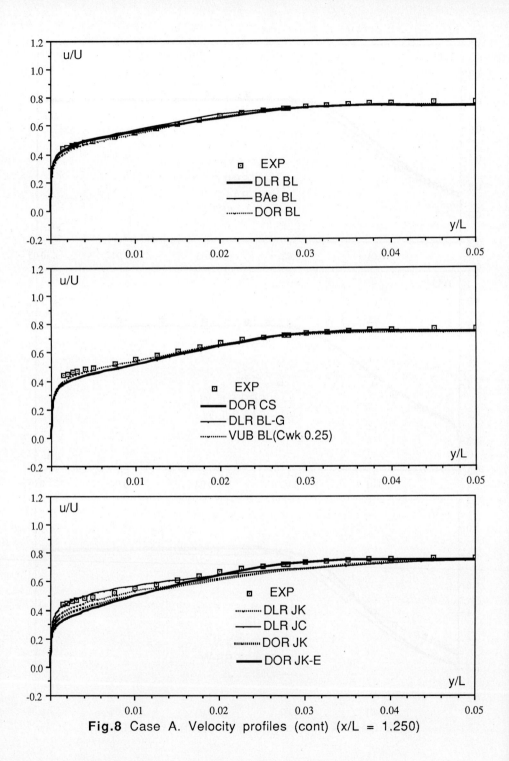

Fig.8 Case A. Velocity profiles (cont) (x/L = 1.250)

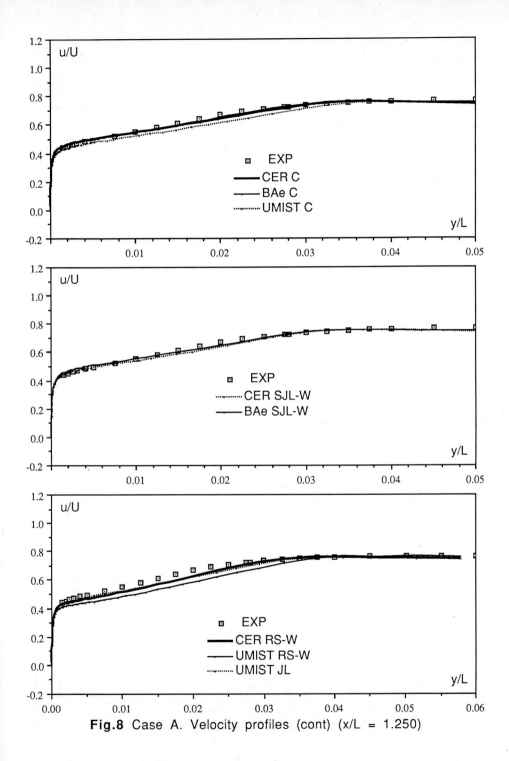

Fig.8 Case A. Velocity profiles (cont) (x/L = 1.250)

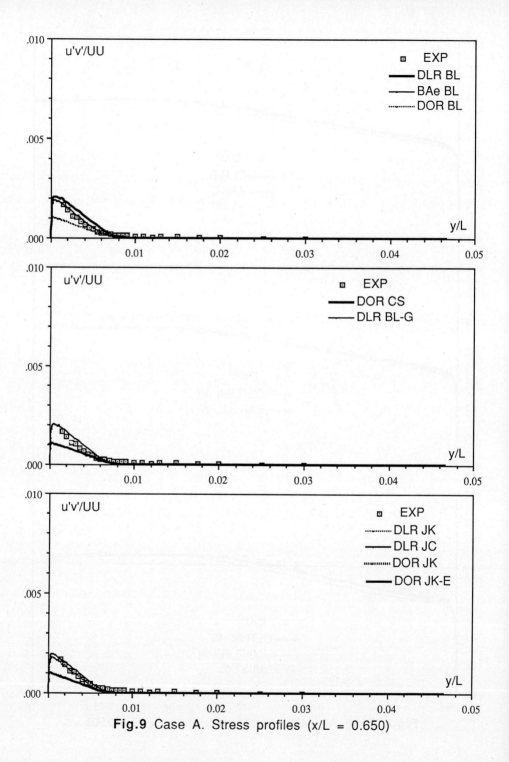

Fig.9 Case A. Stress profiles (x/L = 0.650)

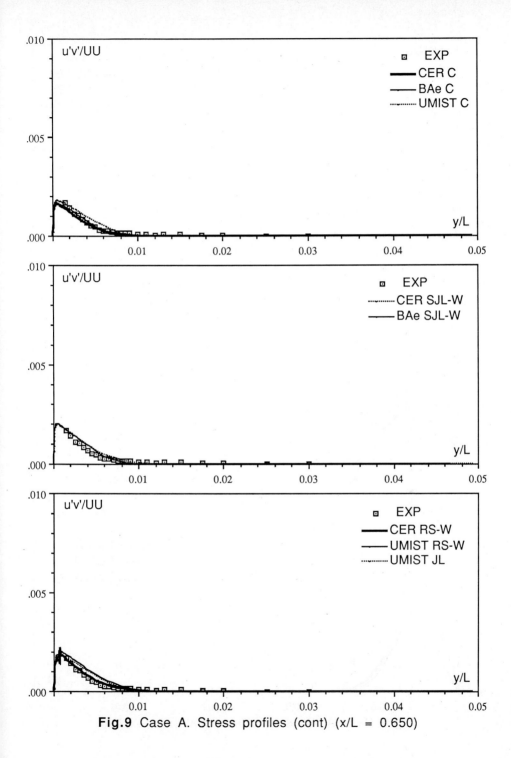

Fig.9 Case A. Stress profiles (cont) (x/L = 0.650)

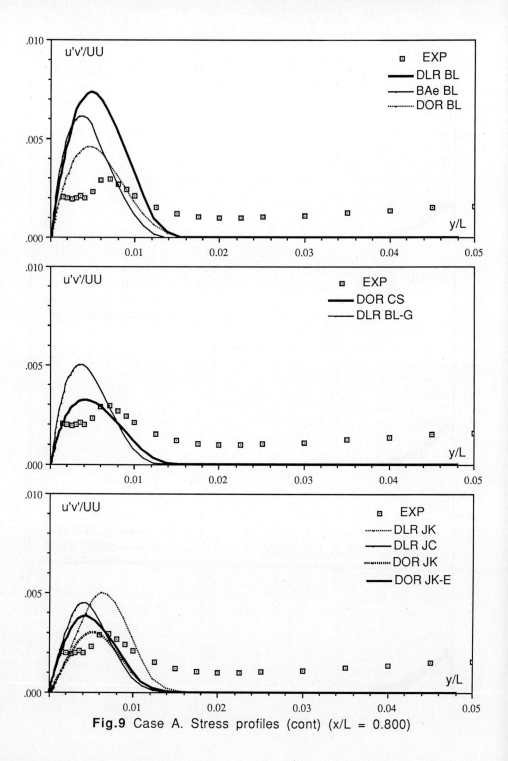

Fig.9 Case A. Stress profiles (cont) (x/L = 0.800)

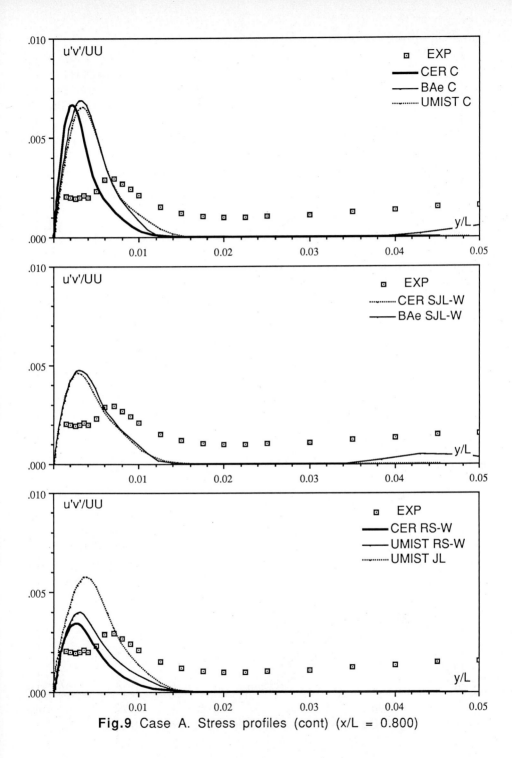

Fig.9 Case A. Stress profiles (cont) (x/L = 0.800)

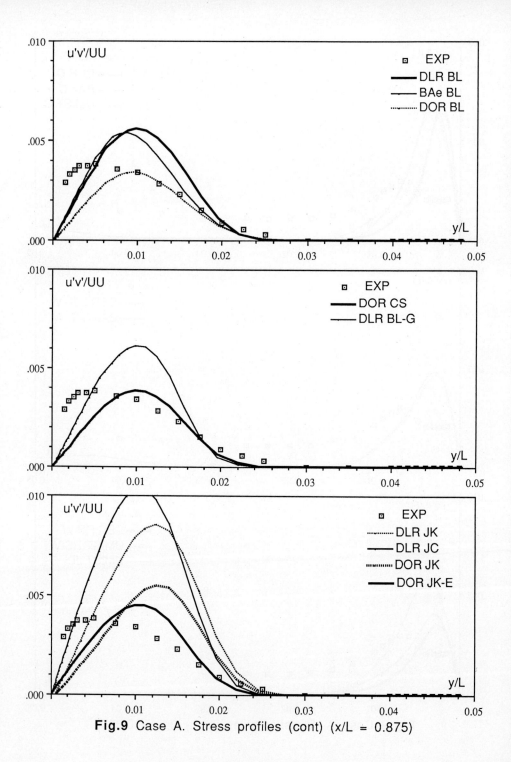

Fig.9 Case A. Stress profiles (cont) (x/L = 0.875)

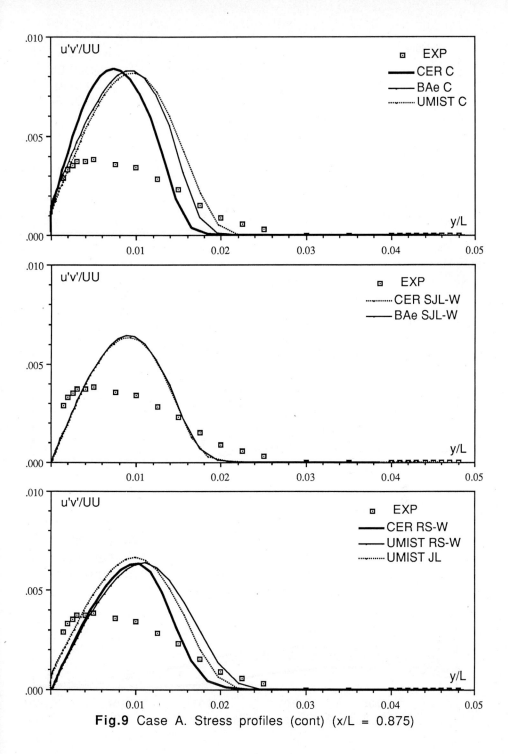

Fig.9 Case A. Stress profiles (cont) (x/L = 0.875)

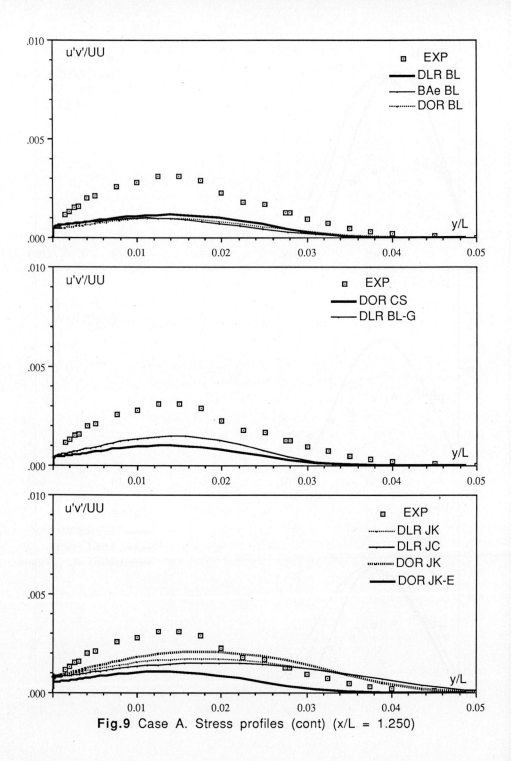

Fig.9 Case A. Stress profiles (cont) (x/L = 1.250)

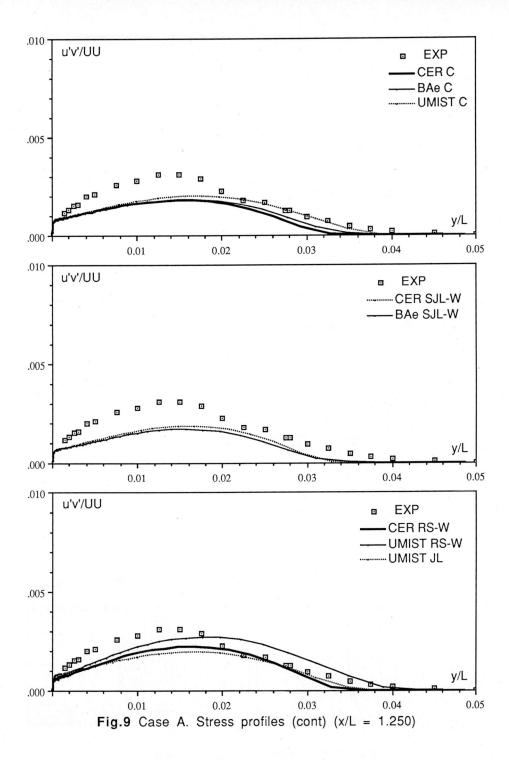

Fig.9 Case A. Stress profiles (cont) (x/L = 1.250)

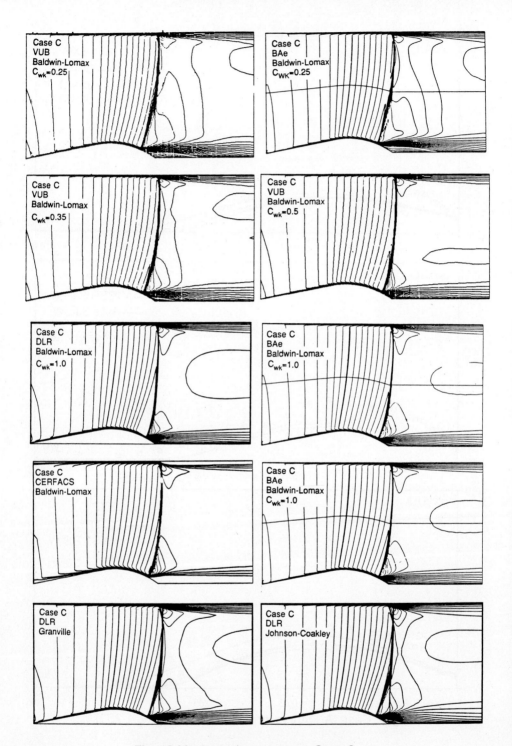

Fig. 10 Mach-number contours, Case C

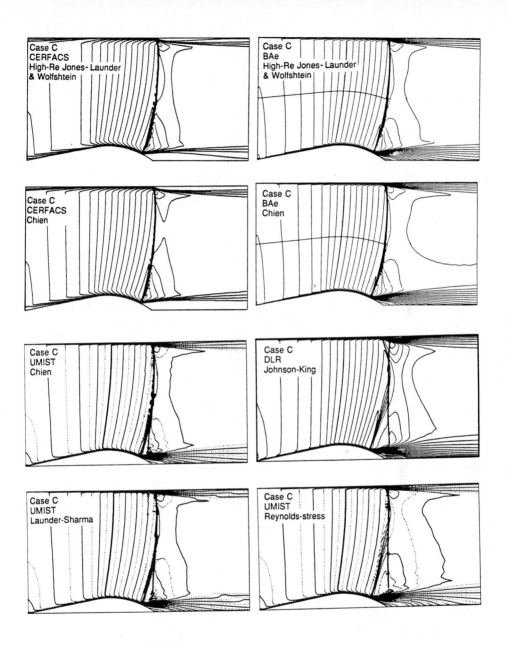

Fig. 10 (cont) Mach-number contours, Case C

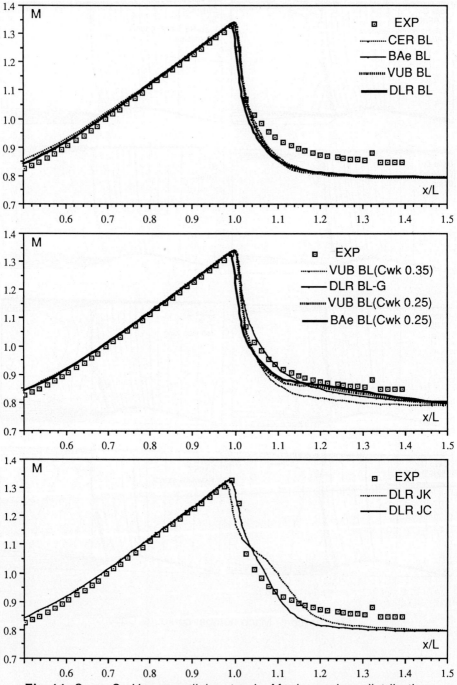

Fig.11 Case C. Upper wall isentropic Mach number distribution

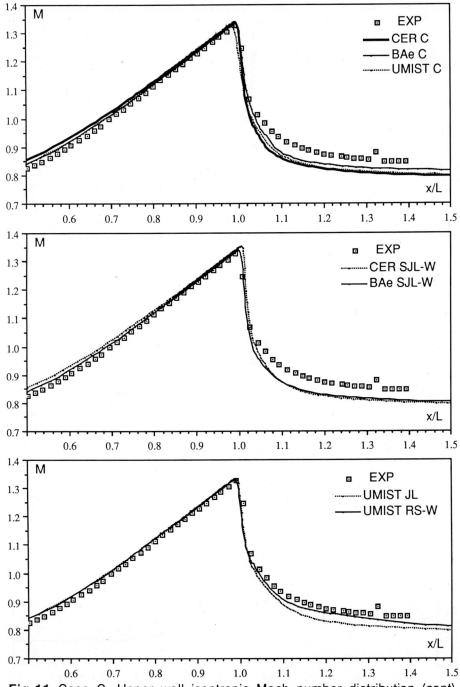

Fig.11 Case C. Upper wall isentropic Mach number distribution (cont)

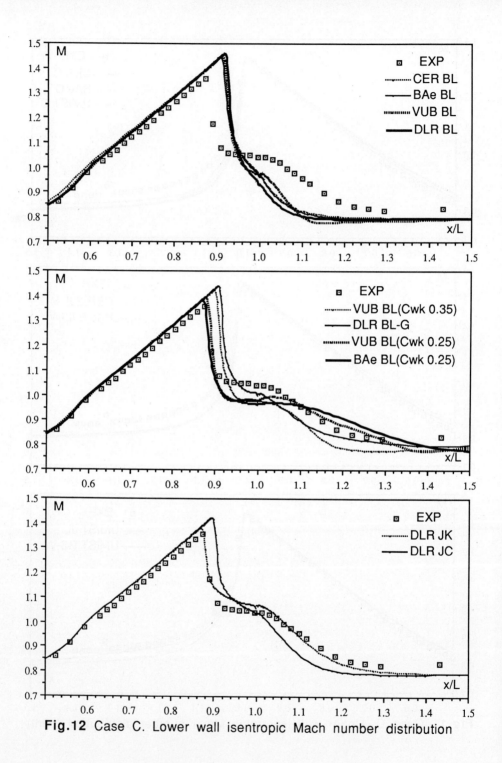

Fig.12 Case C. Lower wall isentropic Mach number distribution

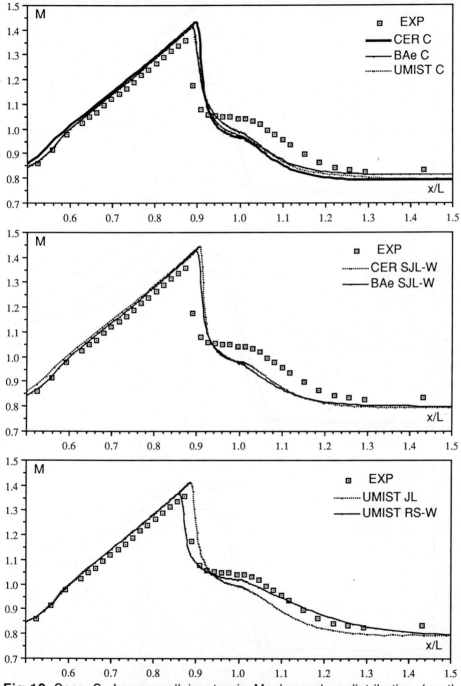

Fig.12 Case C. Lower wall isentropic Mach number distribution (cont)

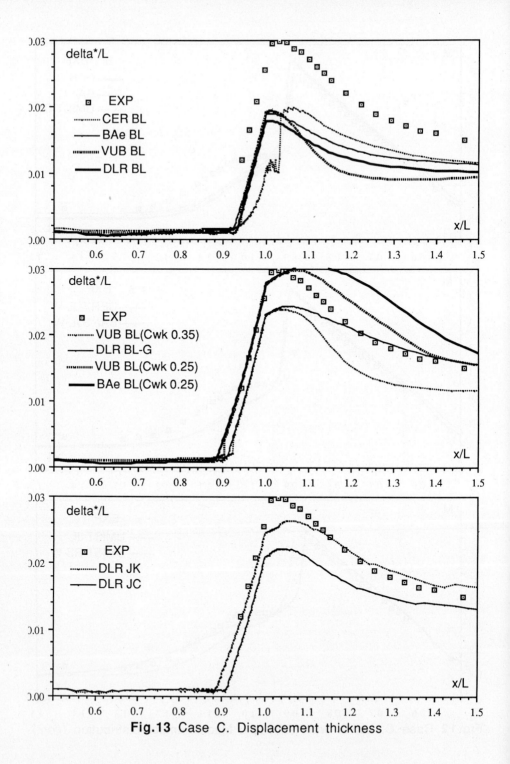

Fig.13 Case C. Displacement thickness

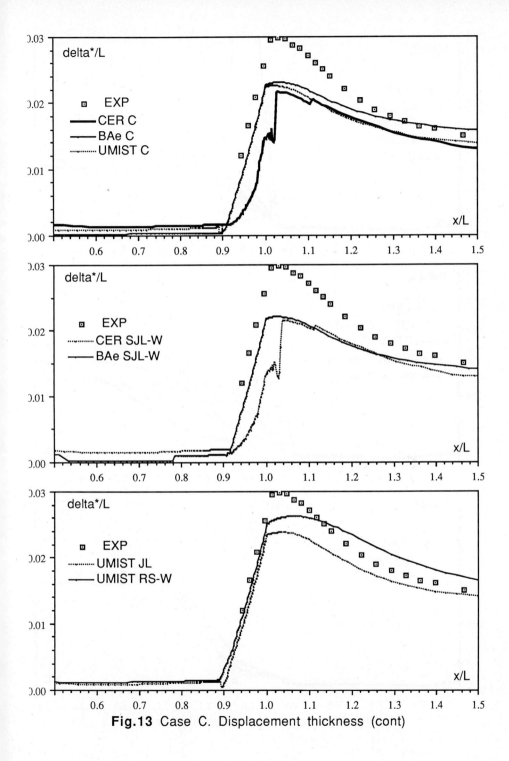

Fig.13 Case C. Displacement thickness (cont)

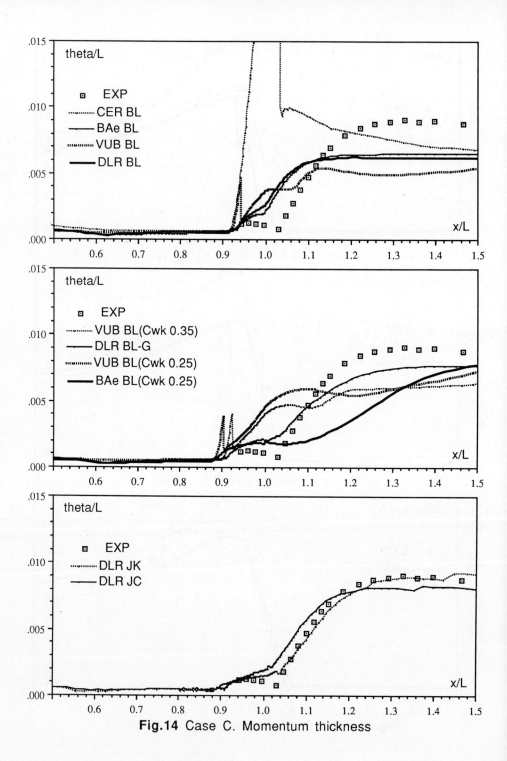

Fig.14 Case C. Momentum thickness

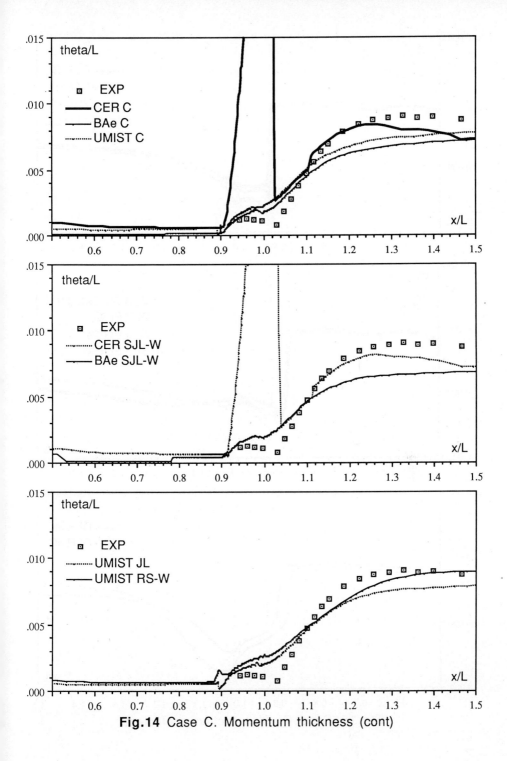

Fig.14 Case C. Momentum thickness (cont)

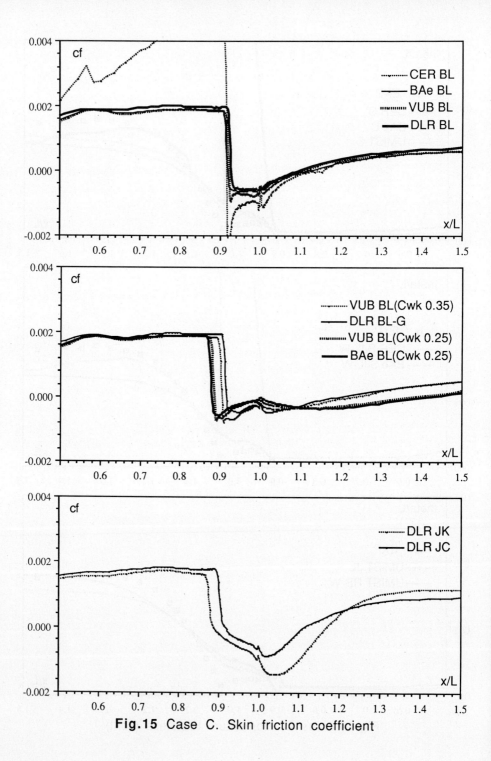

Fig.15 Case C. Skin friction coefficient

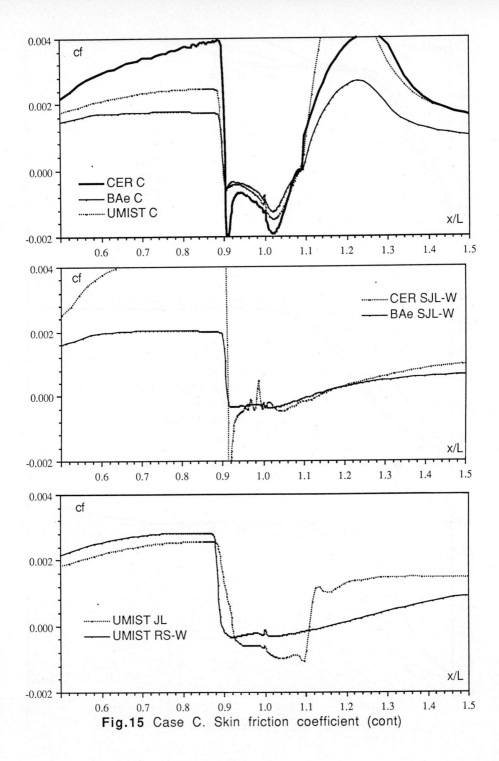

Fig.15 Case C. Skin friction coefficient (cont)

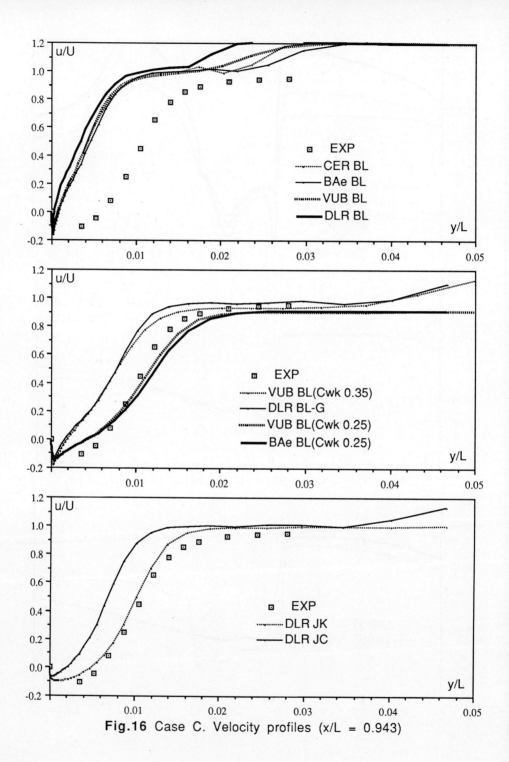

Fig.16 Case C. Velocity profiles (x/L = 0.943)

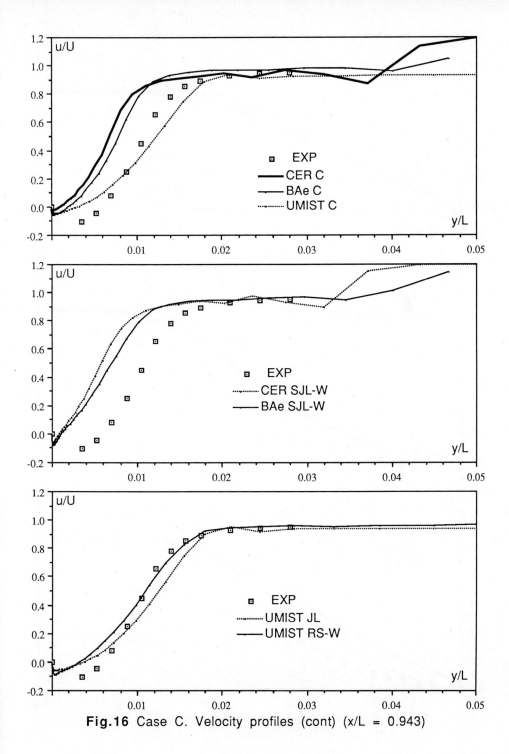

Fig.16 Case C. Velocity profiles (cont) (x/L = 0.943)

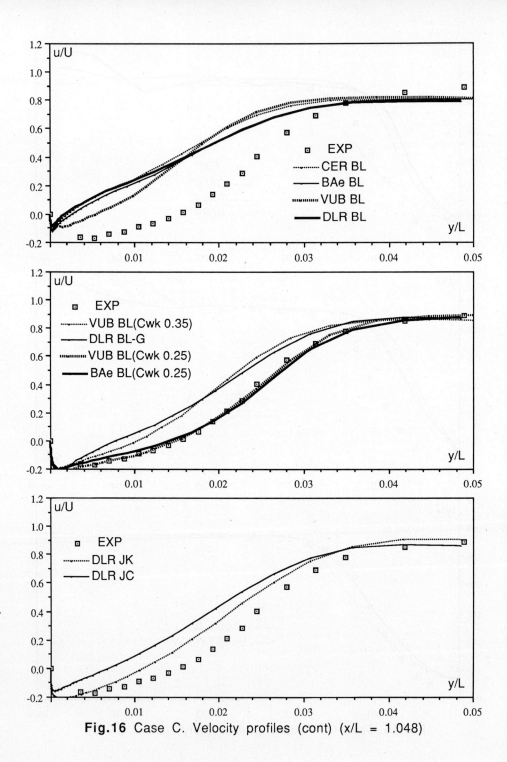

Fig.16 Case C. Velocity profiles (cont) (x/L = 1.048)

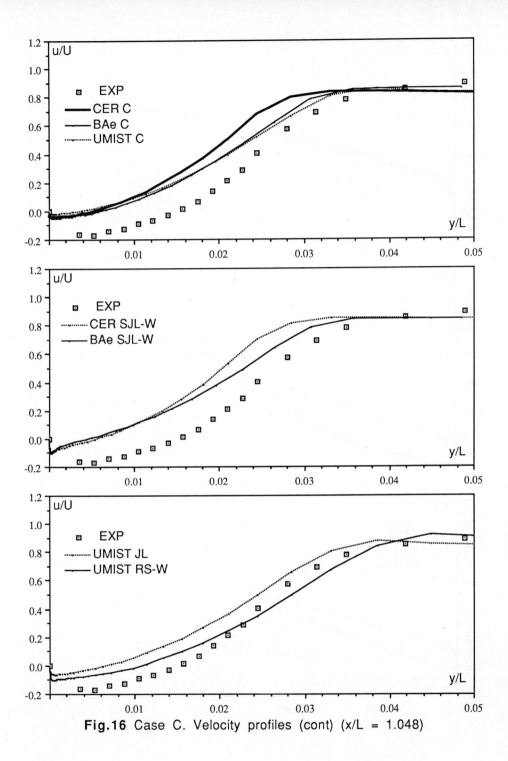

Fig.16 Case C. Velocity profiles (cont) (x/L = 1.048)

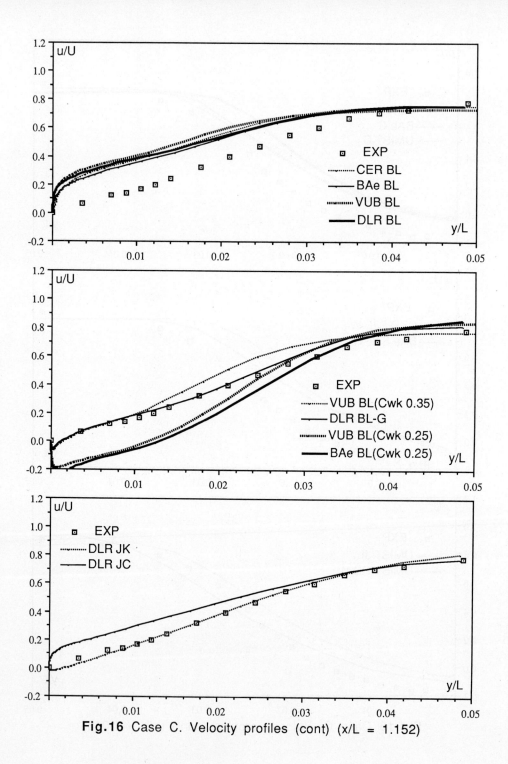

Fig.16 Case C. Velocity profiles (cont) (x/L = 1.152)

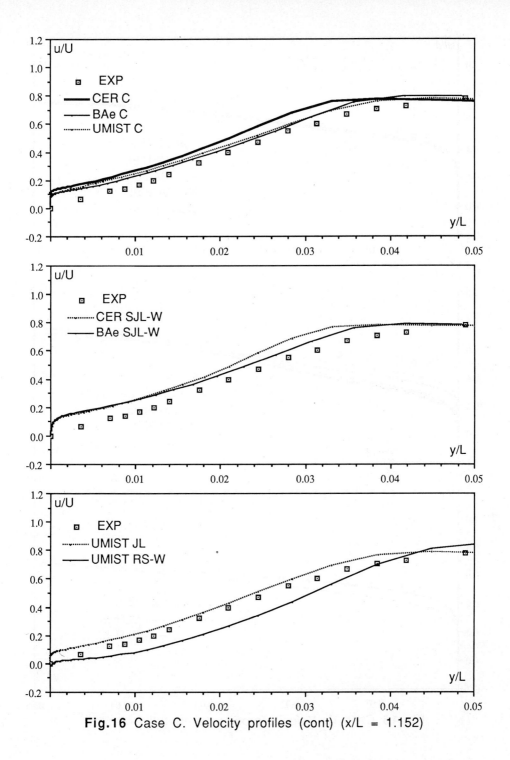

Fig.16 Case C. Velocity profiles (cont) (x/L = 1.152)

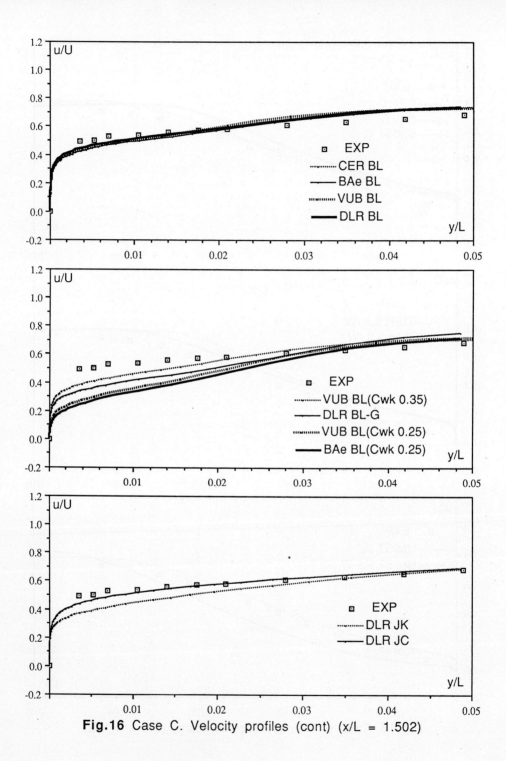

Fig.16 Case C. Velocity profiles (cont) (x/L = 1.502)

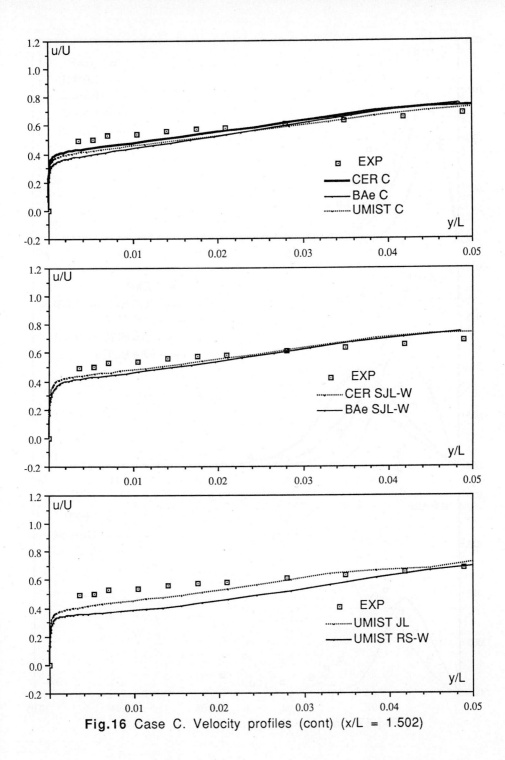

Fig.16 Case C. Velocity profiles (cont) (x/L = 1.502)

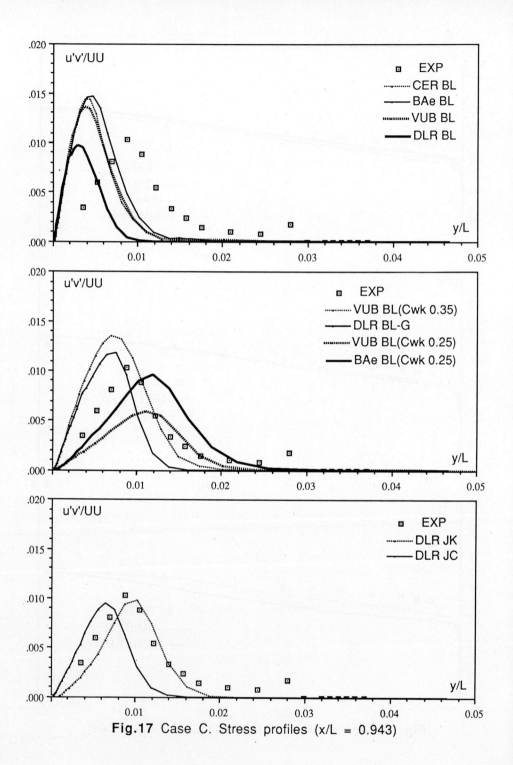

Fig.17 Case C. Stress profiles (x/L = 0.943)

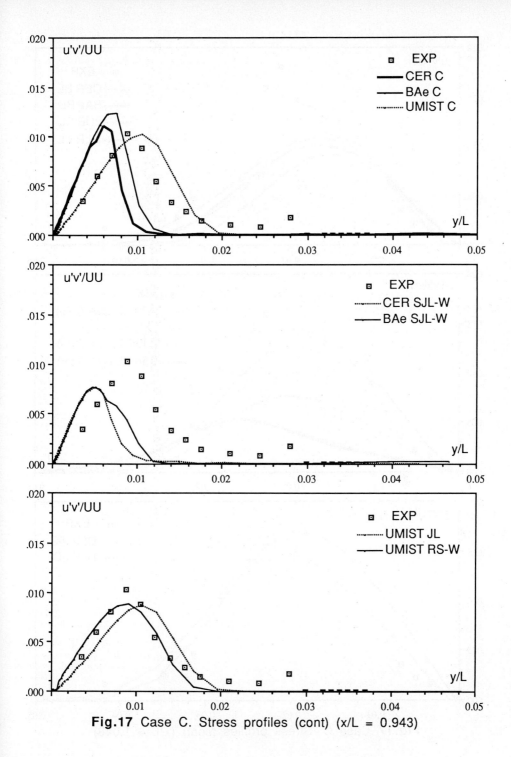

Fig.17 Case C. Stress profiles (cont) (x/L = 0.943)

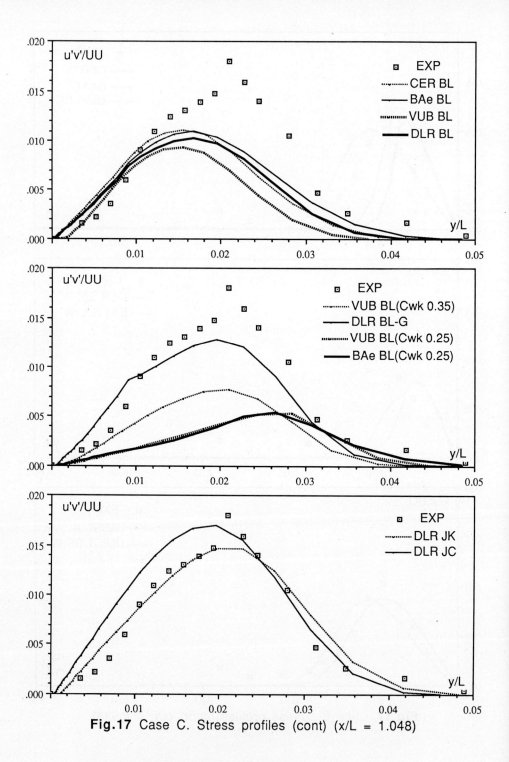

Fig.17 Case C. Stress profiles (cont) (x/L = 1.048)

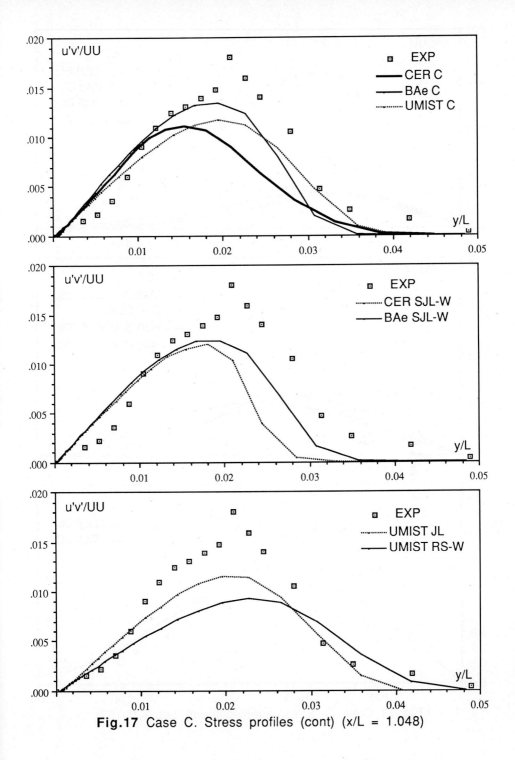

Fig.17 Case C. Stress profiles (cont) (x/L = 1.048)

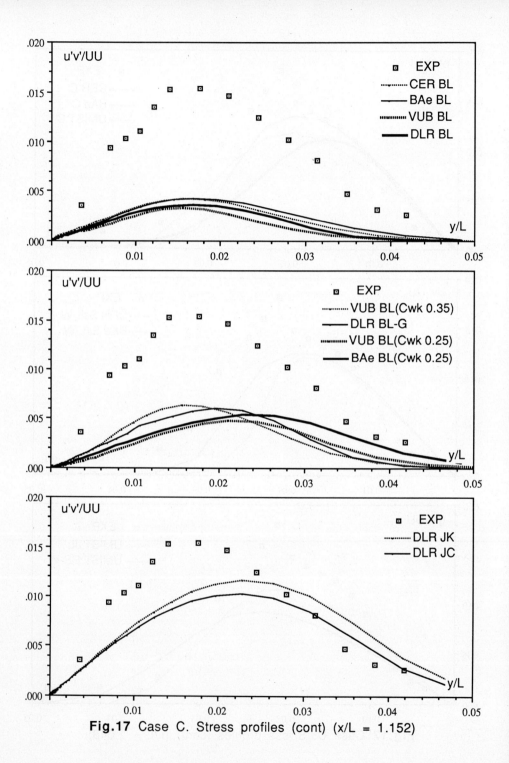

Fig.17 Case C. Stress profiles (cont) (x/L = 1.152)

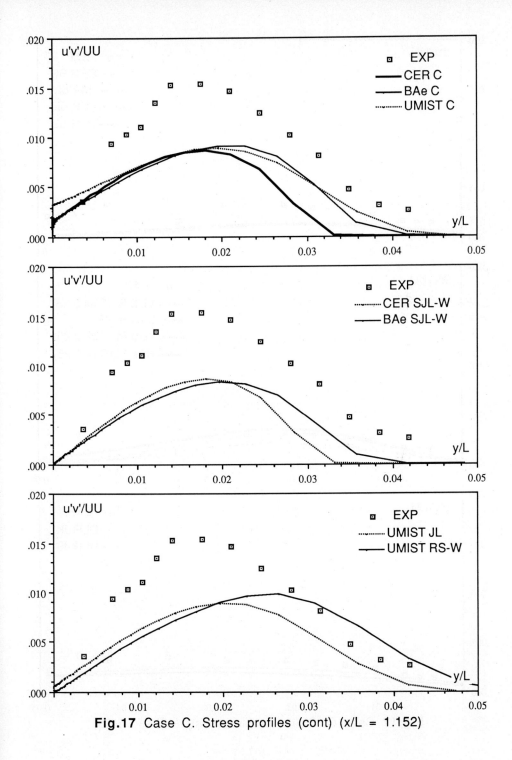

Fig.17 Case C. Stress profiles (cont) (x/L = 1.152)

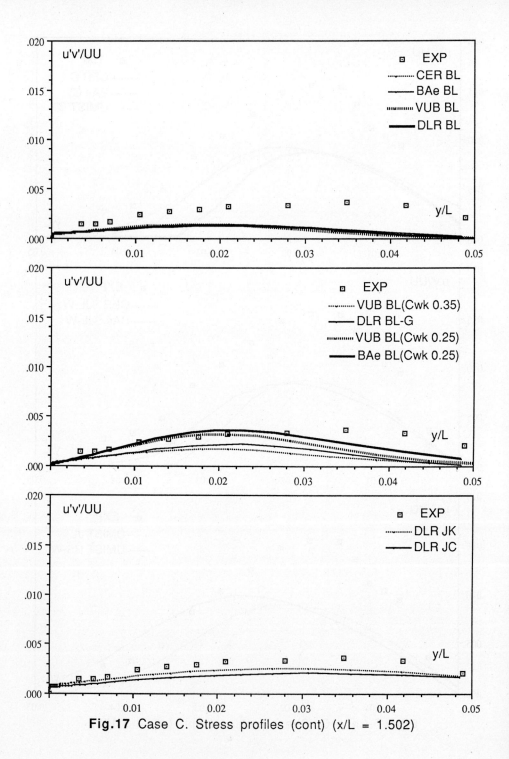

Fig.17 Case C. Stress profiles (cont) (x/L = 1.502)

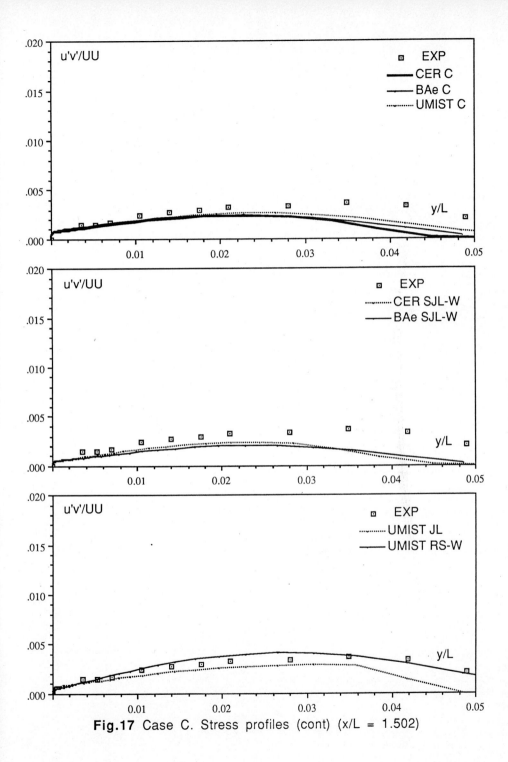

Fig.17 Case C. Stress profiles (cont) (x/L = 1.502)

MAXIMUM LIFT INVESTIGATIONS FOR A-AIRFOIL

5.3

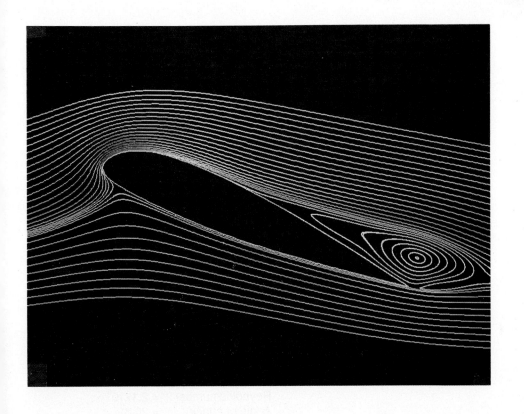

Colour figure - provided by NLR - shows:

**Mach number contours for
2D subsonic flow over an airfoil near maximum-lift (stall) conditions:**

A-Airfoil:

Mach number = 0.15
Reynolds number = 2.1×10^6
Angle of attack = 20.3^0
Transition at x/c = 0.30 (lower surface)
0.05 (upper surface)
Turbulence model BL + GB

5.3 Maximum Lift Investigations for A–Airfoil
Author: **A. Kourta**, CERFACS
Plot Coordinator: **F.J. Brandsma**, NLR

5.3.1 Partners Versus Methods and Models

An overview of the numerical methods and turbulence models used by the partners is presented in Table 1.

5.3.2 Discussion of Results

Different numerical schemes have been used by the partners to compute the A-airfoil at Ma=.15, $Re = 2.1 \times 10^6$ ($\alpha = 7.2^0$, 12.3^0, 13.3^0 and 15.3^0) and $Re = 5.25 \times 10^6$ ($\alpha = 8.08^0$, 13.1^0 and 15.11^0). Several turbulence models have also been tested. Below, these results are compared with experimental values obtained by the F1 and the F2 wind tunnels at ONERA/FAUGA. Figure 1 shows the mandatory grid used.

We begin by comparing the force coefficients C_l and C_d (Table 2 and 3). In general, the lift coefficient is over-predicted and the drag coefficient is under-predicted, regardless of the Reynolds number and the angle of attack. The lift coefficient is over-predicted by all the partners. However, the drag coefficient is over-predicted by SAAB and TUD and under-predicted by BAe, NLR and CERFACS. The values predicted by TUD with an incompressible solver and a Cebeci-Smith model and those predicted by NLR with the Baldwin-Lomax / Goldberg backflow model show good agreement with the experimental data at low angles of attack. However, the results obtained by CERFACS, calculated with the help of an algebraic stress model (RS), demonstrate the best agreement over the range of incidences considered.

The values of the pressure coefficient distribution C_p obtained by the partners (Fig. 2, 10, 11, 24 and 28) are in good agreement with experimental values at low incidence. However, with the exception of CERFACS's computations using RS model, large discrepancies exist at higher angles of attack, these discrepancies increasing with the incidence. In general, no separation region (or a separation that is too weak) is predicted close to the trailing edge. This is apparent in the aft C_p distribution. There is also a significant over-prediction of the leading edge suction at higher angles of attack. There is, in fact a coupling between the significant over-prediction of the suction peak on the upper surface leading edge and the too weak separation at the trailing edge. Too weak separation implies too high circulation which in turn will give a too high suction peak in the nose. The leading edge suction is only correctly estimated if the trailing edge viscous interaction is captured. This could be the main reason for overestimates of C_l by using other models than RS model. Meanwhile, results for the skin friction coefficient distribution C_f compare reasonably with experiments for all angles of incidence and do not indicate any problems with the separation region (Fig. 3, 12, 13, 25 and 29). However, the results obtained by TUD show an over-estimation of this coefficient but generally present quite good agreement.

Boundary layer thickness, displacement thickness (Fig. 4, 14, 15, 26 and 30) and momentum thickness (Fig. 5, 16, 17, 27 and 31) are generally well estimated. This agreement becomes poor near the trailing edge, at the approach of separation and in the separation region. The RS model shows an overprediction of momentum thickness.

With the exception of CERFACS's RS model results, a comparison of the velocity profile (Fig. 6, 7, 18, 19 and 20) clearly indicates the discrepancy between computations and experiments for higher angles of attack. The midchord stations show fairly well-predicted velocity profiles, whereas in the aft stations it would seem that the computation predicts a backflow region that is too thin. Once again, better agreement with experiment is obtained by using the algebraic stess model (RS).

In general, the turbulent shear stress profiles (Fig. 8, 9, 21, 22 and 23) show poor agreement with experimental data, the computed levels being too high close to the airfoil surface in the trailing edge region. Somewhat better agreement with the experiment is obtained by using RS model.

Reynolds number effects are well–predicted for all Reynolds numbers studied.

Grid dependence has also been studied (Table 2 and 3), although no real effects have been seen.

5.3.3 Limits of Applicability

All of the numerical methods used are able to compute the flow around the A-Airfoil at a moderate angle of attack. At higher angles of attack, when separation occurs, it is possible to predict this flow only with a suitable turbulence model. The algebraic stress model yields the best results for the present test cases.

5.3.4 Turbulence Model Validation

Several turbulence models have been used by the partners: algebraic turbulent models (Baldwin-Lomax, Cebeci-Smith, Baldwin-Lomax with the Goldberg backflow model), two layer models ($k - \epsilon$ with Chen-Patel sublayer model, $k - \epsilon$ with Wolfshtein sublayer model); and the algebraic stress model.

The best greement with the experimental values is obtained for low incidences. In the case of separated flows, straight-forward algebraic eddy viscosity models fail to predict realistic mean and turbulent quantities. Improvements can be obtained for the force coefficients by using the Golderg backflow model. Although the $k - \epsilon$ models used in this project gives better results, the separated region close to the trailing edge is still poorly predicted. In conclusion, these turbulence models (algebraic and $k - \epsilon$ models) are inadequate for predicting flows at higher incidence. The results which agree most with the experiments are those obtained by using the algebraic stress model. This turbulence model has been shown to be able to predict stall. The main reason for the superiority of the RS model is believed to be its ability to account for the influence of streamline curvature and the anisotropic character in the normal Reynolds stresses.

It seems that the use of the RS model predicts this better flow than the eddy viscosity model. So, the RS models are more suitable to compute this kind of flow. The RS model approach used in this study (by CERFACS) has some limitations e.g. the eddy viscosity necessary near the wall and the unsatisfactory modelling of the near wall correction in the pressure-strain terms.

5.3.5 Matrix of Distinct Values

The calculated force coefficients are summarized in Table 2 for $\mathrm{Re} = 2.10 \mathrm{x} 10^6$ cases and in Table 3 for $\mathrm{Re} = 5.25 \mathrm{x} 10^6$ cases.

Table 1 Numerical Methods and Turbulence Models

Partners	Numerical Methods			Equations	Turbulence Models
	time stepping	spatial discretization	Convergence Accel.		
BAe	Explicit Runge-Kutta	Cell Centered Finite Volume Artificial Dissipation	Residual Smoothing Multigrid	Averaged N.S. $+k-\epsilon$	Baldwin-Lomax $k-\epsilon$/Wolfshtein
CERFACS	Explicit Runge-Kutta Semi-Implicit ADI	Cell Centered Finite Volume Artificial Dissipation		Averaged N.S. $k-\epsilon$	$k-\epsilon$ / Chien ASM
NLR	Explicit Runge-Kutta	Cell Vertex Finite Volume Artificial Dissipation	Residual Averaging Multigrid	Averaged N.S.	Baldwin-Lomax B-L/Goldberg
SAAB	Explicit Runge-Kutta	Cell Centered Finite Volume Artificial Dissipation	Multigrid	Averaged N.S.	Baldwin-Lomax $k-\epsilon$/Wolfshtein
SAAB/HUT	Implicit Scheme (Factorization)	Cell Centered Finite Volume Van Leer Splitting	Implicit Multigrid	Thin Layer N.S.	Cebeci-Smith
TUD	SIMPLE	Finite Volume Artificial Dissipation	Implicit	Averaged N.S. Incompressible	Cebeci-Smith

Table 2 Force Coefficients (Ma=0.15, Re=2.10x10^6)

Partners	turb. mod.	$\alpha = 7.2^0$		$\alpha = 12.3^0$		$\alpha = 13.3^0$		$\alpha = 15.3^0$	
		C_l	C_d	C_l	C_d	C_l	C_d	C_l	C_d
SAAB	BL	1.04	0.0130	1.56	0.0196	1.75	0.0228	1.85	0.0278
BAe	BL	1.07	0.0118	1.62	0.0156	1.72	0.0181	1.89	0.0197
NLR	BL	0.99	0.0119	1.48	0.0182	1.57	0.0225		
NLR	BL/GB	1.07	0.0130	1.46	0.0171	1.52	0.0185	1.67	0.0259
SAAB/HUT	CS	1.05	0.0125	1.57	0.0187	1.66	0.0207	1.83	0.0256
TUD	CS	1.02	0.0130	1.52	0.0188	1.63	0.0217	1.78	0.0280
CERFACS	$k - \epsilon$/C	1.04	0.0119			1.64	0.0173	1.77	0.0311
BAe	$k - \epsilon$/W	1.07	0.0121	1.60	0.0162	1.69	0.0178	1.87	0.0210
SAAB	$k - \epsilon$/W	1.02	0.0132			1.638	0.0212		
CERFACS	ASM	1.00	0.0120			1.53	0.0208	1.64	0.0227
Expt. F1		1.02	0.0128	1.49	0.0184	1.56	0.0204	1.67	0.0267
Expt. F2		1.02	0.0136			1.52	0.0308	1.29	

Table 3 Force Coefficients (Ma=0.15, Re=5.25x10^6)

Partners	turb. mod.	$\alpha = 8.08^0$		$\alpha = 13.1^0$		$\alpha = 15.11^0$	
		C_l	C_d	C_l	C_d	C_l	C_d
SAAB	BL	1.137	0.0125	1.651	0.0206	1.827	0.0265
BAe	BL	1.18	0.0127	1.72	0.0176	1.92	0.0200
NLR	BL/GB	1.18	0.0083	1.67	0.0172	1.82	0.0209
SAAB/HUT	CS	1.15	0.0124	1.66	0.0186	1.85	0.0227
TUD	CS	1.13	0.0125	1.62	0.0169	1.79	0.0230
BAe	$k - \epsilon$/W	1.16	0.0131	1.69	0.0183	1.89	0.0210
CERFACS	$k - \epsilon$/C	1.15	0.0108	1.64	0.0205	1.80	0.0193
Expt. F1		1.12	0.0127	1.59	0.0187	1.72	0.0241

Table 4 Number of contributions submitted by the partners

Partners	Ma=0.15, Re=2.10$10^6$				Ma=0.15, Re=5.25$10^6$		
	$\alpha = 7.2^0$	$\alpha = 12.3^0$	$\alpha = 13.3^0$	$\alpha = 15.3^0$	$\alpha = 8.08^0$	$\alpha = 13.1^0$	$\alpha = 15.1^0$
SAAB	3	2	3	2	2	2	2
BAe	2	2	3	2	2	2	2
NLR	2	2	2	1	1	1	1
TUD	1	1	1	1	1	1	1
CERFACS	2	0	2	2			

Table 5 Grids used

Partners	grids	size	$y^+(1)$
BAe	BAe	256x64	≤2
	NLR	384x64	≤3
CERFACS	CERFACS	353x65	≤1
	NLR	384x64	
NLR	NLR	384x64	
SAAB	SAAB	384x64	
		384x128	≤1
TUD	NLR	384x64	

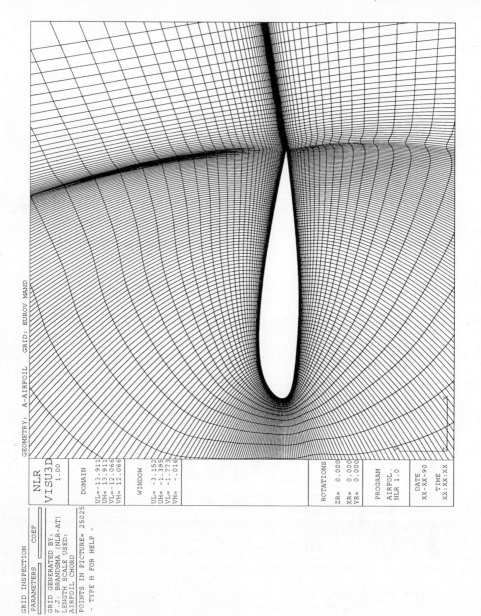

Fig. 1 Mandatory grid

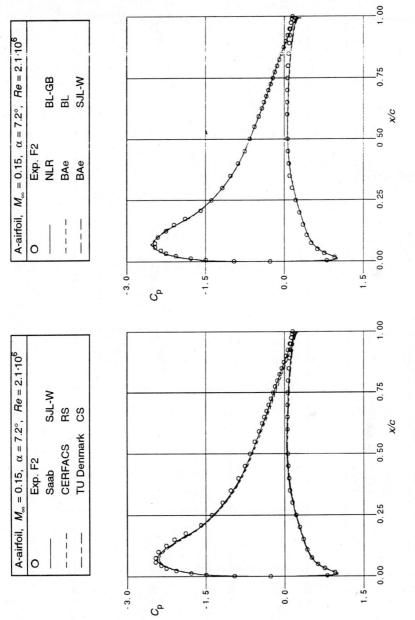

Fig. 2 Pressure coefficient distribution ($Re = 2.10 \times 10^6$, $\alpha = 7.2^0$)

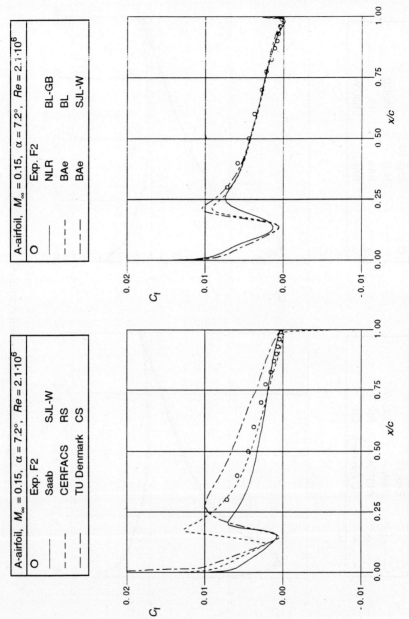

Fig. 3 Skin friction distribution ($Re = 2.10 \times 10^6$, $\alpha = 7.2°$)

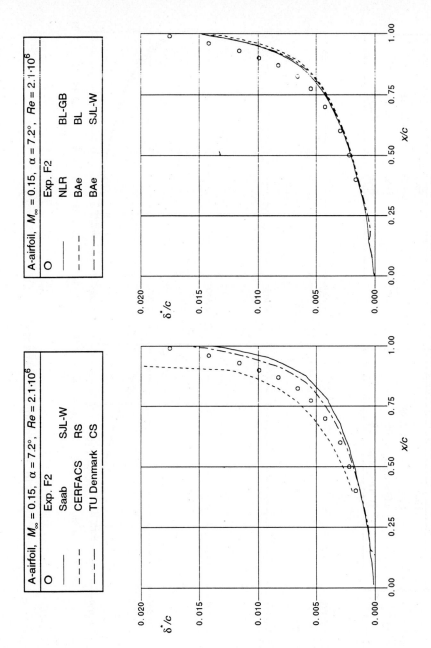

Fig. 4 Displacement thickness ($Re = 2.10 \times 10^6$, $\alpha = 7.2°$)

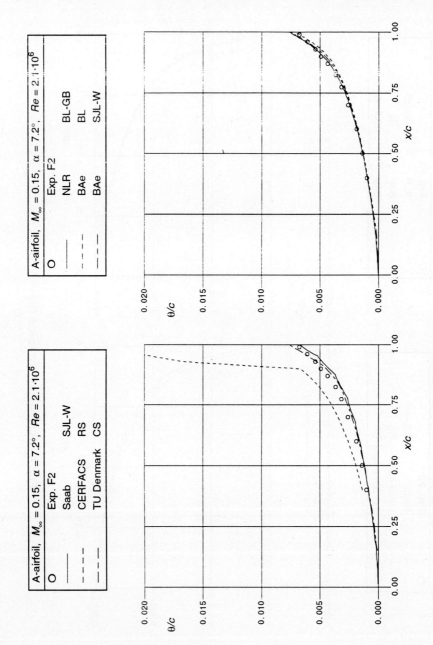

Fig. 5 Momentum thickness ($Re = 2.10 \times 10^6$, $\alpha = 7.2^0$)

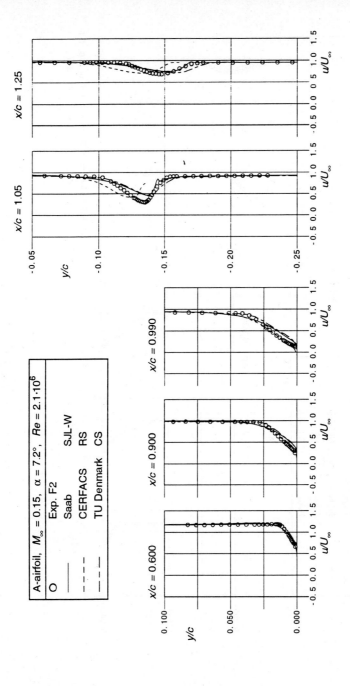

Fig. 6 Velocity profiles (Re = 2.10×10^6, $\alpha = 7.2^0$)

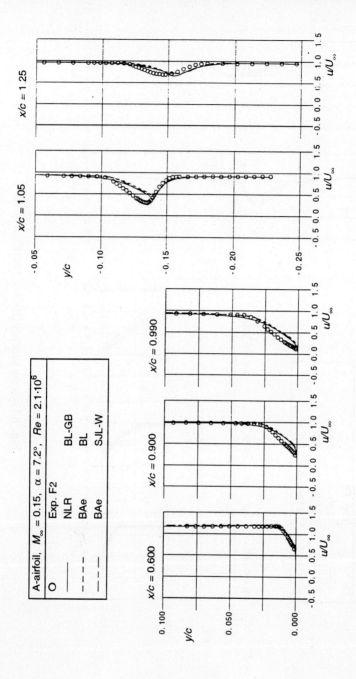

Fig. 7 Velocity profiles (Re = 2.10×10^6, $\alpha = 7.2^0$)

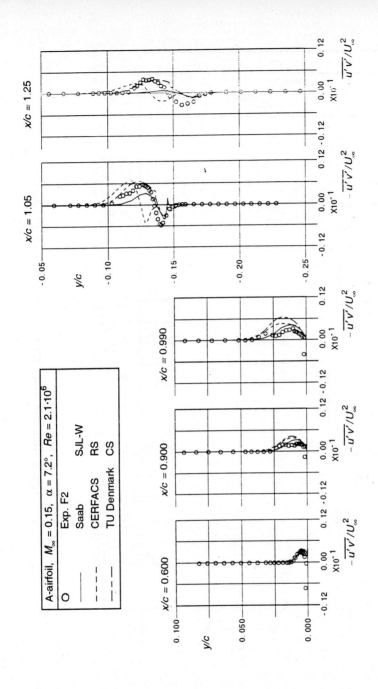

Fig. 8 Turbulent shear stress ($Re = 2.10 \times 10^6$, $\alpha = 7.2^0$)

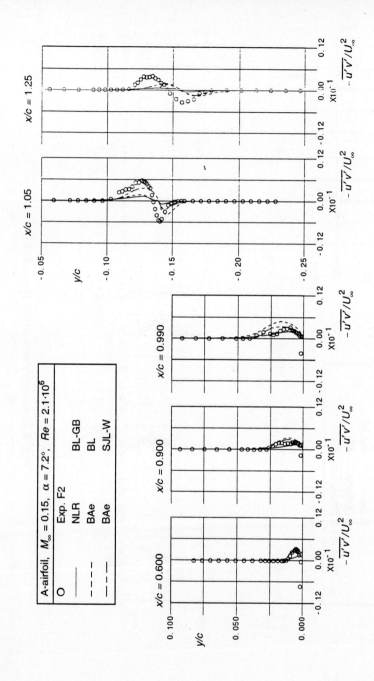

Fig. 9 Turbulent shear stress ($Re = 2.10 \times 10^6$, $\alpha = 7.2^0$)

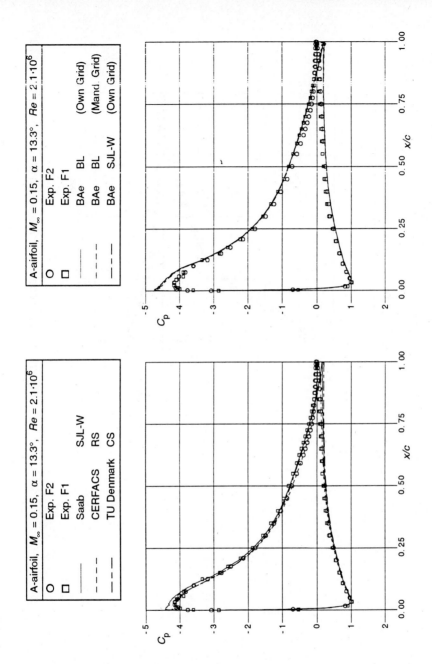

Fig. 10 Pressure coefficient distribution ($Re = 2.10 \times 10^6$, $\alpha = 13.3^0$)

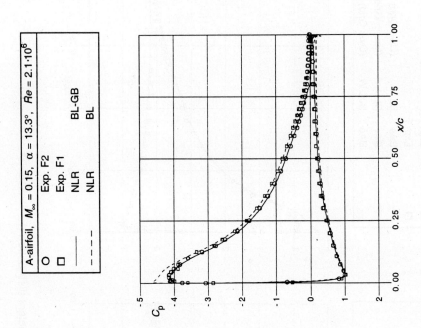

Fig. 11 Pressure coefficient distribution (Re = 2.10×10^6, $\alpha = 13.3^0$)

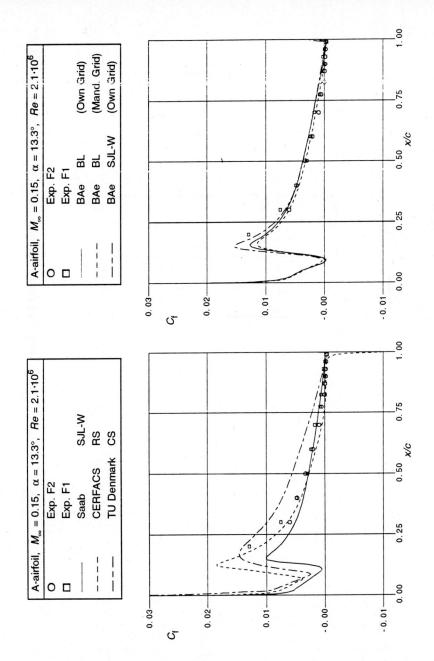

Fig. 12 Skin friction distribution ($Re = 2.10 \times 10^6$, $\alpha = 13.3^0$)

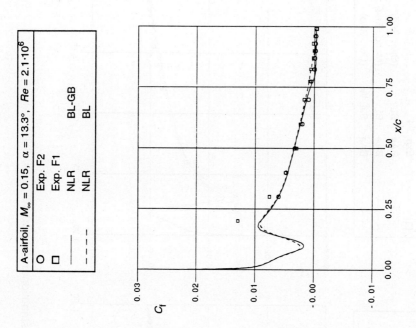

Fig. 13 Skin friction distribution (Re = 2.10×10^6, $\alpha = 13.3^0$)

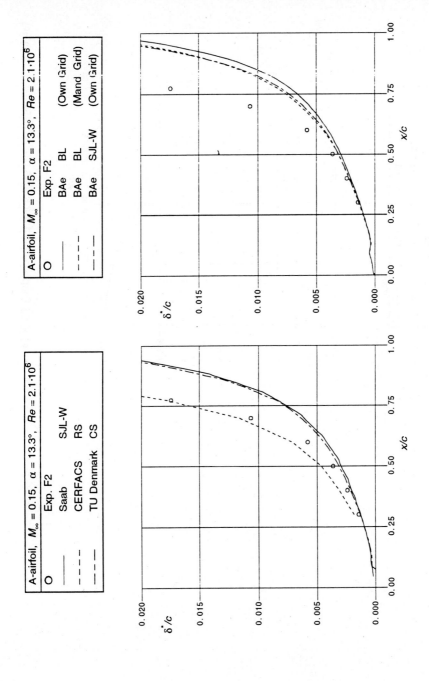

Fig. 14 Displacement thickness ($Re = 2.10 \times 10^6$, $\alpha = 13.3^0$)

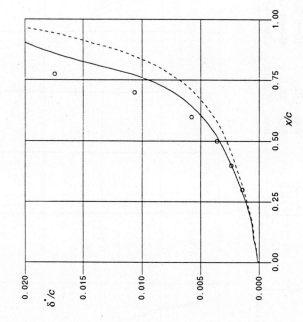

Fig. 15 Displacement thickness (Re = 2.10×10^6, $\alpha = 13.3^0$)

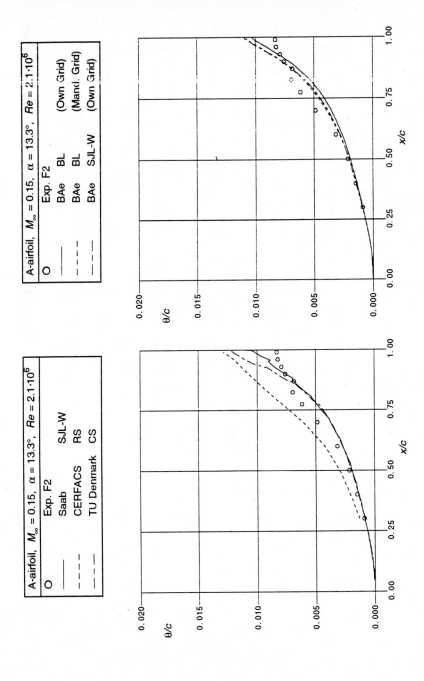

Fig. 16 Momentum thickness ($Re = 2.10 \times 10^6$, $\alpha = 13.3°$)

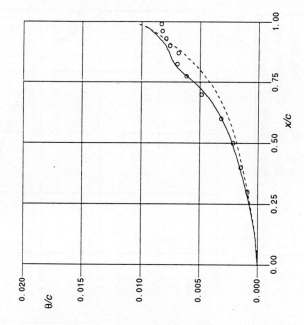

Fig. 17 Momentum thickness ($Re = 2.10 \times 10^6$, $\alpha = 13.3°$)

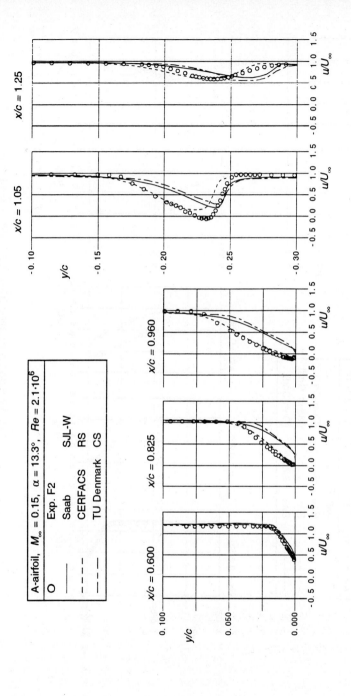

Fig. 18 Velocity profiles ($Re = 2.10 \times 10^6$, $\alpha = 13.3°0$)

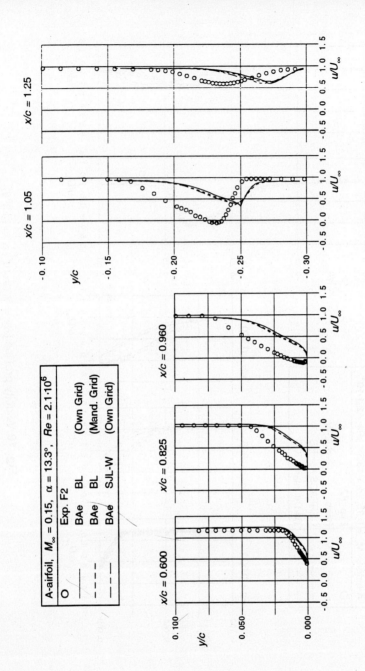

Fig. 19 Velocity profiles ($Re = 2.10 \times 10^6$, $\alpha = 13.3°0$)

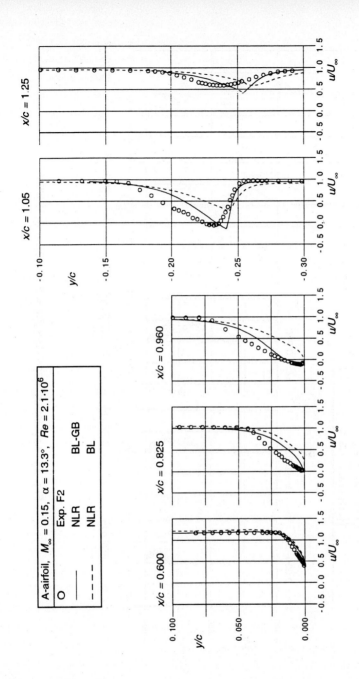

Fig. 20 Velocity profiles ($Re = 2.10 \times 10^6$, $\alpha = 13.3^0$)

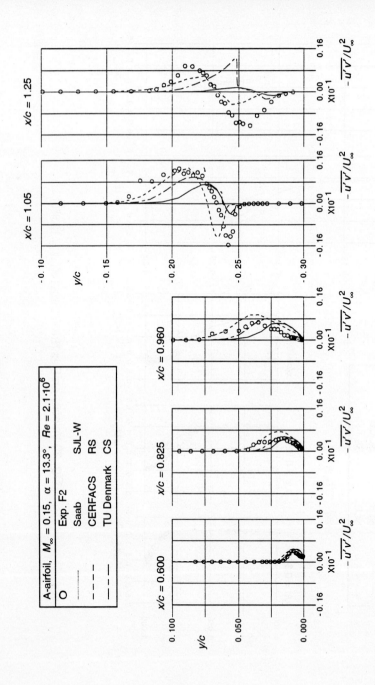

Fig. 21 Turbulent shear stress (Re = 2.10×10^6, $\alpha = 13.3^\circ$)

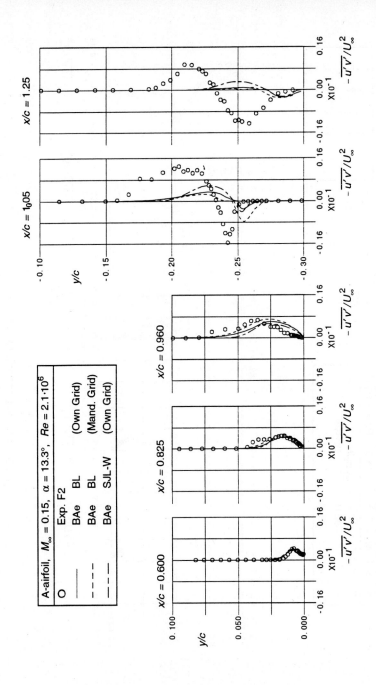

Fig. 22 Turbulent shear stress (Re = 2.10×10^6, $\alpha = 13.3°$)

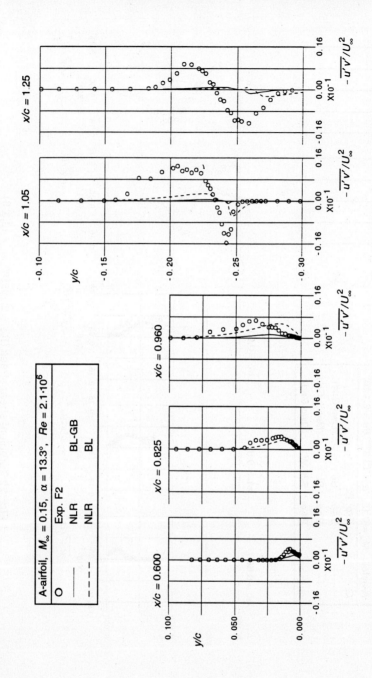

Fig. 23 Turbulent shear stress ($Re = 2.10 \times 10^6$, $\alpha = 13.3^0$)

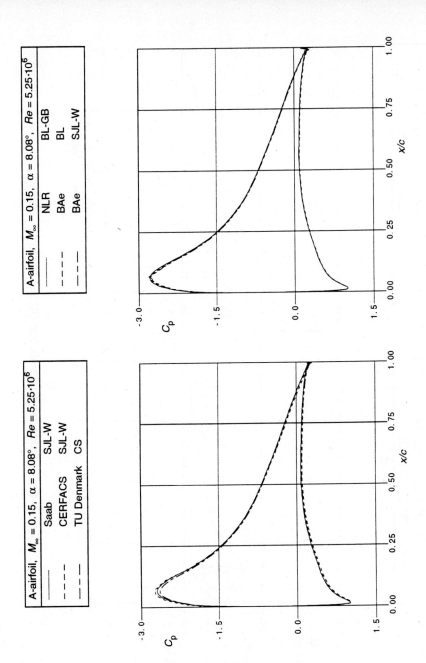

Fig. 24 Pressure coefficient distribution (Re = 5.25×10^6, $\alpha = 8.08°$)

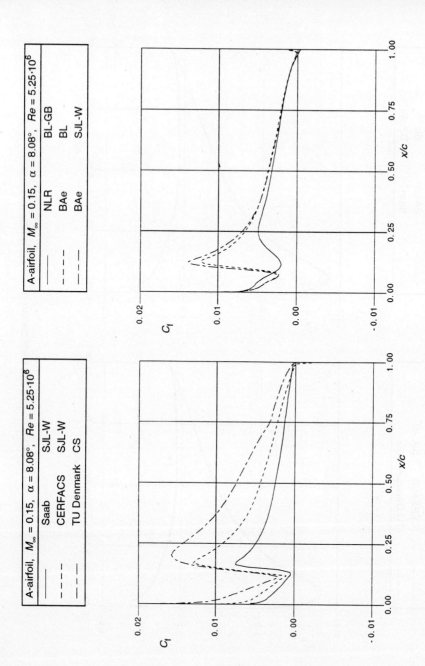

Fig. 25 Skin friction distribution ($Re = 5.25 \times 10^6$, $\alpha = 8.08°$)

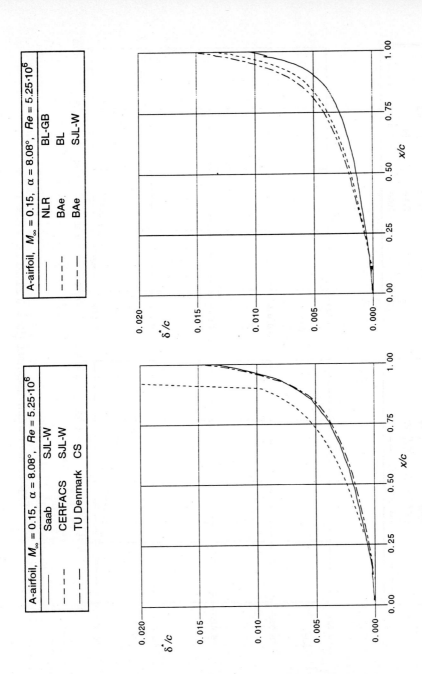

Fig. 26 Displacement thickness ($Re = 5.25 \times 10^6$, $\alpha = 8.08°$)

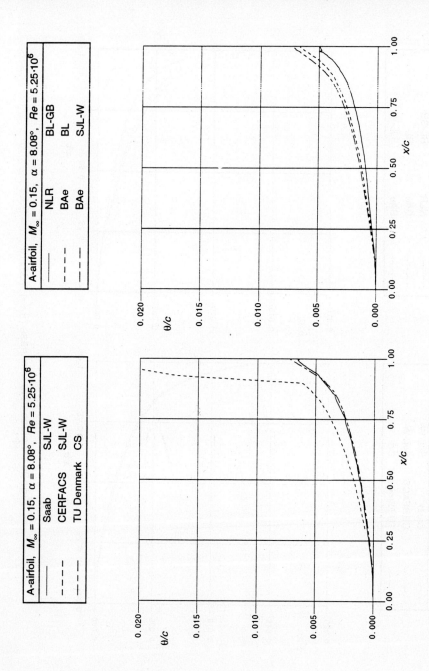

Fig. 27 Momentum thickness ($Re = 5.25 \times 10^6$, $\alpha = 8.08^0$)

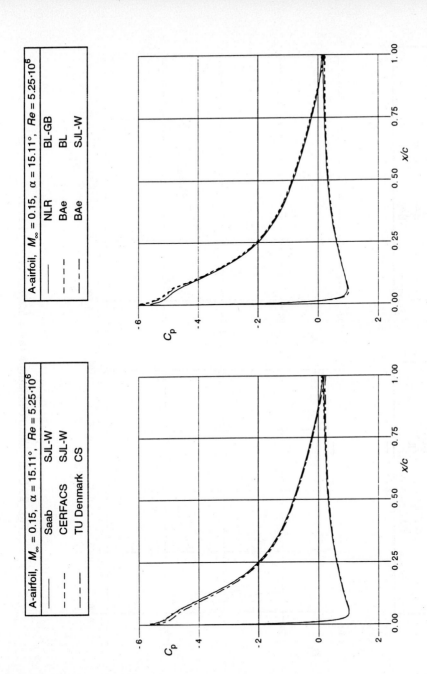

Fig. 28 Pressure coefficient distribution ($Re = 5.25 \times 10^6$, $\alpha = 15.11°$)

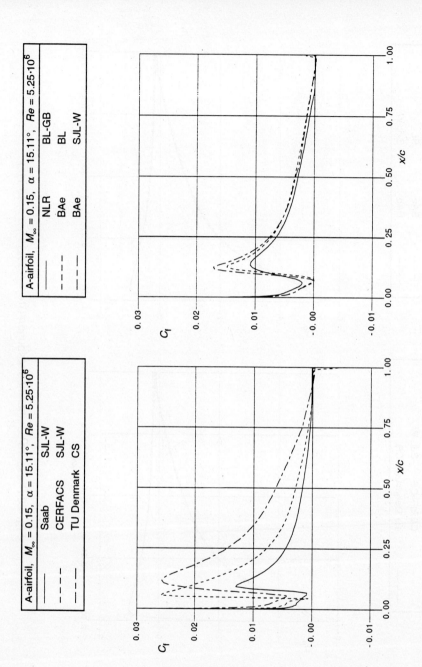

Fig. 29 Skin friction distribution (Re = 5.25×10^6, $\alpha = 15.11°$)

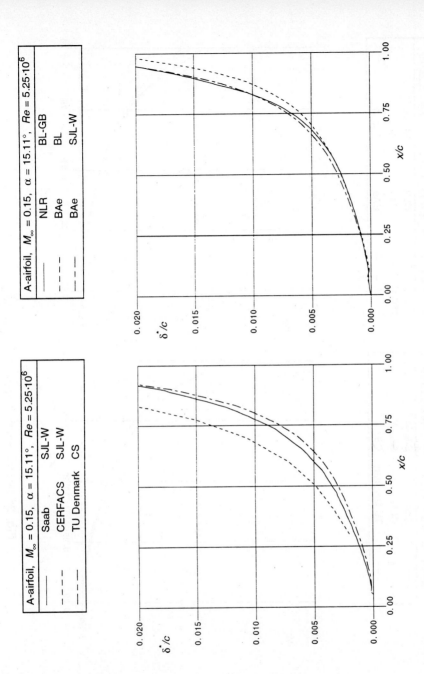

Fig. 30 Displacement thickness ($Re = 5.25 \times 10^6$, $\alpha = 15.11°$)

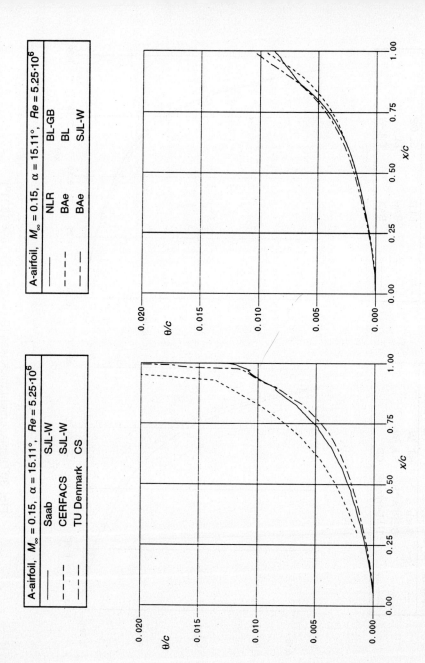

Fig. 31 Momentum thickness ($Re = 5.25 \times 10^6$, $\alpha = 15.11°$)

HIGH-LIFT INVESTIGATIONS FOR NLR 7301 FLAPPED AIRFOIL

5.4

Colour figure - provided by SAAB - shows:

**Mach number contours for
2D subsonic flow over a two-element airfoil at high angle of attack:**

NLR 7301 flapped airfoil: Mach number = 0.185
Reynolds number = 2.51×10^6
Angle of attack = 13.1^0
Transition on wing at x/c on
lower / upper surface = 0.710 / 0.035
Transition on flap at x/c on
lower / upper surface = none / 1.073
Turbulence model: SJL-W

5.4 High–Lift Investigations for NLR7301 Flapped Airfoil

Author: **F.J. Brandsma**, NLR
Plot Coordinator: **T. Larsson**, SAAB

5.4.1 Introduction

In the design of airplanes, construction of efficient high-lift devices for take-off and landing conditions is a very important issue. In order to obtain the high-lift necessary during take-off and landing conditions, wings usually are adjusted with trailing-edge flaps and/or leading-edge slats. In the design of these configurations CFD plays an increasing role. Especially the capability of CFD methods to perform simulations for full-scale free flight Reynolds numbers (which is not possible in wind-tunnel experiments), make CFD methods an important tool to support the design of high-lift devices which usually relies on extensive wind tunnel experiments. Because of the complicated nature of the flows around these configurations, where interaction between several shear layers (boundary layers and wakes) occur, it may be clear that calculation methods based on the (Reynolds-averaged) Navier-Stokes equations are expected to be necessary in order to predict the characteristics of high-lift devices accurately.

In the present work, several Navier-Stokes methods employed by the different partners in the EUROVAL project, are validated against experimental results for the NLR7301/Flap configuration described below. The objective of this validation excercise is to get insight in the predictive capability of Navier-Stokes methods for high-lift flows, and the information provided by this excercise may be used to improve the methods in order to come to more efficient tools for the design of high-lift devices. Important issues of the current Navier-Stokes methods with respect to predicting high-lift flows are turbulence modelling and also the generation of computational grids. Both aspects are covered in the present work.

5.4.2 Partners and Methods

Five partners contributed to this task: BAe, Saab, Cerfacs, Dornier, and CFD-Norway, where it should be mentioned that the contribution of Saab includes also some work carried out at the Helsinki University of Technology (HUT). In Table 1, an overview of the numerical methods and the different turbulence models employed by these partners is presented. It should be mentioned here that all the methods listed in Table 1 use structured computational grids and allow these grids to consist of several blocks (in order to generate a suitable structured Navier-Stokes grid for a multi-element airfoil configuration it is in practice unavoidable to consider a blocked grid). Full descriptions of the methods used by each partner are given in chapter 4. The descriptions of the standard versions of the different turbulence models can be found in chapter 3, and here the same abbreviations for the different models are adopted, being

    BL   Balwin&Lomax (algebraic)
    CS   Cebeci&Smith (algebraic)

C Chien (k-ε)
JL Jones&Launder / Launder&Sharma (k-ε)
SJL "Standard" High-Reynolds number Jones&Launder (k-ε)
W Wolfshtein (One Equation Low-Reynolds number)
LB Lam&Bremhorst (k-ε)

Table 1 Overview of contributors to the NLR7301 Flapped Airfoil application and their methods

Partner	Method	Turbulence Model
BAe (J.J. Benton)	Time-marching; Cell-centred FV; Central differencing; 4-stage Runge-Kutta; $2^{nd}/4^{th}$-order artificial dissipation; Residual smoothing; W-cycle multigrid.	C
Saab (B. Arlinger, T. Larsson)	Time-marching; Cell-centred FV; Central differencing; 5-stage Runge-Kutta; $2^{nd}/4^{th}$-order artificial dissipation; Multigrid.	BL, SJL-W
HUT (J. Hoffren)	Thin Layer Navier-Stokes; Time-marching; Cell-centred FV; Van Leer splitting (+ central diff. for viscous terms); Implicit relaxation; V-cycle multigrid.	CS
CERFACS (A. Kourta)	Time-marching; Cell-centred FV; Steger&Warming 2nd order flux splitting (+ central diff. for viscous terms); Explicit MacCormak; Local time-stepping.	BL, JL
Dornier (W. Fritz)	Time-marching; Cell-centred FV; Central differencing; multi-stage Runge-Kutta (3-stage mostly used); $2^{nd}/4^{th}$-order artificial dissipation; Residual smoothing; V- or W-cycle multigrid.	BL, LB
CFD-Norway (I. Øye)	Thin Layer Navier-Stokes; Time-marching; Cell-centred FV; Central differencing; 3-stage Runge-Kutta; $2^{nd}/4^{th}$-order Artificial dissipation; Local time-stepping.	BL, C

As can be seen in Table 1, all contributors employ a finite volume (FV) method, where four of the six methods mentioned employ central differencing combined with artificial dissipation. The fact that there is a kernel of similar methods, makes it possible to focus on differences in turbulence modelling and/or grids being used. On the other hand, there is still sufficient variation in the methods to make comparisons of the results from the point of view of the numerical schemes (Thin Layer Navier-Stokes vs. full Navier-Stokes, upwind vs. central differencing). However, it should already be mentioned here, that the discrepancies resulting from the differences in (implementation of) turbulence models may be much larger than the differences due to the variations in the numerical schemes employed.

5.4.3 Test Cases

The main objective of the NLR7301/Flap high-lift experiments (Van den Berg, 1979) has been to obtain sufficiently detailed experimental data for purpose of validation of CFD methods. The wing with flap configuration has been derived from an early supercritical section, NLR7301, from which a flap has been 'cut out'. Care has been taken to keep the flows under consideration relatively simple by avoiding separations. To realize this, a moderate flap angle, $\delta=20°$, has been used, whereas the shroud in the main section (where the flap has been cut out) has been reshaped such that a smooth surface resulted towards the trailing edge (hence the flap can not actually be retracted anymore). The experiments cover two values of the gap widths between main section and flap, of which only the larger value of 2.6% c has been considered in the present validation excercise (c is the basic "flap-retracted" airfoil chord). The overlap of main wing section and flap is 5.3% c. The configuration is depicted in Figure 1 where also the static-pressure holes and the boundary layer measuring stations are indicated. The experiments have been performed for a Reynolds number (based on the basic-airfoil chord) of $Re_c=2.51 \cdot 10^6$ and at free stream Mach numbers close to $M_\infty=0.185$. In the 1979 experiments, besides measurements of complete polars, three angles of incidence have been investigated in more detail, of which two, $\alpha=6°$ and $\alpha=13.1°$ (the latter being close to maximum lift), have been selected as the two mandatory test cases for the present validation excercise.

Experimental pressure distributions for these cases are available (Fig. 1b) as well as a limited number of skin friction values, and boundary layer quantities such as displacement- and momentum thicknesses as well as velocity profiles at the stations depicted in Figure 1a. The experimental results confirm that the flow is attached everywhere except for a thin laminar separation bubble near the leading edge of the main wing section being responsible for the transition from laminar to turbulent flow there. The flow remains laminar for a large part of the lower surface of the main wing section for both angles of incidence. For the gap of 2.6% c under consideration the wake of the main wing section passes entirely above the flap and the boundary layer at the flap lower surface remains laminar for both angles of incidence. In the boundary layer on the upper surface of the flap transition from laminar to turbulent flow occurs due to the unfavourable pressure gradient, however, for both angles of incidence no strong interaction between the wake of the main wing section and the flap boundary layer could be observed for the considered value of the gap up to the flap wake where the first interactions are visible. The (approximate) transition locations observed in the experiments are for both conditions presented in Table 2.

Table 2 Experimentally observed transition locations for the NLR7301/flap configuration at $M_\infty=0.185$, $Re_c=2.51\cdot10^6$ (from Van den Berg, 1979)

	$\alpha = 6°$		$\alpha = 13.1°$	
	Main	Flap	Main	Flap
upper	x/c = 0.04	x/c = 1.10	x/c = 0.035	x/c = 1.073
lower	x/c = 0.65	none	x/c = 0.71	none

In order to gain better insight in the behaviour of the flow above the flap, in 1989 the same configuration (with the 2.6%c gap) has been the subject of an experimental investigation in which for a number of flap boundary layer- and wake stations (see Fig. 2) hot wire measurements have been performed and for two angles of attack ($\alpha=6°$ and $\alpha=13.1°$) experimental Reynold's stress profiles have been obtained (Gooden and van Lent, 1989). This experimental work has been carried out under contract to the Netherlands Agency for Aerospace Programs (NIVR) and the data report mentioned above is not freely available. However, permission of the NIVR has been obtained to use the experimental hot wire measurements for one of the two angles of incidence, $\alpha=13.1°$ (the condition near maximum lift), for purpose of validation within in the framework of the present EUROVAL project. For this purpose a special (reduced) version of the data report has been made available to the EUROVAL partners (Gooden, 1991). For this case the calculated Reynold's stress profiles by the different partners are compared with the experimental data.

5.4.4 Computational Grids

Three partners, BAe, CFD-Norway, and Dornier, constructed computational grids for the NLR7301 Flapped Airfoil configuration and circulated these grids among the other partners.

The grid constructed by BAe (BAE grid) consists of 37 blocks. The grid has been constructed in two steps. An elliptic grid generator is used to construct a C-type multi-block grid which is suitable fo Euler calculations, satisfying orthogonality conditions at the surfaces, the far field boundaries and specified fixed control lines (some of the interior block-boundaries). Grid spacing is specified along the boundaries mentioned above. The far field boundary is placed at approximately ten chords away from the configuration. Furthermore, care has been taken that the topology is such that (for the final Navier-Stokes grid) the viscous regions are confined to a well defined set of innermost blocks. The initial Euler grid consists of 16448 cells and has actually been used by BAe to produce a solution of the Euler equations for both cases (see section 5.4.5). In the second step, the innermost C-grid blocks around the main wing section and the flap, and their wake extensions are taken out and replaced with blocks for which algebraic grids have been generated taking care of the very fine cells required near the walls for a Navier-Stokes equations. The outer blocks are kept unchanged. The resulting grid, depicted in Figure 3, consists of 28288 cells. The computational results on this grid show that the first grid line over the surface the law-of-the wall coordinate is near unity, $y^+ \approx 1$, over the largest part of the surface, and is less than 2 everywhere (except near the small separation region on the main airfoil). The grid has been used for one or more calculations by BAe, HUT, CERFACS, Dornier, and CFD-Norway. Some of the partners reconstructed the grid by taking some blocks together (in order to reduce the number of blocks for purpose of efficiency of their solvers), however, keeping the grid point distribution unchanged. For the higher angle of incidence ($\alpha=13.1°$) some of the partners (CFD-Norway and Dornier) reported initially difficulties to obtain converged solutions on the BAe grid and for that reason these partners constructed alternative grids, and distributed them among the other partners.

The grid constructed by CFD-Norway (CFD-N grid) consists of 5 blocks and a total number of 26172 grid cells (see Fig. 4). Also here, care has been taken that the grid spacing near the surfaces is sufficiently fine, meaning that along the first grid line the condition $y^+ \approx 1$ is satisfied to the best possible extent. The main difference with the BAe grid is the distribution of grid points in the wake regions (compare Figs. 3b and 4b), where the CFD-Norway grid seems to spread out more rapidly than in the BAe grid. The results for the higer angle of incidence, $\alpha=13.1°$ (see section 5.4.5) show that the wake distribution in the CFD-Norway grid is less suitable for this flow condition than the BAe grid is. The CFD-Norway grid has been used (for one or more calculations) by Saab and CFD-Norway.

In order to overcome the problems Dornier found with the BAe grid for the $\alpha=13.1°$ condition, they constructed for that case a more or less "flow adapted" grid (DORNIER grid). On the basis of a former flow solution, an elliptic grid generator is used where the source terms of the partial differential equations are adapted to the total pressure loss in that flow solution. The resulting grid for that case, of which some details are presented in Fig. 5, consists of a total number of 49392 cells. Compared with the BAe grid, also the Dornier grid shows a more rapidly smoothed out grid density in the wake regions. However, the dense grid spacing is maintained over a larger part of the wake region than is the case for the BAe grid. The Dornier grid has been used only by Dornier itself for the $\alpha=13.1°$ case.

5.4.5 Discussion of Results

Overview of Contributions

An overview of the contributions of the different partners can be found in Table 3 for the $\alpha=6°$ case and in Table 4 for the $\alpha=13.1°$ case. In these tables the calculated lift and drag coefficients can be found, where the turbulence model and the grid employed to produce the results are also indicated. The turbulence model and the grid employed are also inidicated in the Figure legends which have the form: Partner [Turbulence model, Grid]. This notation will also be adopted in the text from now on when referring to specific results.

Table 3 Overview of contributions to the NLR7301/Flap case at $M_\infty=0.185$, $\alpha=6°$, and $Re_c=2.51\cdot10^6$, and the calculated lift- and drag coefficients

Partner	Turbulence Model	Grid	C_L Main	C_L Flap	C_L Total	C_D Main	C_D Flap	C_D Total
BAE	*)	BAE *)	2.294	0.386	2.680	-0.1372	0.1402	0.0030
	C	BAE	2.109	0.348	2.457	-0.0929	0.1253	0.0323
Saab	BL	CFD-N	2.084	0.347	2.431	-0.0890	0.1240	0.0349
HUT	CS	BAE	-	-	2.417	-	-	0.0343
CERFACS	BL	BAE	1.990	0.310	2.300	-0.0800	0.1160	0.0360
	JL	BAE	2.018	0.324	2.242	-0.0810	0.1190	0.0380
Dornier	BL	BAE	-	-	2.359	-	-	0.0381
	LB	BAE	-	-	2.342	-	-	0.0375
Exp.	-	-	-	-	2.416	-	-	0.0229

*) Euler calculation on BAE-Euler grid

Table 4 Overview of contributions to the NLR7301/Flap case at $M_\infty=0.185$, $\underline{\alpha=13.1°}$, and $Re_c=2.51\cdot10^6$, and the calculated lift- and drag coefficients

Partner	Turbulence Model	Grid	C_L			C_D		
			Main	Flap	Total	Main	Flap	Total
BAE	*)	BAE *)	3.274	0.383	3.657	-0.1828	0.1898	0.0070
	C	BAE	2.852	0.313	3.165	-0.0945	0.1550	0.0604
Saab	SJL-W	CFD-N	2.843	0.316	3.159	-0.0880	0.1540	0.0658
HUT	CS	BAE	-	-	3.193	-	-	0.0613
CERFACS	BL	BAE	2.665	0.284	2.949	-0.0470	0.1530	0.1060
	JL	BAE	2.531	0.270	2.801	-0.0040	0.1420	0.1380
Dornier	BL	DORNIER	2.960	0.355	3.315	-0.0819	0.1730	0.0911
	LB	DORNIER	2.921	0.341	3.262	-0.0918	0.1686	0.0758
CFD-Norway	BL	BAE	2.903	0.301	3.204	-0.0880	0.1540	0.0660
	BL	CFD-N	2.836	0.311	3.148	-0.0920	0.1520	0.0600
	C	BAE	2.794	0.297	3.091	-0.0820	0.1490	0.0670
	C	CFD-N	2.758	0.306	3.064	-0.0810	0.1490	0.0680
Exp.	-	-	-	-	3.141	-	-	0.0445

*) Euler calculation on BAE-Euler grid

The set of Figures for the α=6° case consists of
- Comparison of calculated and experimental pressure distributions for the complete configuration (Fig. 6)
- Comparison of calculated and experimental pressure distributions on the flap (Fig. 7)
- Comparison of calculated and experimental surface skin friction coefficients for the complete configuration (Fig. 8)
- Comparison of calculated and experimental surface skin friction coefficients on the flap (Fig. 9)
- Comparison of calculated and experimental displacement thicknesses and momentum thicknesses for the complete configuration (Fig. 10)

The set of Figures for the α=13.1° case consists of
- Comparison of calculated and experimental pressure distributions for the complete configuration (Fig. 11)
- Comparison of calculated and experimental pressure distributions on the flap (Fig. 12)
- Comparison of calculated and experimental surface skin friction coefficients for the complete configuration (Fig. 13)
- Comparison of calculated and experimental surface skin friction coefficients on the flap (Fig. 14)
- Comparison of calculated and experimental displacement thicknesses for the complete configuration (Fig. 15)
- Comparison of calculated and experimental momentum thicknesses for the complete configuration (Fig. 16)
- Comparison of calculated and experimental velocity profiles (Fig. 17). The comparison is made for the stations where also hot-wire measurments have been available. These are the main airfoil wake (flap boundary layer) stations 8, 12, 13, and 14 (see Fig. 2), and the flap wake station 16 (see Fig. 2)
- Comparison of calculated and experimental Reynolds stress profiles (Fig. 18) at the stations mentioned above
- Contourline plots of the total pressure loss ($1-p_t/p_{t0}$) near the complete configuration as calculated by the different contributors (Fig. 19)
- Results of the Euler calculation carried out by BAe, consisting of pressure distribution for the complete configuration and for the flap

Discussion of results for the α=6° case

For this case all partners predicted reasonable values for the lift coefficients as can be seen in Table 3, where the results of HUT, BAe (NS result) and Saab come close to the experimental value of 2.416. This observation can also be made for the pressure distributions (Figs. 6 and 7) where the partners mentioned above show the best agreement with the experimental results. The rather low lift coefficients predicted by CERFACS seems to be correlated to their numerical scheme. It is shown in Figs. 6 and 7 that for both turbulence models they use, the suction peak on the main airfoil is poorly predicted, whereas on the flap upper surface the pressure is rather overpredicted, although the results with the JL (k-ε) model seem much better than the BL results on this points. Apparently the Steger-Warming flux splitter provides an excessive amount of dissipation, which is also reflected in the skin friction distributions (Figs. 8 and 9) where the CERFACS results seem far too low (especially on the main upper surface) compared to the results of the other partners, which do agree better with the experiment. The other upwind method (HUT), which employs Van Leer splitting, seems to be better suited for Navier-Stokes calculations. The partners using the central difference / artificial dissipation approach, all use mechanisms to reduce the amount of artificial dissipation in the boundary layer. From the skin friction distributions it is also seen that all partners predicted a (small) laminar separation bubble close to the experimentally observed location. From the point of view of turbulence modelling only CERFACS showed improved results using the JL model compared to the (algebraic) BL model results. The results of BAE with the C model are close to the BL model results of Saab and the CS model results of HUT, and no argument in favour of either k-ε or algebraic models is found.

All partners largely overpredict the value of the drag coefficient as can be seen in Table 3. This issue will be discussed in more detail for the α=13.1° case, where the same kind of overprediction is found, and where more detailed information is available on the possible explanations for this large discrepance.

Also the discussion on the influence of the computational grid is postponed to the α=13.1° case where more material is available to judge upon the quality of the grids.

Discussion of results for the α=13.1° case

Also for this case most of the partners found reasonable values for the lift coefficients compared to the experimental value (Table 4). Extremely close to the experimental lift coefficient is the result of Saab, whereas the largest deviations are to be found in the CERFACS results and also the Dornier results. Looking into pressure distributions and skin friction distributions, it may be concluded that the "averaged" deviation from the experimental values has become larger compared with the α=6° case, as could be expected because of the fact the the α=13.1° case should be close to maximum lift. In particular the flow in the slot region and in the upper surface boundary layer and wake of the flap becomes more difficult to resolve by the considered turbulence models for this case, as will be demonstrated below.

The variation in turbulence models, numerical schemes, and grids used for the results included in Table 4, opens several viewpoints to look at the solutions in more detail. Although effects of each of these aspects (turbulence modelling, numerical scheme, grid) may strictly not be treated separately, an attempt will be made to draw some conclusions about details of the solutions from these three different viewpoints.

First of all, the conclusions about the numerical schemes drawn for the α=6° case, seem to be confirmed by the α=13.1° results. The Steger&Warming flux splitting seems again not too well suited for Navier-Stokes calculations for the present application. The suction peak is again not well resolved (Fig. 11), and the flap pressure distribution shows some large deviations from the experiments. The already suspected excessive diffusion of the scheme is also reflected in the very low values predicted for the skin friction (Figs. 13 and 14) especially on the flap. With respect to skin friction distribution a similar behaviour is found in the Dornier results, however, it is expected that this is more due to turbulence modelling and/or to the computational grid than to the numerical scheme employed. It is also interesting to search for difference in the solutions due to the fact that some of the partners used Thin Layer Navier-Stokes (TLNS) methods (HUT and CFD-Norway). Detailed results are only available for CFD-Norway. Focussing on solutions based on more or less the same grids and turbulence models, one could compare the CFD-Norway [C, BAE] results with the BAe [C, BAE] results in Figs. 12-19, as well as the CFD-Norway [C, CFD-N] results with the Saab [SJL-W, CFD-N] results. Both comparisons lead to the conclusion that on this point the

Saab and BAe results mentioned above are much closer to the experiment than the mentioned CFD-Norway results. The most clarifying pictures on this point are those of the flap pressure distributions (Fig. 12) and the velocity profiles (Fig. 17), from which it may be observed that the TLNS approach gives more diffusive solutions (and worse comparison with the experiments) than the full Navier-Stokes methods

From the viewpoint of computational grids, the results are rather inconclusive. It is observed by the partners employing the CFD-N grid that the distribution of grid points is not satisfactory in the wake regions. However, the CFD-Norway results produced with the BL turbulence model on the CFD-N grid show on many points (flap pressure, velocity profiles, and Reynolds stress profiles) better agreement with the experiments than those on the BAe grid. The C model results of CFD-Norway on the other hand, show especially for the velocity profiles much better results on the BAe grid than on the CFD-N grid. Comparing on the same basis the results of Saab [SJL-W, CFD-N] with results of BAe [C, BAE] (both results produce the best over-all agreement with the experiment), no firm conclusions can be drawn with respect to a possible better quality of either the BAE grid or the CFD-Norway grid other than that in the near wake of the main wing section the results on the BAE grid seem indeed to be slightly better than the results on the CFD-N grid (compare the velocity and Reynolds stress profiles for station 8 and 12, and also the pressure distributions near the flap leading edge). It is seen also from the total pressure contour plots presented in Fig. 19, that for the calculations on the CFD-Norway grid (CFD-N [BL, CFD-N], CFD-N [C, CFD-N], and Saab [SJL-W, CFD-N]) the wake decays more rapidly than is the case for the calculation on the BAe grid (CFD-N [BL, BAE], CFD-N [C, BAE], and BAe [C, BAE]). However, there is no hard experimental evidence to decide which situation is the best. Concerning the Dornier grid, also hardly any conclusions can be drawn because of the fact that only Dornier has used that grid. On several points the Dornier results seem to fail to predict the wake flow behind the main airfoil (see e.g. the velocity and Reynolds stress profiles, as well as the flap pressure distributions) resulting in a far too thin wake (see also the total pressure loss contours). The fact that on these points the results do not seem to be very much improved by changing from an algebraic to a k-ε model does indicate some negative effect of the grid used. This conclusion is partly contradicted by the fact that their results with respect to displacement thickness and momentum thickness did indeed improve by changing the turbulence model.

Important differences between the solutions (and deviations from the experiments) do finally result from the turbulence modelling. The most obvious difference in turbulence modelling is the fact that both algebraic as well as k-ε models have been employed. The most important region to focus on is the wake of the main wing section passing over the flap. The experimental velocity profiles (Fig. 17) show there, as mentioned before, hardly any interaction between the wake and the flap boundary layer up to station 13. For stations 14 and 16 (the flap wake), however, the merging of the shear layers is visible. It is also for these stations that all results (independent from the turbulence model) show the largest deviations from the experimental velocity profiles. Looking to the velocity profiles there are two observations which show typical shortcomings of both the algebraic and the k-ε model (in the formulations as applied here). First of all, partners who used both the algebraic BL model as well as a version of the k-ε model, show that at the stations 8, 12, and 13, the velocity profiles are better predicted by the simple BL model than with a comparable calculation (same grid) with their version of the k-ε model. For these stations the k-ε models show typically a decay of the main wing section wake which is much faster than in the experiment, showing a too high rate of dissipation produced by these models. However, for the stations where the interaction between the shear layers become important (14 and 16), the k-ε models produce more realistic velocity profiles than the BL model does, probably due to the fact that history effects are completely missing in the BL model. The best results on this point stem from BAe and Saab. With respect to the Reynolds stress profiles the results from CERFACS and Dornier, both using the BL model as well as a k-ε model, seem for both turbulence models to be very unrealistic. On this point the possible defects in numerical scheme and/or computational grid seem to make it impossible to draw any conclusions with respect to turbulence modelling for these partners. The results of BAe, Saab, and CDF-Norway, however, show that the Reynolds stress profiles are in general better predicted using k-ε models than with the BL model. The high stress levels at the stations 13 and 14 as observed in the experiment could hardly be predicted by any of the methods (except maybe the odd result for station 14 produced by CFD-Norway using the BL model on the CFD-N grid). It is assumed that the high level of turbulent stress at these stations is due to the destabilizing effect of the wake curvature (due to the flap flow) on the turbulence intensity in wake and boundary layer. The results show that this effect is not adequately represented in the BL model as well as in the versions of the k-ε model employed here. Finally, it should be concluded here that most partners (except may be for CFD-Norway) found slightly improved overall agreement with the experiments applying k-ε models than was the case with the algebraic models, although the difference are not that big that a definitive choice for the k-ε (in the versions used here) models in order to handle this type of flows can be justified based on the material presented for the present NLR73/flap application.

Finally, the last item to be discussed here is the fact that just like for the $\alpha=6°$ case the calculated drag coefficients are largely overpredicted compared to the experiment. Among several explanations, the arguments of Saab that also in the experiment the pressure drag resulting from integrating the pressure coefficient along the surfaces would be larger than the experimental value obtained from a wake traverse seems to be reasonable at first glance. However, a wake traverse in the Navier-Stokes results would produce rather inaccurate results and makes it impossible to check whether the discrepancy is really due to the difference in drag calculation. Furthermore, there are some arguments for believing that the Navier-Stokes methods might be, despite all attempts to reduce artificial dissipation inside the boundary layers, still too dissipative. First of all it should be mentioned that the NLR7301/Flap testcases have also been studied in the framework of a GARTEUR Action Group in which several viscid/inviscid interaction methods are applied to these cases (Van den Berg, 1983). These methods consistently underpredict the experimental drag values where some of these methods predicted (despite the fact that they are based on boundary layer methods) just like the Navier-Stokes methods in the present work, quite realistic pressure distributions for these cases. The assumption that too much dissipation is present in the results produced by the Navier-Stokes methods is also partly supported by the observations from the Euler results of BAe of which the pressure distribution is presented in Fig. 20. Although the grid spacing of the Euler grid in normal direction is different from the Navier-Stokes grid, the non zero drag found in that calculation (C_D=0.0070) indicates that a numerical boundary layer is present. Presumably, the relatively coarse grid in streamwise direction near the main airfoil leading edge (which also is used for the Navier-Stokes grid) gives rise to excessive entropy production in that region, which has actually been reported by BAE for their Euler results. Hence, it is expected that part of this spurious entropy production may also have been included in the Navier-Stokes equations. Local grid refinement and/or more adequate formulations of artificial dissipation is certainly an issue to be addressed in future work on validation of Navier-Stokes methods for high-lift devices.

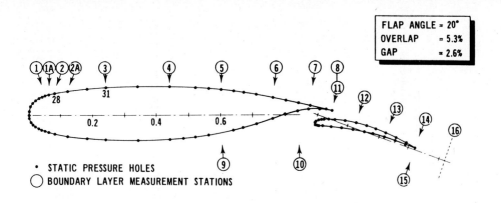

a) Positions of static pressure holes and boundary layer measurement station

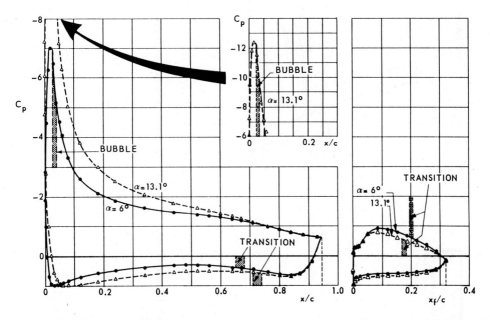

b) Measured pressure distributions, transition regions, and location of laminar separation bubbles at $\alpha=6°$ and $\alpha=13.1°$

Fig. 1 NLR7301/Flap measurements at $M_\infty=0.185$ and $Re_c=2.51 \cdot 10^6$ (from Van den Berg, 1979)

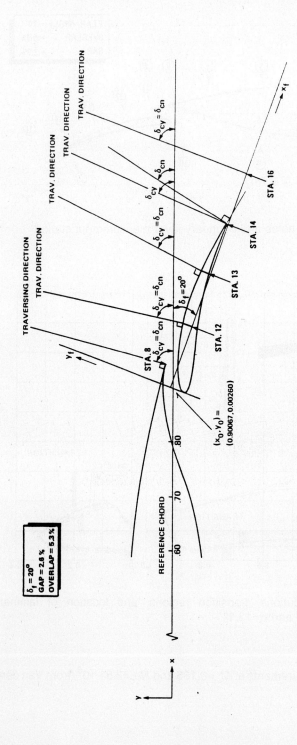

Fig. 2 NLR7301/Flap measurements; Positions of hot-wire measurement stations and definition of direction angles (from Gooden, 1991)

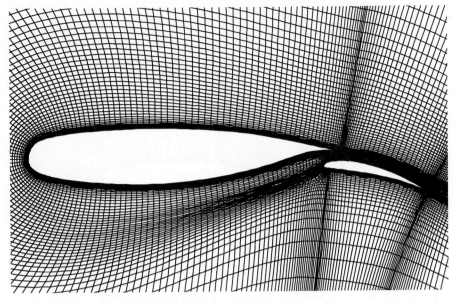
a) Region near the complete configuration

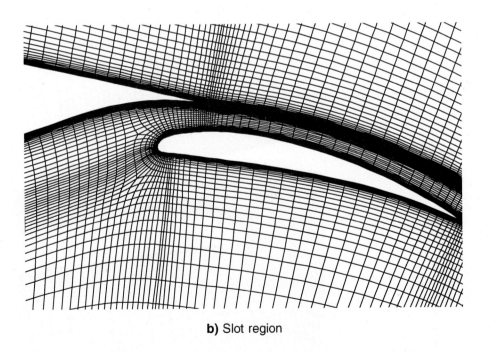
b) Slot region

Fig. 3 BAe grid for the NLR7301/Flap (2.6%c gap) configuration

a) Region near the complete configuration

b) Slot region

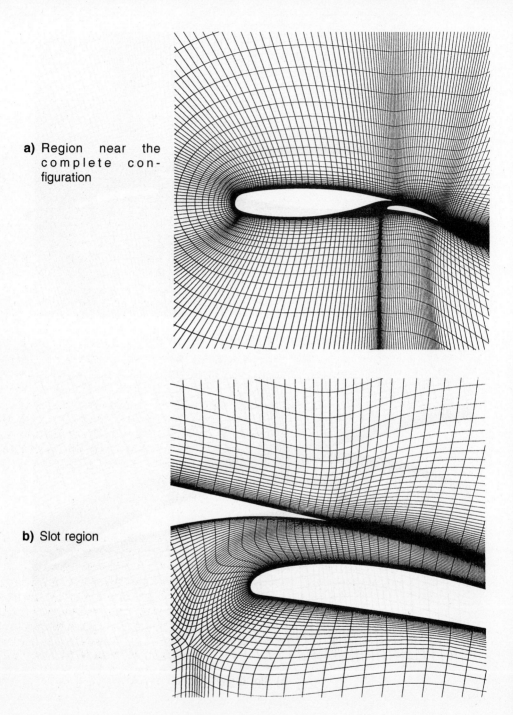

Fig. 4 CFD-Norway grid for the NLR7301/Flap (2.6%c gap) configuration

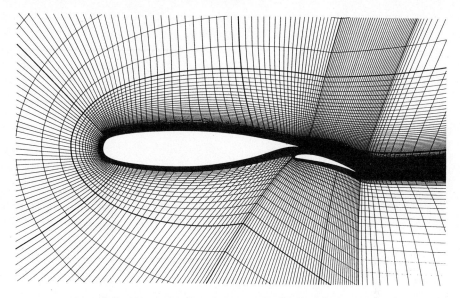

a) Region near the complete configuration

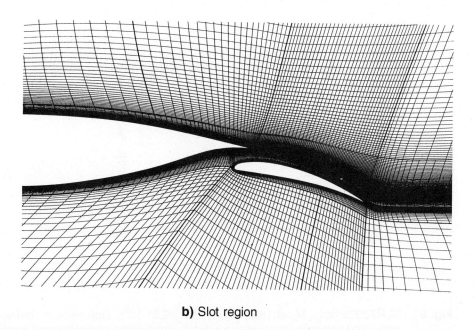

b) Slot region

Fig. 5 Dornier grid for the NLR7301/Flap (2.6%c gap) configuration (adapted to the flow at M_∞=0.185, α=13.1°, Re_c=2.51·10^6)

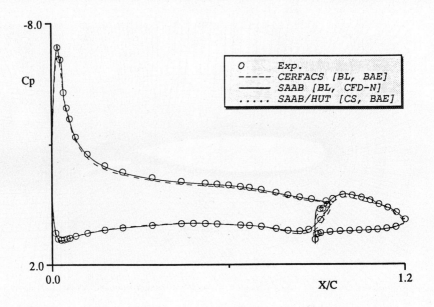

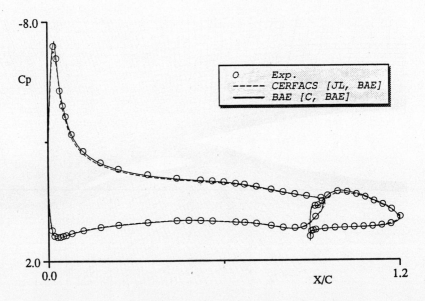

Fig. 6 NLR7301/Flap, M_∞=0.185, α=6°, Re_c=2.51·10^6; Comparison between measured and calculated pressure distributions on the complete configuration

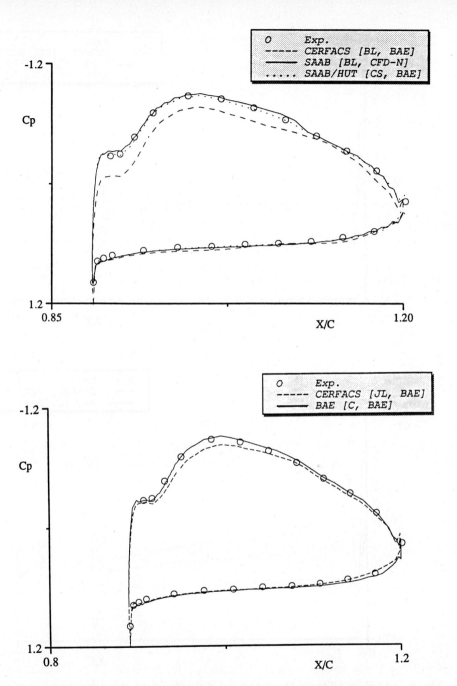

Fig. 7 NLR7301/Flap, M_∞=0.185, α=6°, Re_c=2.51·10^6; Comparison between measured and calculated pressure distributions on the flap

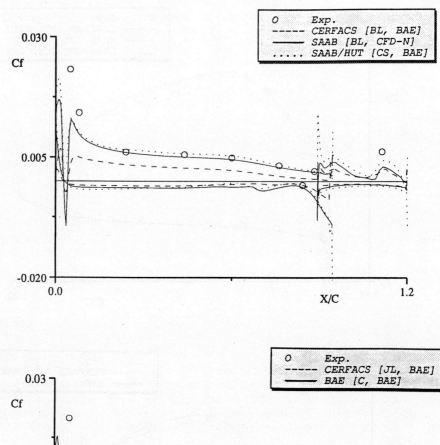

Fig. 8 NLR7301/Flap, M_∞=0.185, α=6°, Re_c=2.51·10^6; Comparison between measured and calculated surface skin friction coefficients on the complete configuration

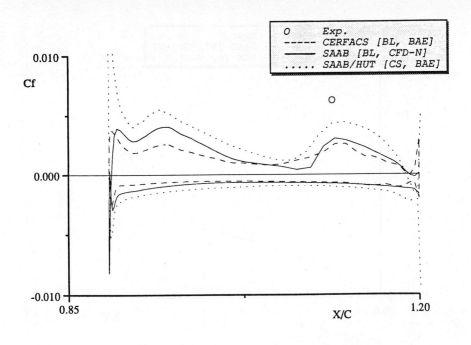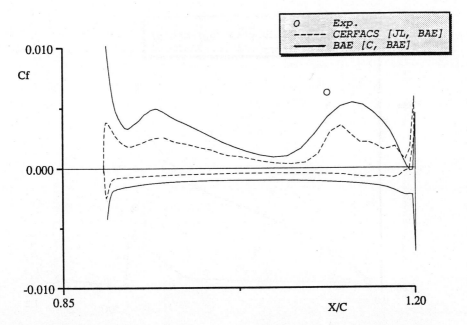

Fig. 9 NLR7301/Flap, $M_\infty=0.185$, $\alpha=6°$, $Re_c=2.51\cdot10^6$; Comparison between measured and calculated surface skin friction coefficients on the flap

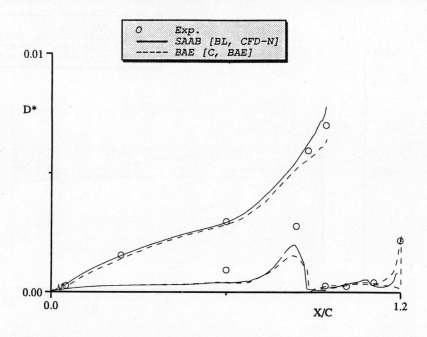
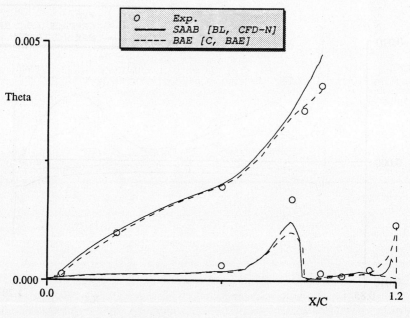

Fig. 10 NLR7301/Flap, $M_\infty=0.185$, $\alpha=6°$, $Re_c=2.51\cdot 10^6$; Comparison between measured and calculated displacement thicknesses and momentum thicknesses on the complete configuration

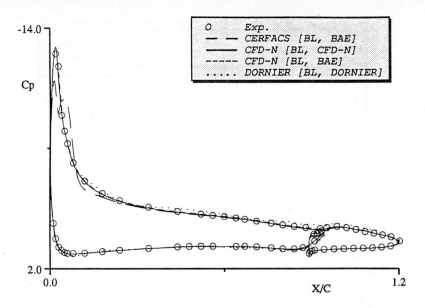

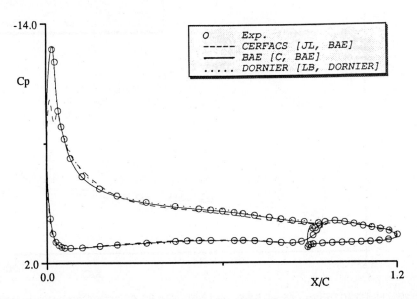

Fig. 11 NLR7301/Flap, M_∞=0.185, α=13.1°, Re_c=2.51·10^6; Comparison between measured and calculated pressure distributions on the complete configuration

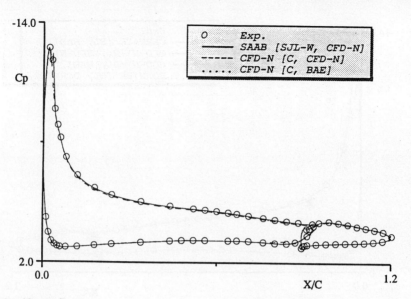

Fig. 11 (Continued)

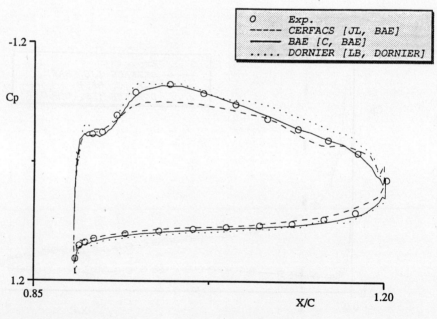

Fig. 12 NLR7301/Flap, $M_\infty=0.185$, $\alpha=13.1°$, $Re_c=2.51·10^6$; Comparison between measured and calculated pressure distributions on the flap

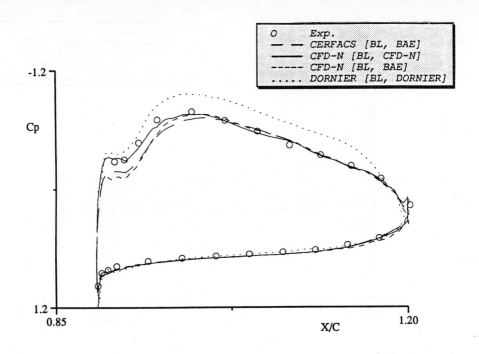

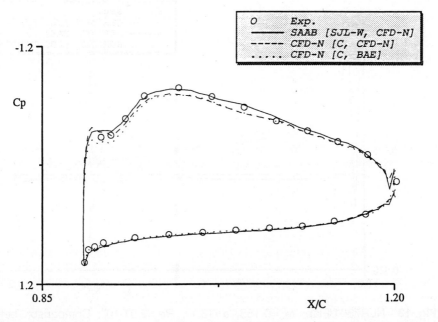

Fig. 12 (Continued)

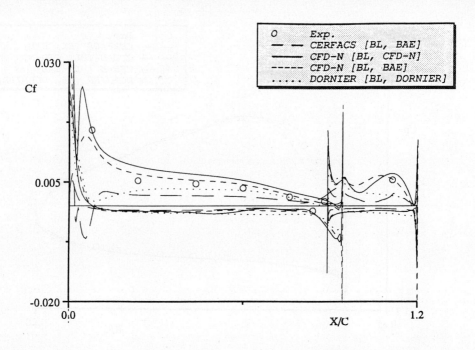

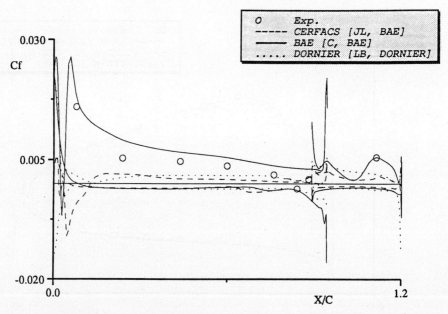

Fig. 13 NLR7301/Flap, M_∞=0.185, α=13.1°, Re_c=2.51·10^6; Comparison between measured and calculated surface skin friction coefficients on the complete configuration

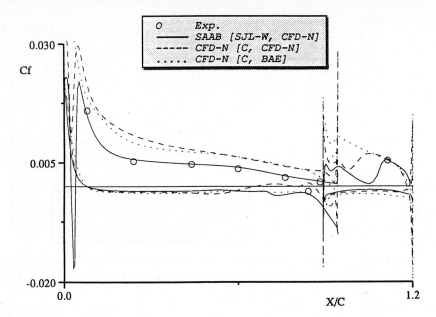

Fig. 13 (Continued)

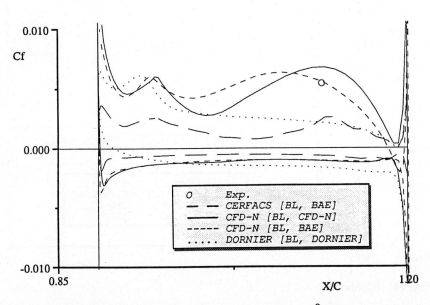

Fig. 14 NLR7301/Flap, $M_\infty=0.185$, $\alpha=13.1°$, $Re_c=2.51\cdot10^6$; Comparison between measured and calculated surface skin friction coefficients on the flap

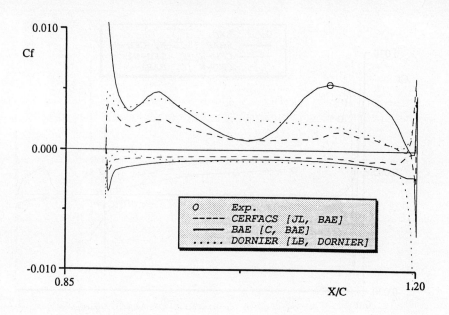

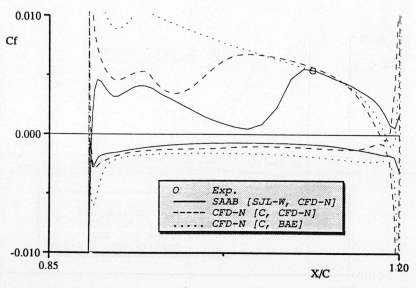

Fig. 14 (Continued)

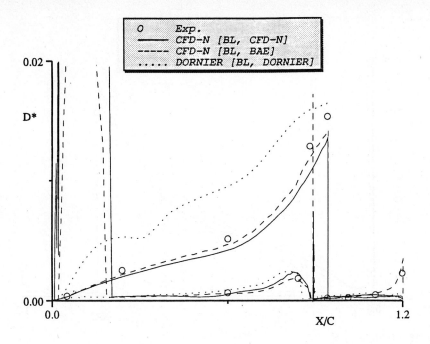
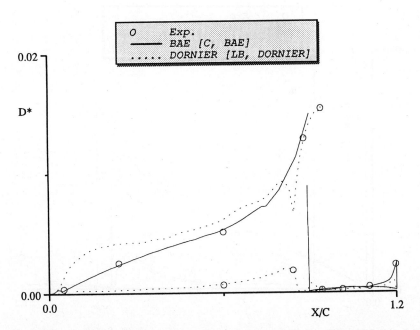

Fig. 15 NLR7301/Flap, M_∞=0.185, α=13.1°, Re_c=2.51·10^6; Comparison between measured and calculated displacement thicknesses (complete configuration)

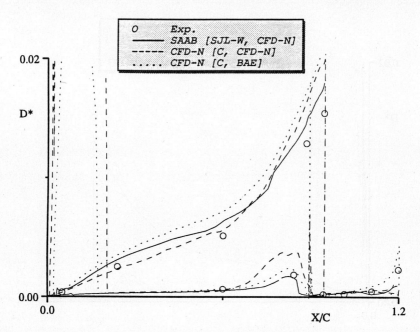

Fig. 15 (Continued)

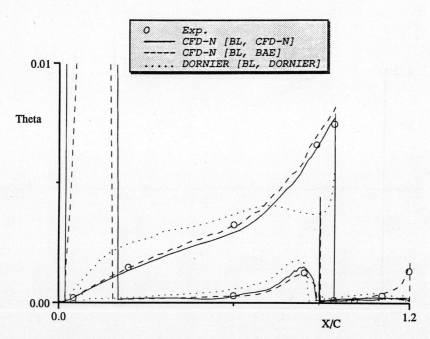

Fig. 16 NLR7301/Flap, M_∞=0.185, α=13.1°, Re_c=2.51·10^6; Comparison between measured and calculated momentum thicknesses (complete configuration)

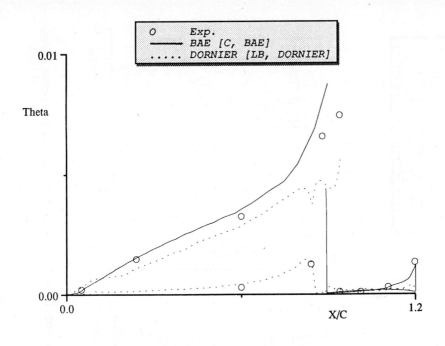
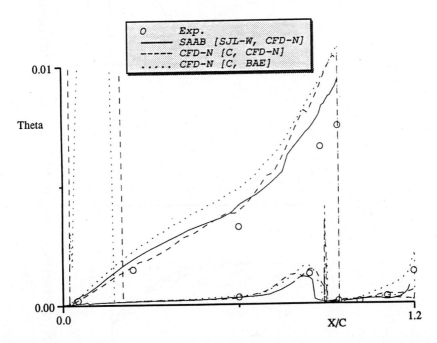

Fig. 16 (Continued)

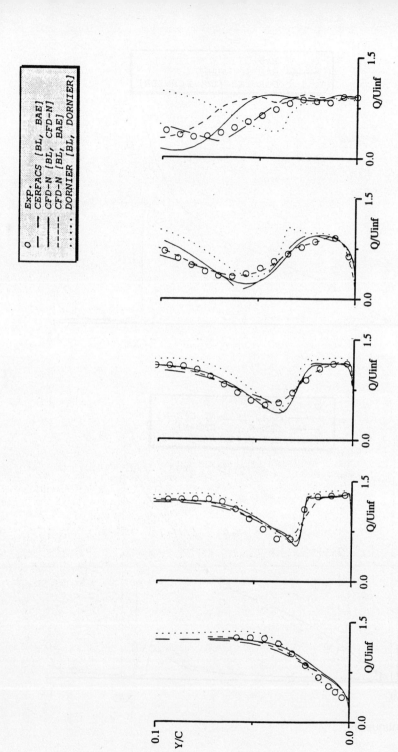

Fig. 17 NLR7301/Flap, $M_\infty=0.185$, $\alpha=13.1°$, $Re_c=2.51\cdot10^6$; Comparison between measured and calculated velocity profiles

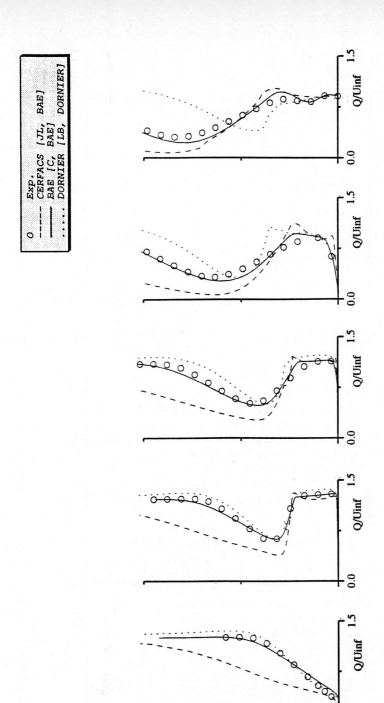

Fig. 17 (continued) NLR7301/Flap, $M_\infty=0.185$, $\alpha=13.1°$, $Re_c=2.51\cdot10^6$; Comparison between measured and calculated velocity profiles

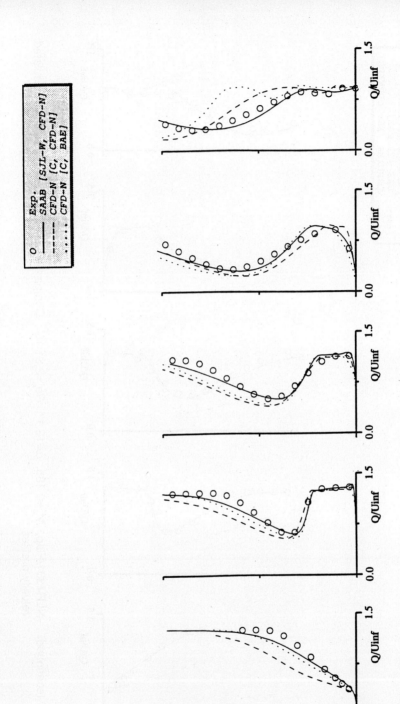

Fig. 17 (continued) NLR7301/Flap, $M_\infty=0.185$, $\alpha=13.1°$, $Re_c=2.51\cdot10^6$, Comparison between measured and calculated velocity profiles

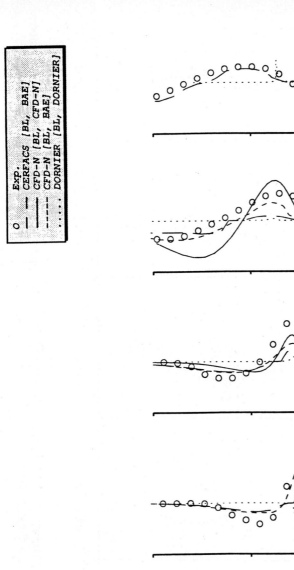

Fig. 18 NLR7301/Flap, $M_\infty=0.185$, $\alpha=13.1°$, $Re_c=2.51\cdot10^6$; Comparison between measured and calculated Reynolds stress profiles

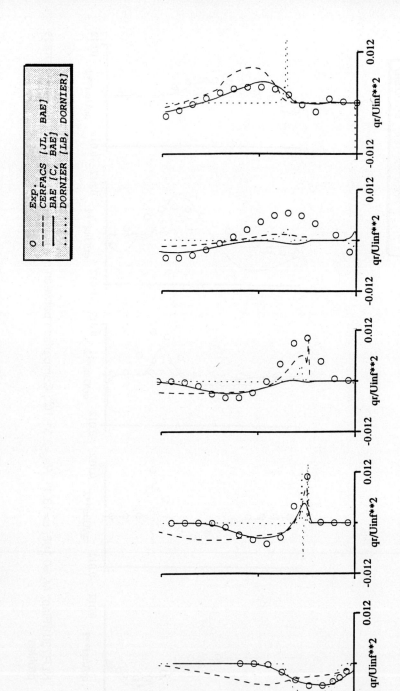

Fig. 18 (continued) NLR7301/Flap, $M_\infty=0.185$, $\alpha=13.1°$, $Re_c=2.51\cdot 10^6$; Comparison between measured and calculated Reynolds stress profiles

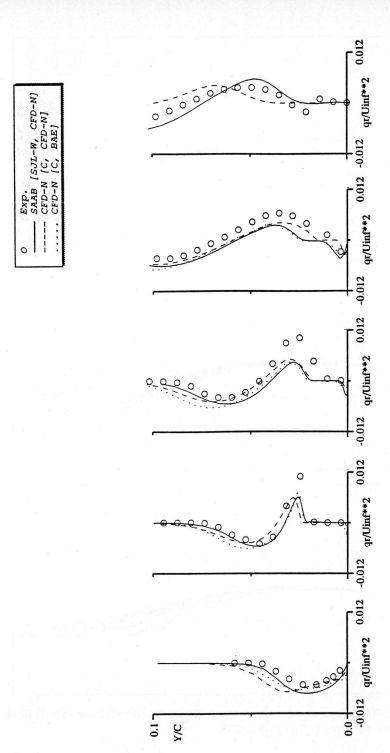

Fig. 18 (continued) NLR7301/Flap, $M_\infty=0.185$, $\alpha=13.1°$, $Re_c=2.51\cdot 10^6$. Comparison between measured and calculated Reynolds stress profiles

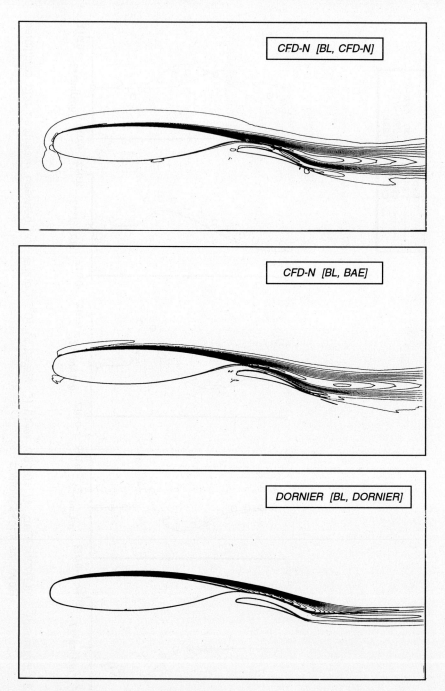

Fig. 19 NLR730/Flap, M_∞=0.185, α=13.1°, Re_c=2.51·10^6; Calculated total pressure loss contours ($\Delta[1-p_t/p_{t0}]$=0.002)

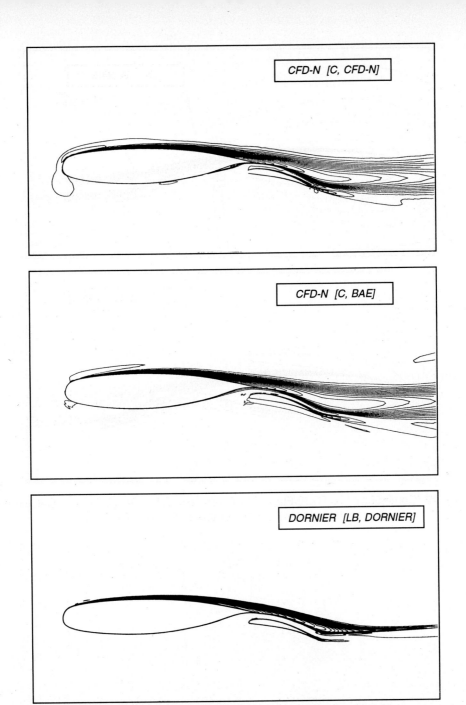

Fig. 19 (Continued) NLR730/Flap, $M_\infty=0.185$, $\alpha=13.1°$, $Re_c=2.51\cdot10^6$; Calculated total pressure loss contours ($\Delta[1-p_t/p_{t0}]=0.002$)

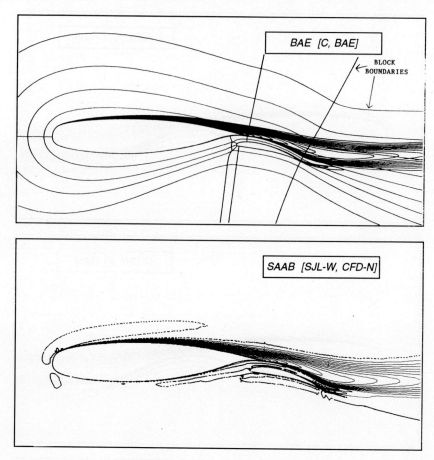

Fig. 19 (Continued)   NLR730/Flap, M_∞=0.185, α=13.1°, Re_c=2.51·10^6; Calculated total pressure loss contours ($\Delta[1-p_t/p_{t0}]$=0.002)

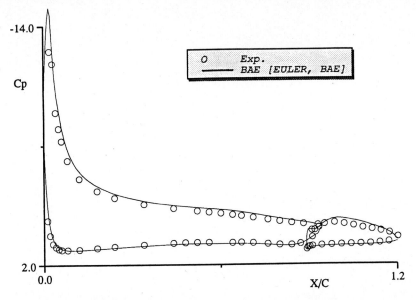

a) Pressure distribution on complete configuration

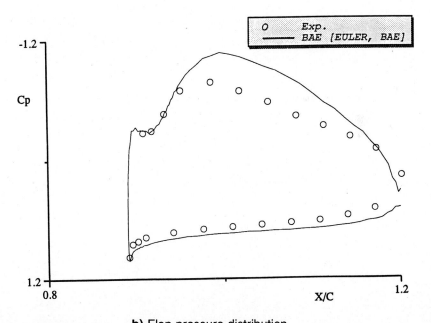

b) Flap pressure distribution
Fig. 20 NLR7301/Flap, M_∞=0.185, α=13.1°; Experiment vs. Euler calculation

TWO-DIMENSIONAL BOUNDARY LAYERS

5.5

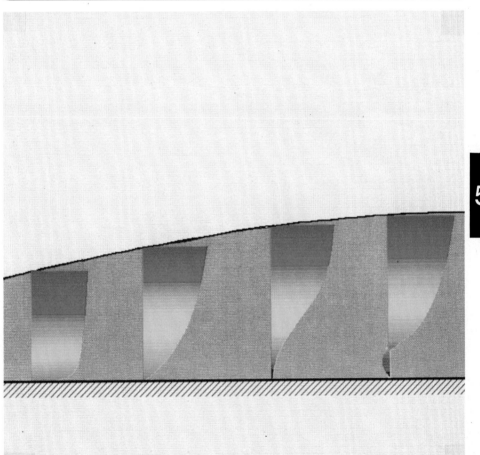

Colour figure - provided by Dornier - shows:
Separating boundary layer due to an adverse pressure gradient.

5.5 Two-Dimensional Boundary Layers

Author: **W. Haase**, Dornier
Plot Coordinator: **W. Haase**, Dornier

5.5.1 Introduction

The evaluation of turbulence models using Navier-Stokes equations has proven to be time consuming and difficult with respect to the occurrence of artificial viscosity which often swamps the real viscosity. Therefore, a validation process on using boundary layer equations is highly desirable, due to the fact that these equations can be solved by numerical techniques which are free from numerical viscosity. Hence, results gained from such a validation process are not altered in any way by the occurrence of non-physical viscosities.

5.5.2 Test Cases

In the following, Table 1 presents the testcases which have been used for the validation process and the partners involved. It should be noted at this point, that both DA and RAE joined the current task in the second year of the EUROVAL project on a voluntary basis, i.e. their additional effort was appreciated as an upgrade to the project and exhibits something of the cooperative spirit in the complete EUROVAL group.

Table 1 Partners and Selected Test Cases

Test Case		Partner		
		Dornier	RAE	Deutsche Airbus
Stanford '68:	IDENT-1300	X	X	
	IDENT-1400	X	X	X
	IDENT-2200	X	X	
	IDENT-2400	X	X	X
	IDENT-2600	X	X	

These five Stanford '68 test cases, taken into account for the validation of turbulence models, are in fact well known and served already in the past as test cases for the validation of various boundary layer methods. The chosen test cases cover the flow range from equilibrium to non-equilibrium flows, thus forming a proper basis for a comparison of the performance of different turbulence models. A brief description of the test cases reads:

    IDENT-1300: Ludwieg-Tillmann, dp/dx<0
    IDENT-1400: Wieghardt, flat plate (dp/dx=0) flow
    IDENT-2200: Clauser I, retarded equilibrium boundary layer
    IDENT-2400: Bradshaw, retarded relaxing flow
    IDENT-2600: Bradshaw & Ferriss, retarded equilibrium boundary layer

5.5.3 Turbulence Models

Up to five different turbulence models (from a total of six) have been applied to the aforementioned test cases by the different partners and the results thereof serves as a first data base which has to be extended in a future effort with respect to other turbulence models and an increased number of test cases. In Table 2, the turbulence models used by the partners are presented.

Table 2 Partners and Turbulence Models Applied

Turbulence Model		Partner		
		Dornier	RAE	Deutsche Airbus
Baldwin-Lomax	(BL)	X		
Cebeci-Smith	(CS)	X	X	X
Johnson-King	(JK)	X	X	X
Johnson-Coakley	(JC)		X	
Horton	(HO)	X		
Hassid-Poreh	(HP)	X		

The first two models involved in the validation process, Baldwin-Lomax and Cebeci-Smith, are algebraic turbulence models. The next three, Johnson-King, Johnson-Coakley and Horton, are 1/2-equation models while the Hassid-Poreh model denotes a one-equation turbulence model. Abbreviations for the models, BL for Baldwin-Lomax, CS for Cebeci-Smith, JK for Johnson-King, JC for Johnson-Coakley, HO for Horton and HP for Hassid-Poreh are used wherever appropriate.

All standard versions of the applied turbulence models are documented in section 3. Important modification to the original versions are presented in the contributor's chapter 4, and they need only to be referenced in the current chapter.

5.5.4 Discussion of Results

5.5.4.1 Stanford IDENT-1300

The investigated flow of test case IDENT-1300 is accelerating due to a moderate adverse pressure gradient and is close to equilibrium, at least up to 3.5m. The boundary layer edge velocity, U_e, and its first derivative in streamwise direction, dU_e/dx, is depicted in Figs. 1 and 2, respectively. Although the velocity distribution in Fig. 1 shows a smooth behaviour, the gradient thereof, Fig. 2, exhibits some "wiggles". The oscillations in the x-direction, particularly that one at x=1.5m, cause some wavy behaviour in the results, especially for the form parameter and the skin friction. Although it is possible, to avoid such a wavy input in principle, the presented velocity distribution is based on the originally given values, (Kline et al, 1968), and those values have been used as a mandatory input for the computations of the involved partners. The turbulence models employed for the current test case are identified in the corresponding diagrams.

Figs. 3a and 3b depict the shape factors calculated by Dornier (Fig. 3a) and RAE (Fig.3b). It is obvious from both figures that the Johnson-King model is slightly off, while all the other models involved behave quite well. This discrepancy, appearing in nearly all following computations, needs to be investigated further.

Skin friction results are presented in Figs. 4a and 4b. Again the Johnson-King model, being quite sensitive to the starting conditions, is off compared with the measurements. On the other hand, the CS model, the model by Horton and the HP model behave reasonably. The Johnson-Coakley modification of the Johnson-King model, with an inner layer eddy viscosity formulation based on a new velocity scale, compare section 3.2.3, improves the boundary layer length scales' accuracy, however the wall shear stress is still insufficient.

Moreover, a reasonable accuracy with the experimental findings should be achieved comparing the momentum loss thickness for the different turbulence models. Starting with the same initial Re_θ-value, the momentum loss thickness distributions calculated by Dornier are given in Fig. 5a, the corresponding RAE results in Fig. 5b, respectively. The observed scatter, employing different turbulence models, is in its extent comparable to measurement inaccuracies.

More information about the performance of different turbulence models can be gained by checking boundary layer profile properties at various stations. In the following, profile results are presented at x=4.332m. According to the comments given by Kline et al (1968), the flow at the selected position is near to equilibrium but not exactly in such a status. A set of four different distributions is presented in Figs. 6a (Dornier results) and 6b (RAE results) to Figs. 9a and 9b, respectively.

Fig. 6 depicts the velocity profile at x=4.332m. A reasonable comparison is achieved, although the eddy viscosity distributions, Fig. 7, differ significantly. From the latter figures, it is obvious that an intermittency function is in use in all turbulence models. However, the different shape of the Baldwin-Lomax eddy viscosity profile is caused by an inefficient intermittency formulation in the outer region (Stock and Haase, 1989). The significant gap between the Hassid-Poreh and the Johnson-King model, the JK model offers approximately 50% less eddy viscosity than the HP model, leads to an underestimation of momentum, compare Fig. 6 for the JK model, and consequently to an underestimation in skin friction, Figs. 4a and 4b. Additionally, Reynolds stress profiles, Fig. 8, are presented together with dimensionless velocity $u^+=u/u_\tau$ versus the dimensionless wall distance (Reynolds number) $z^+=zu_\tau/\nu$, Fig. 9, with u_τ being the friction velocity.

5.5.4.2 Stanford IDENT-1400

The Stanford test case IDENT-1400, flow along a flat plate at constant pressure, represents a flow case being in a totally equilibrium status. Therefore, it may be assumed that the differences in the results applying different turbulence models appears to be negligible.

The boundary layer length scales are given in Figs. 10a and 10b to Figs. 12a and 12b for the form parameter, skin friction and momentum loss thickness. Rather obvious is the coincidence of computation and measurement for the Hassid-Poreh model, although the scatter between the other models is small, too. Apart from the Baldwin-Lomax model which underpredicts the experimental form parameter, Fig. 10a, all other models exhibit a slight overprediction of H and underpredict skin friction, Fig. 11a and 11b, by approximately 5%. The momentum loss thickness distributions, Figs. 12a and 12b, depict an underestimation for nearly all models, apart from Hassid-Poreh and the RAE calculations for the Johnson-King model, Fig. 12b. The latter model, however, exhibits for the IDENT-1400 test case a slightly improved performance over CS and HO. The Johnson-Coakley model on the other hand, seems to be less accurate in predicting equilibrium flows, compare Figs. 10b, 11b, and 12b.

At station x=4.98m, results for the computed velocity profiles, Fig. 13a and 13b, eddy viscosity profiles in Figs. 14a and 14b as well as the dimensionless velocity profiles in Figs. 15a and 15b are presented. The different eddy viscosity distributions for HP, BL and (all together) the other models, Fig. 14a, lead consequently to deviations in the outer layer Reynolds stress distributions.

The computations at x=4.98m, performed by DA, indicate different eddy viscosity profiles for CS and JK models. The smaller eddies produced by the JK model, Fig. 14b, result directly in a better representation of the velocity profile for the CS model, Fig. 13b, although the eddy viscosity for the latter model exhibits a non-smooth transition from the inner- to the outer-layer model.

To draw a conclusion from the IDENT-1400 results, the comparison between experimental findings and computations for the Hassid-Poreh model is supposed to be excellent. Both 1/2-equation models, JK and HO, behave very similar, in contrast to all other test cases. A possible reason for that may be due to the specific test case of a flat plate flow. From the Dornier method it turned out that the σ-values, equs. (JK8) and (HO3), for both models (Johnson-King and Horton), being a measure for non-equilibrium, are close to unity with values of 1.04 for Johnson-King and 1.01 for the Horton model, apart from a small initial maximum of about 1.1 for both models.

5.5.4.3 Stanford IDENT-2200

The flow, investigated as IDENT 2200, is a retarded equilibrium boundary layer flow, particularly a flow with a mild positive pressure gradient as it can be taken from the boundary layer edge velocity presented in Fig. 16 and its x-dervative in Fig. 17.

Based on that x-wise velocity distribution, Figs. 18a and 18b to Fig. 20a and 20b present form parameter, skin friction and momentum loss thickness distributions. Concerning the momentum loss thickness distributions in Fig. 20a (Dornier calculations), the computational results for all turbulence models coincide perfectly, but, at the same time, overpredicting θ with increasing x-values. The Johnson-King model, however, is quite off. The latter is also true for skin friction, Figs. 19a and 19b, and the form parameter distributions in Figs. 18a and 18b. A reason may be, that the JK model produces σ-values being greater than unity throughout the computational domain while the HO model shows a crossover at approximately x=14ft from values being greater than one to those being less than unity.

The diversification of the results leads to the assumption that the currently investigated flow case seems to be a challenge for all models and, hence, it is difficult to argue about comparisons with the experimental findings. Additionally, it is interesting to note that the Cebeci-Smith and the Horton turbulence models give rather similar answers.

The JK model, as mentioned to be very sensitive to initial settings and often overpredicting non-equilibrium influences, exhibits, particularly for the skin friction, Figs. 19a and 19b, with a strong initial increase in skin friction, a behaviour being contradictory to the other models. The Dornier results, Fig. 19a, and the RAE results, Fig. 19b, depict an almost identical qualitative flow behaviour. The change from JK to the JC model, however, avoids that drastic initial c_f rise, Fig. 19b, and switches the skin friction from over- to underprediction by nearly the same amount of gap to the experiments.

In the following, wall-normal (profile) distributions for the employed turbulence models are compared at x=18.58ft, a position where the momentum loss thickness is still in a reasonable agreement with the measurements. Fig. 21 shows the computed velocity profiles. All models, apart from the Baldwin-Lomax model, underestimate the measured velocity profile, the Johnson-King model is again even more off. Although the JK model predicts the profile shape quite well, the underestimation of momentum in the near wall region may be caused by an overprediction of the eddy viscosity near the wall, Fig. 22, by 15-20%. It should be noted at this point, that the Navier-Stokes results for airfoil flows show the same tendency for decelerated flow, if one compares the computed velocity profiles caused by the Johnson-King model near the trailing edge region on the upper surface of the RAE2822, see section 5.1.

Additionally, for all turbulence models involved, Reynolds stress distributions are presented in Fig. 23, together with the dimensionless velocity profiles at x=18.58ft in Fig. 24.

5.5.4.4 Stanford IDENT-2400

The IDENT-2400 test case describes an initial equilibrium boundary layer in moderate positive pressure gradient. The pressure gradient abruptly decreases to zero at x=5ft and the flow relaxes towards a new equilibrium as it is illustrated in Fig. 25 for the U_e distribution and its x-derivative in Fig. 26.

The computed results for form parameter, skin friction and momentum loss thickness are presented together with the experimental values in Figs. 27a and 27b to Figs. 29a and 29b. Again, in the "a"-series of these figures, results performed by Dornier are shown, whereas in the "b"-series results computed by DA and RAE are presented. The momentum loss thickness, θ, normally being used as an indicator for the accuracy of a numerical approach, shows clearly that the Cebeci-Smith, the Horton and the Hassid-Poreh model are advantageous over Baldwin-Lomax and Johnson-King, Figs. 29a and 29b. Skin friction, Figs. 28a and 28b, and form parameter distributions, Figs. 27a and 27b, show a large scatter for the various turbulence models. Apart from the DA results, presenting a clear advantage for the JK model, all other skin friction results show an overprediction for the JK model and an underprediction for all other models employed. This coincides with the results of IDENT-2200, a test case which also denotes a flow type with a moderate positive pressure gradient. Using the Johnson-Coakley modification to the JK model, leads, as already pointed out for test case IDENT-2200, to an improvement of the results as it can be taken from Figs. 27b, 28b and 29b, respectively.

It was found from the experiments that the last profile, given at approximately 8ft, has not yet reached the equilibrium state, as for example the skin friction, Figs. 28a and 28b, has not begun to decrease again, as it should be corresponding to the zero-pressure gradient. This may be underlined by the σ-distributions of both JK and HO models being in good qualitative agreement with respect to their maximum σ-values of approximately σ=1.5 at the x=8ft position.

Regarding the fact that profile shapes for velocity, eddy viscosity and dimensionless velocity upstream of x=5ft can be qualitatively compared with those presented for test case IDENT-2200, profiles are presented in the following only for the x=7.92ft position which already lies in the zero pressure gradient area.

A careful check of the profile results, depicted in Figs. 30a and 30b to Figs. 32a and 32b, demonstrate again the good overall-performance of the Hassid-Poreh model and a reasonable behaviour of the CS and HO model. The BL model, although the intermittency is poor, exhibits also reasonable results. The JK model, overpredicts - as for all decelerated flows - the turbulent viscosity, Figs. 31a and 31b, which may lead directly to an underpredition of the kinetic energy in the near wall area.

5.5.4.5 Stanford IDENT-2600

The last test case chosen for the validation of turbulence models on a boundary layer basis is characterised by an equilibrium boundary layer in moderate pressure gradient. This flow is ahead of the first profile station of IDENT-2400 equivalent to that case.

Once again, and although the scatter for the form parameter, Figs. 33a and 33b, skin friction, Figs. 34a and 34b, and momentum loss thickness, Figs. 35a and 35b, is rather large, the Johnson-King model is even worse compared to the experimental findings. Changing to the Johnson-Coakley version, however, improves the situation, see results performed by RAE in Figs. 33b, 34b and 35b. The overprediction of skin friction, Fig. 34b, for the JK model is converted into an underprediction for the JC model with results being very close to the CS calculations. Both algebraic models, Baldwin-Lomax and Cebeci-Smith, as well as the Horton model produce reasonable results for the momentum loss distribution, Figs. 35a and 35b, with a clear advantage for the Baldwin-Lomax model.

For x=6.92ft, velocity profiles, eddy viscosity, Reynolds stress and u^+ distributions are presented in Figs. 36 to 39. Undoubtedly, the Baldwin-Lomax model shows the best performance but overpredicts the velocity profile slightly. However, that overprediction is not the most important feature, it is rather the shape of the velocity profile which does only exhibit - in contrast to the experimental findings - an only "weak" inflection point. This might correspond to the higher eddy viscosity in the near wall region together with an increased maximum value, and farther outside to the lack of a proper represenation of intermittency.

The velocity profile according to JK, see Fig. 36, depicts a clear minimum for momentum which may be correlated with an overprediction of eddy viscosity in an area closer to the wall. This, however, is only a local viewpoint and does not take into account the flow situation upstream of that station.

5.5.5 Conclusions

To conclude from the previous discussion, it turns out that a proper and precise evaluation of the advantages and disadvantages of the investigated turbulence models needs to be carried out cautiously. Moreover, the validation of turbulence models in the described way should be accomplished by an excellent understanding of the individual models in both their physical and numerical representations.

The answers gained from the current initiative, can be directed in a way that it proves generally necessary to maximize the understanding of the nature of the individual models. Consequently, this includes the establishment of a data base - accessible by other interested users - which must depend on standardized turbulence models, i.e. models must be used firstly in their original form, and secondly, properly described and reliable modifications may be introduced. It is unambigious, that such an approach is urgently necessary and needs an extended effort. Due to the fact that a single boundary layer calculation is rather cheap compared to Navier-Stokes computations, several tests and re-calculations can be easily taken into account. It is obvious that the validation of turbulence models on a boundary layer basis gives an excellent tool at hand for applying those (already "boundary-layer-tested") turbulence models in Navier-Stokes methods more efficiently and with the best possible background knowledge.

A conclusion from the computations presented for the six different turbulence models may be drawn as follows:

- Wherever applied, the one-equation model by Hassid-Poreh shows the best overall agreement with the experimental findings.
- The best algebraic model is still the Cebeci-Smith model which is in most cases comparable to the (1/2)-equation model of Horton.
- In all cases, the Horton model is superior to the Johnson-King model.
- The Baldwin-Lomax model underpredicts the momentum loss thickness in all test cases with pressure gradient. The intermittency of that turbulence model is in most cases ineffective (inside the boundary layer).
- The (1/2)-equation Johnson-King model shows some disadvantages for the investigated flows, being all not far away from equilibrium. For decelerated flow situations, all boundary layer results exhibit the same tendency, i.e. an underestimation of the near wall velocity profile.
- The (1/2)-equation Johnson-Coakley model, being a modification to the JK model, exhibits a considerable improvement over the original Johnson-King model.
- It is worthwhile to mention that all investigated models offer a good representation of flows with a zero pressure gradient.

The present results, although - as mentioned - not in a final stage and with the strong demand for continuation, clearly demonstrate the capability of boundary layer methods to validate different turbulence models in the absence of numerical dissipation.

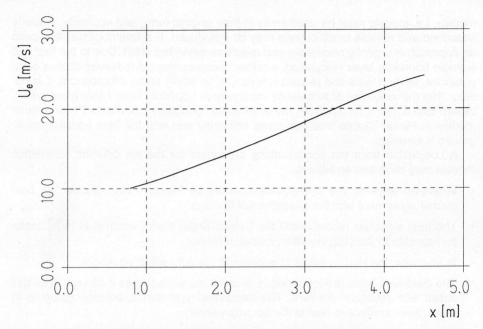

Fig. 1 IDENT-1300: Velocity distribution at boundary layer edge

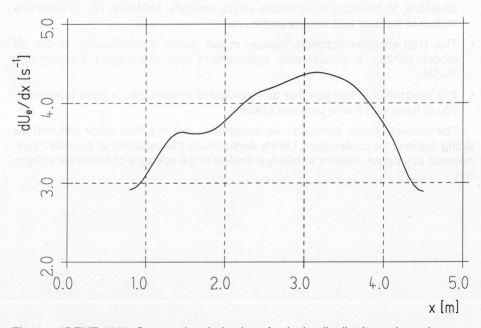

Fig. 2 IDENT-1300: Streamwise derivative of velocity distribution at boundary layer edge

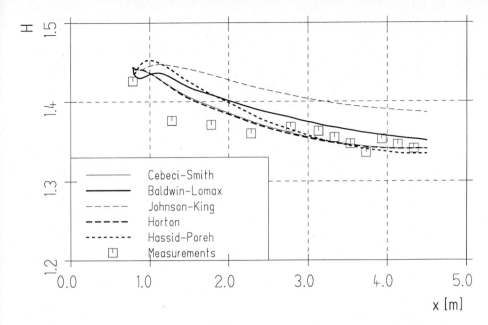

Fig. 3a IDENT-1300: Form parameter distributions calculated by Dornier

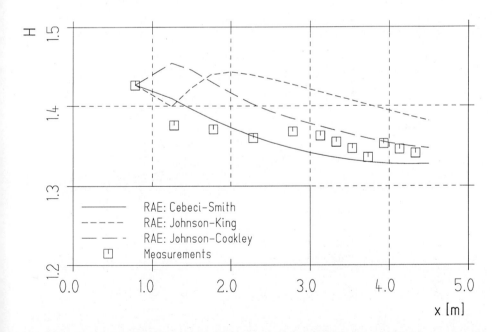

Fig. 3b IDENT-1300: Form parameter distributions calculated by RAE

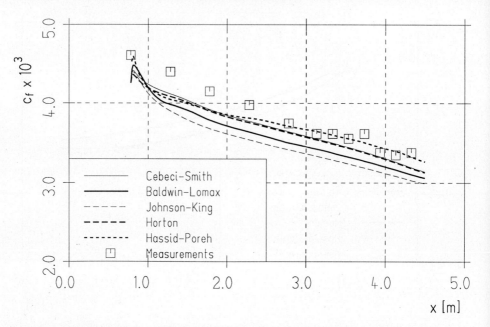

Fig. 4a IDENT-1300: Skin friction distributions calculated by Dornier

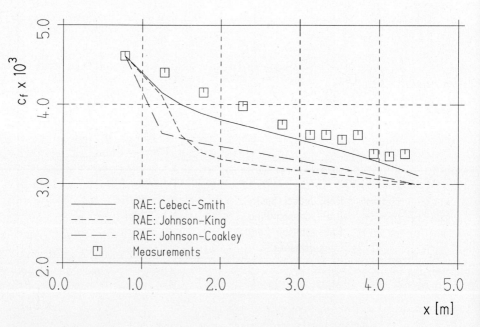

Fig. 4b IDENT-1300: Skin friction distributions calculated by RAE

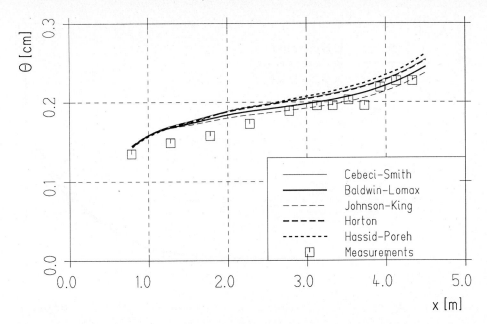

Fig. 5a IDENT-1300: Momentum loss thickness distributions calculated by Dornier

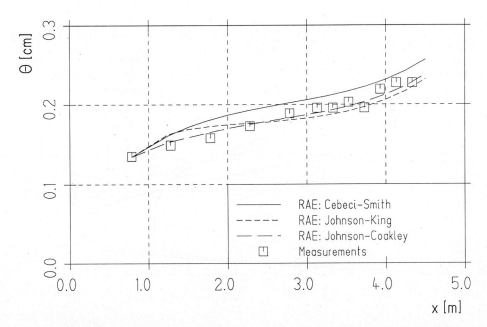

Fig. 5b IDENT-1300: Momentum loss thickness distributions calculated by RAE

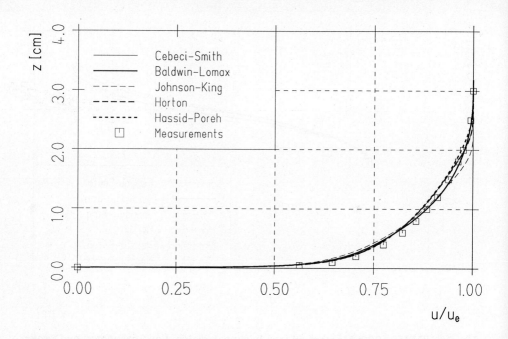

Fig. 6 IDENT-1300: Velocity Profiles at x = 4.332 m calculated by Dornier

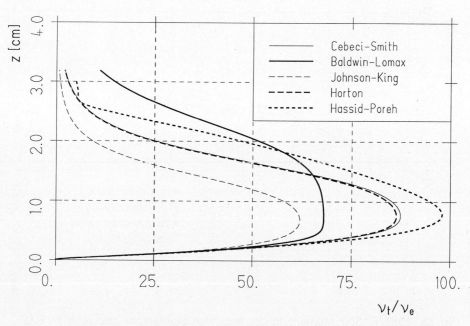

Fig. 7 IDENT-1300: Eddy-viscosity profiles at x = 4.332 m calculated by Dornier

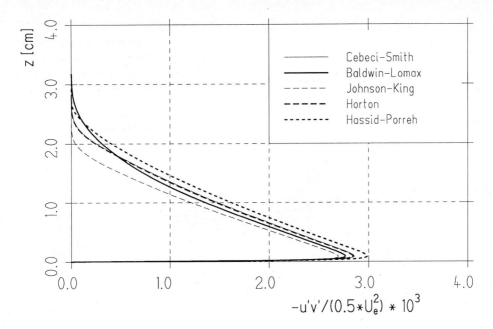

Fig. 8 IDENT-1300: Reynolds-stress profiles at x = 4.332 m calculated by Dornier

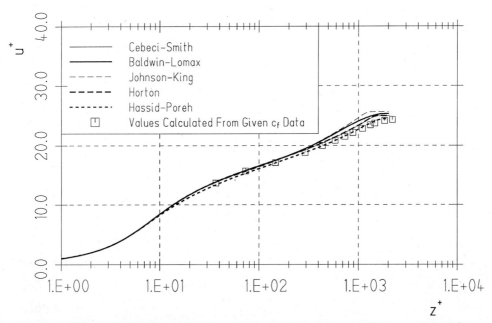

Fig. 9 IDENT-1300: Dimensionless-velocity profiles at x = 4.332 m calculated by Dornier

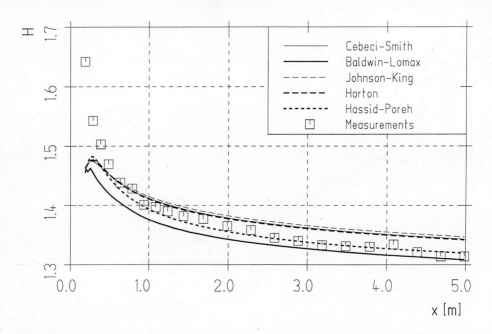

Fig. 10a IDENT-1400: Form parameter distributions calculated by Dornier

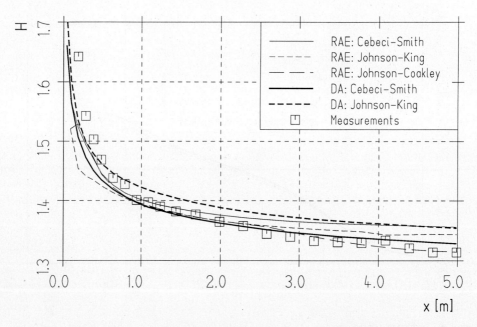

Fig. 10b IDENT-1400: Form parameter distributions calculated by RAE and DA

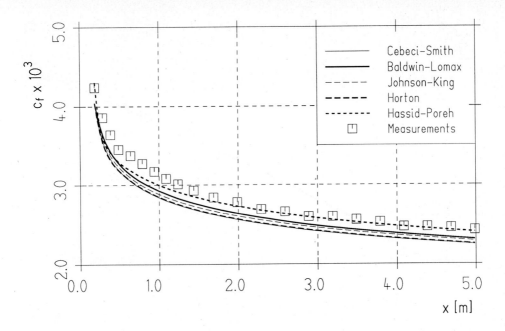

Fig. 11a IDENT-1400: Skin friction distributions calculated by Dornier

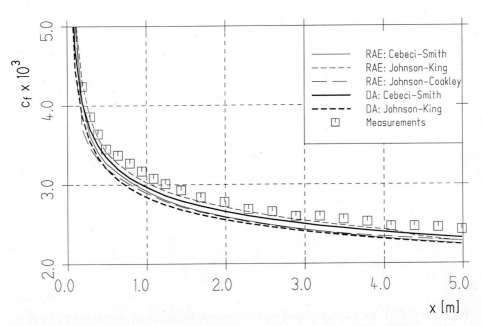

Fig. 11b IDENT-1400: Skin friction distributions calculated by RAE and DA

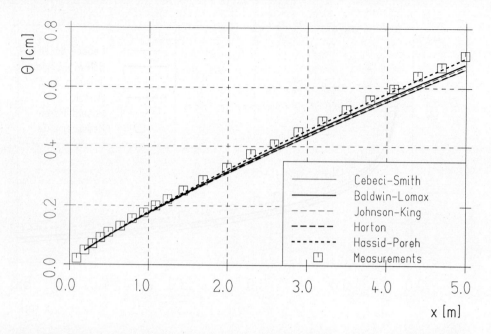

Fig. 12a IDENT-1400: Momentum loss thickness distributions calculated by Dornier

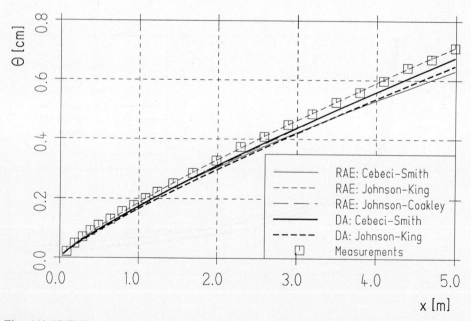

Fig. 12b IDENT-1400: Momentum loss thickness distributions calculated by RAE and DA

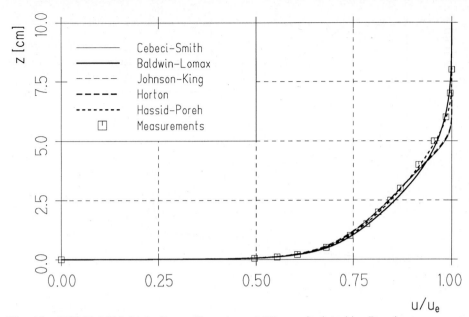

Fig. 13a IDENT-1400: Velocity profiles at x = 4.98 m calculated by Dornier

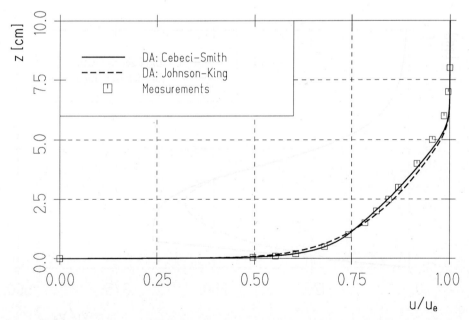

Fig. 13b IDENT-1400: Velocity profiles at x = 4.98 m calculated by DA

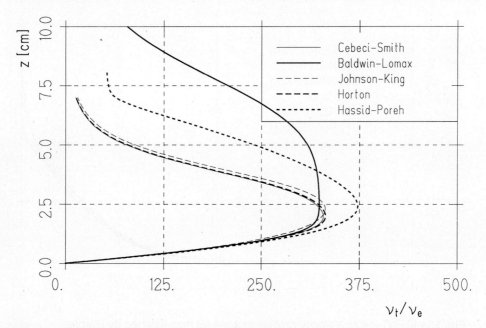

Fig. 14a IDENT-1400: Eddy-viscosity profiles at x = 4.98 m calculated by Dornier

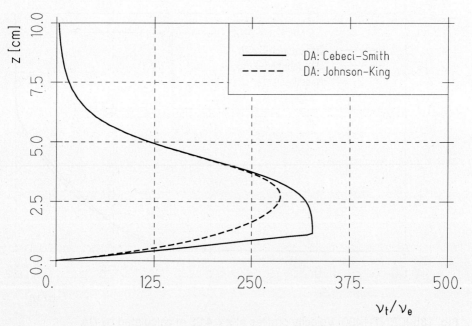

Fig. 14b IDENT-1400: Eddy-viscosity profiles at x = 4.98 m calculated by DA

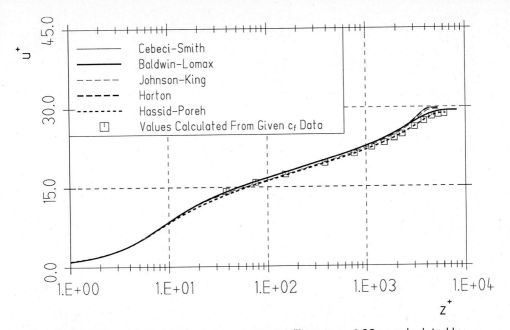

Fig. 15a IDENT-1400: Dimensionless-velocity profiles at x = 4.98 m calculated by Dornier

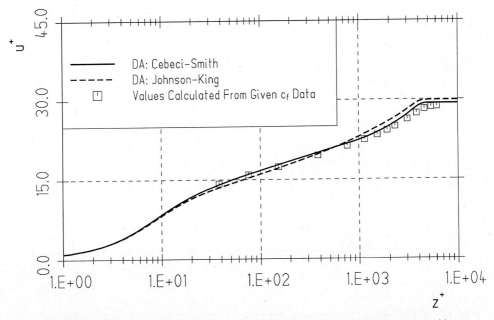

Fig. 15b IDENT-1400: Dimensionless-velocity profiles at x = 4.98 m calculated by DA

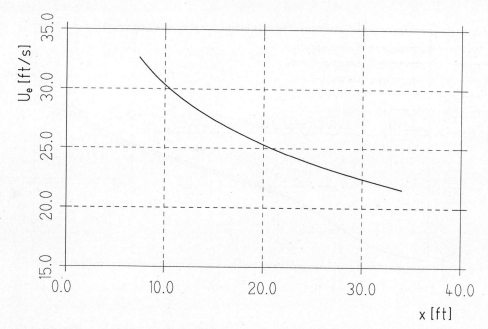

Fig. 16 IDENT-2200: Velocity distribution at boundary layer edge

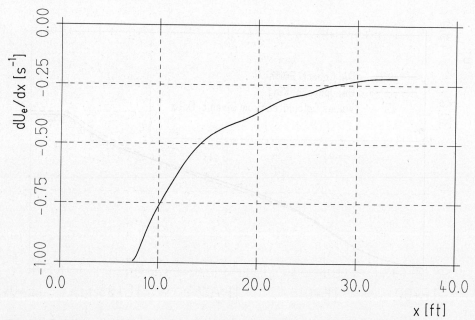

Fig. 17 IDENT-2200: Streamwise derivative of velocity distribution at boundary layer edge

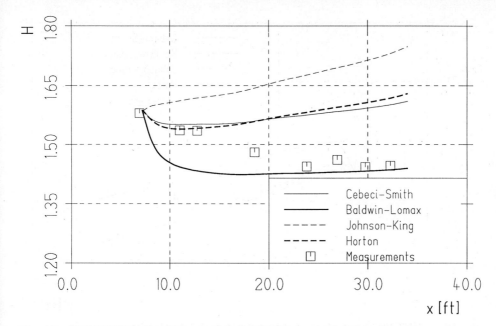

Fig. 18a IDENT-2200: Form parameter distributions calculated by Dornier

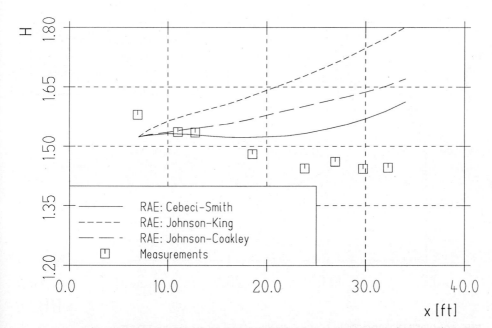

Fig. 18b IDENT-2200: Form parameter distributions calculated by RAE

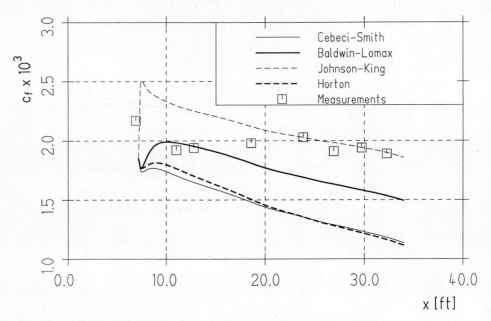

Fig. 19a IDENT-2200: Skin friction distributions calculated by Dornier

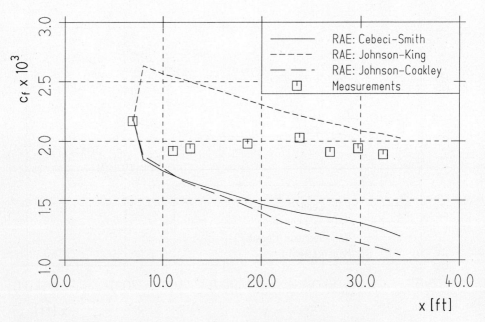

Fig. 19b IDENT-2200: Skin friction distributions calculated by RAE

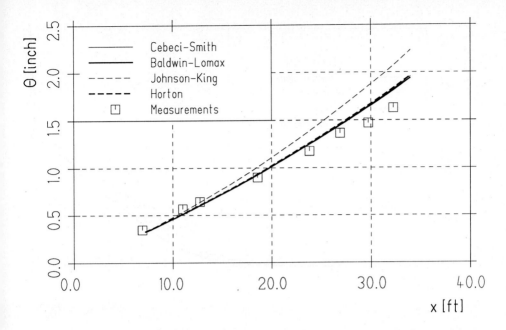

Fig. 20a IDENT-2200: Momentum loss thickness distributions calculated by Dornier

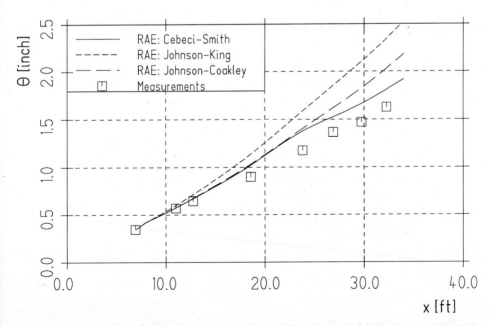

Fig. 20b IDENT-2200: Momentum loss thickness distributions calculated by RAE

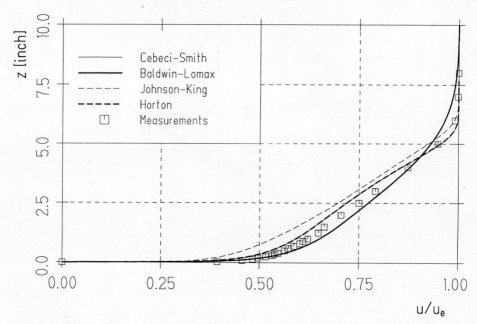

Fig. 21 IDENT-2200: Velocity Profiles at x = 18.58 ft calculated by Dornier

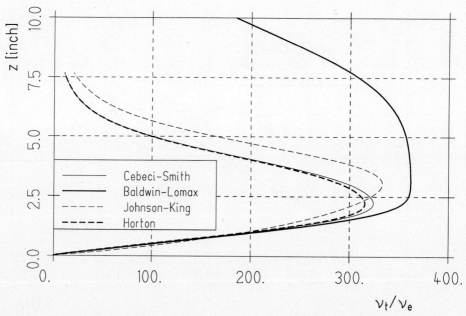

Fig. 22 IDENT-2200: Eddy-viscosity profiles at x = 18.58 ft calculated by Dornier

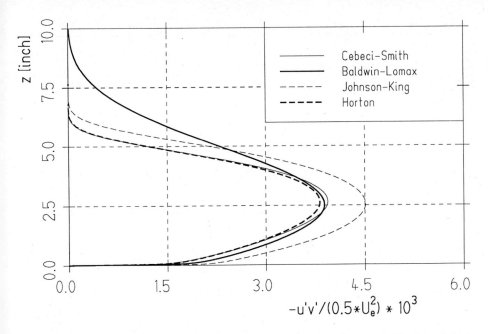

Fig. 23 IDENT-2200: Reynolds-stress profiles at x = 18.58 ft calculated by Dornier

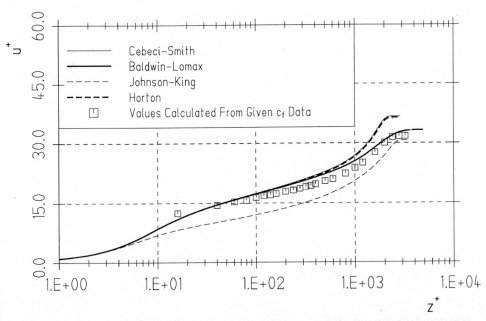

Fig. 24 IDENT-2200: Dimensionless-velocity profiles at x = 18.58 ft calculated by Dornier

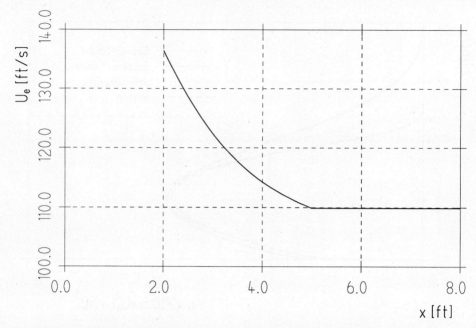

Fig. 25 IDENT-2400: Velocity distribution at boundary layer edge

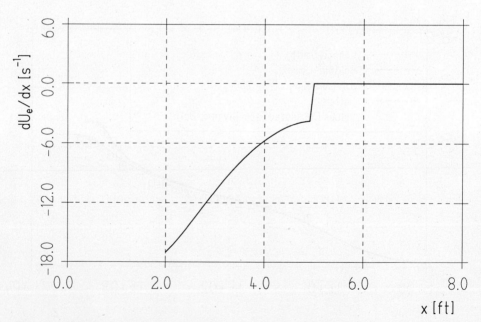

Fig. 26 IDENT-2400: Streamwise derivative of velocity distibution at boundary layer edge

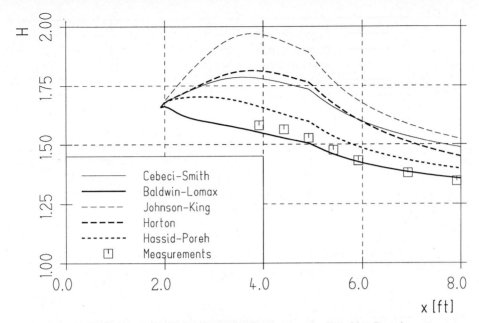

Fig. 27a IDENT-2400: Form parameter distributions calculated by Dornier

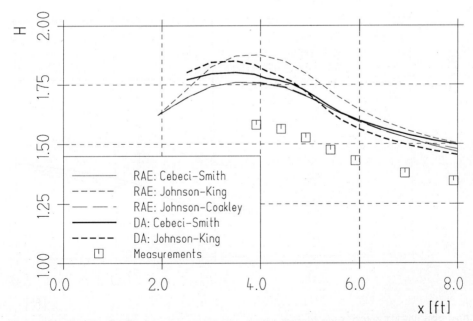

Fig. 27b IDENT-2400: Form parameter distributions calculated by RAE and DA

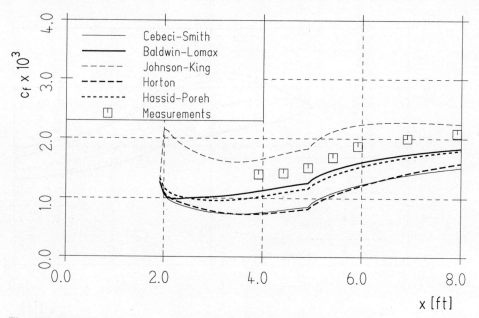

Fig. 28a IDENT-2400: Skin friction distributions calculated by Dornier

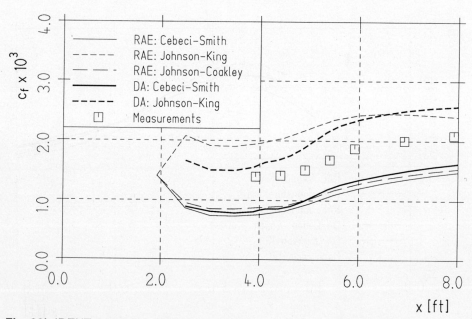

Fig. 28b IDENT-2400: Skin friction distributions calculated by RAE and DA

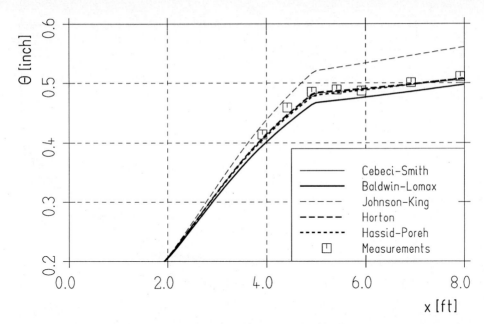

Fig. 29a IDENT-2400: Momentum loss thickness distributions calculated by Dornier

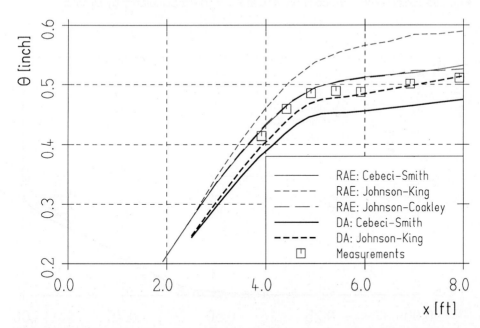

Fig. 29b IDENT-2400: Momentum loss thickness distributions calculated by RAE and DA

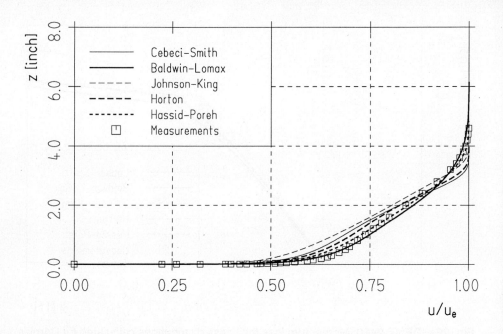

Fig. 30a IDENT-2400: Velocity profiles at x = 7.917 ft calculated by Dornier

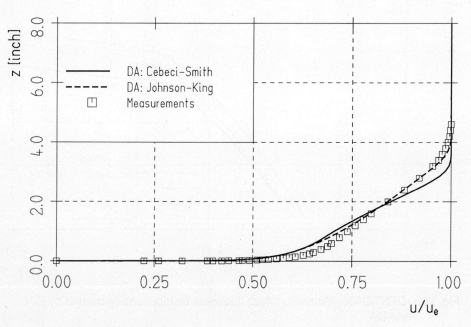

Fig. 30b IDENT-2400: Velocity profiles at x = 7.917 ft calculated by DA

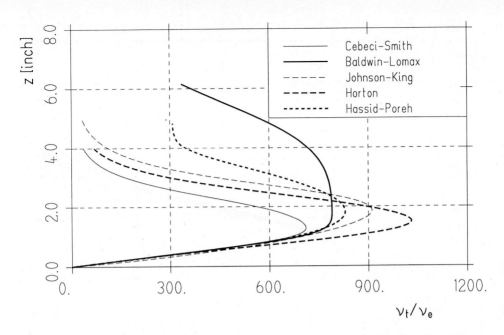

Fig. 31a IDENT-2400: Eddy-viscosity profiles at x = 7.917 ft calculated by Dornier

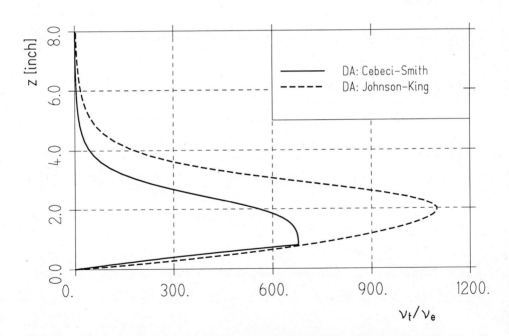

Fig. 31b IDENT-2400: Eddy-viscosity profiles at x = 7.917 ft calculated by DA

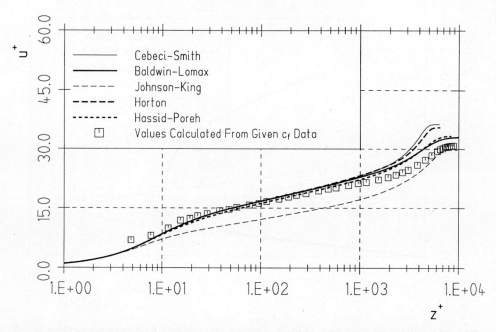

Fig. 32a IDENT-2400: Dimensionless-velocity profiles at x = 7.917 ft calculated by Dornier

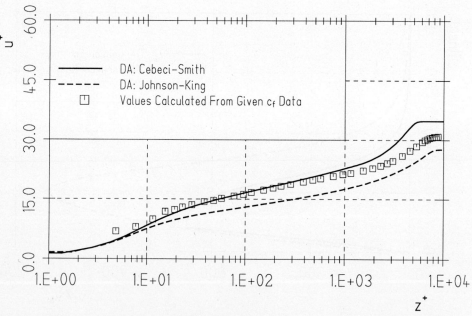

Fig. 32b IDENT-2400: Dimensionless-velocity profiles at x = 7.917 ft calculated by DA

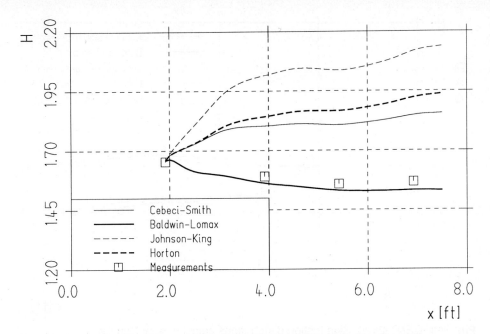

Fig. 33a IDENT-2600: Form parameter distributions calculated by Dornier

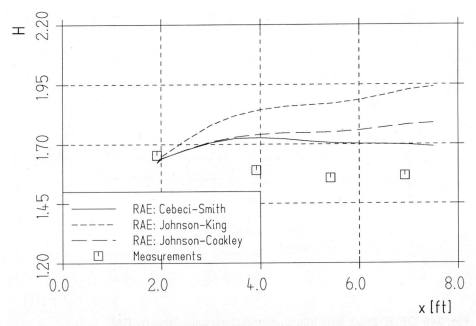

Fig. 33b IDENT-2600: Form parameter distributions calculated by RAE

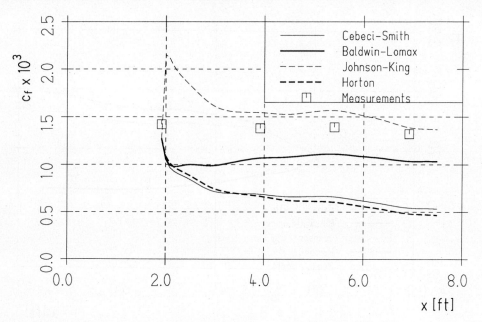

Fig. 34a IDENT-2600: Skin friction distributions calculated by Dornier

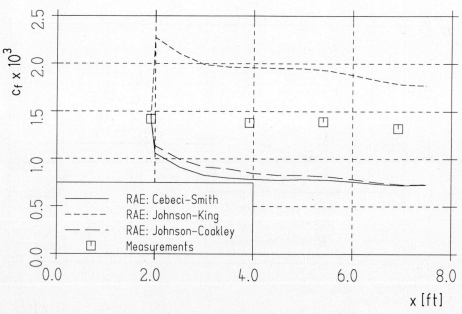

Fig. 34b IDENT-2600: Skin friction distributions calculated by RAE

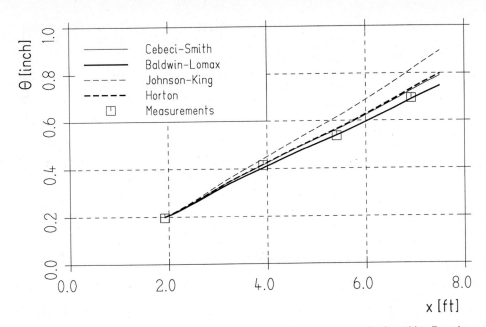

Fig. 35a IDENT-2600: Momentum loss thickness distributions calculated by Dornier

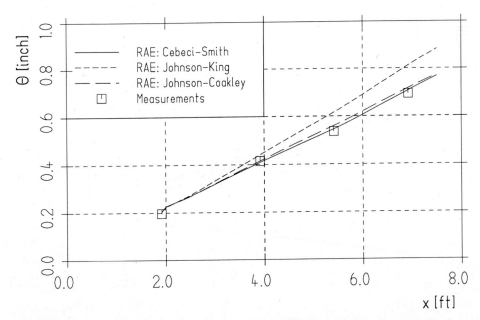

Fig. 35b IDENT-2600: Momentum loss thickness distributions calculated by RAE

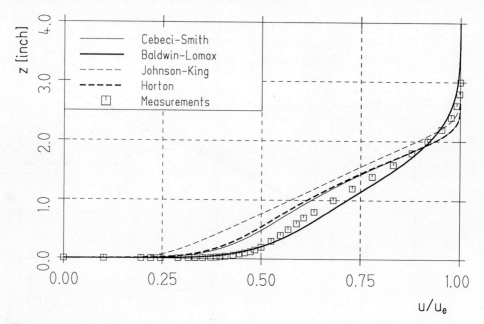

Fig. 36 IDENT-2600: Velocity profiles at x = 3.917 ft calculated by Dornier

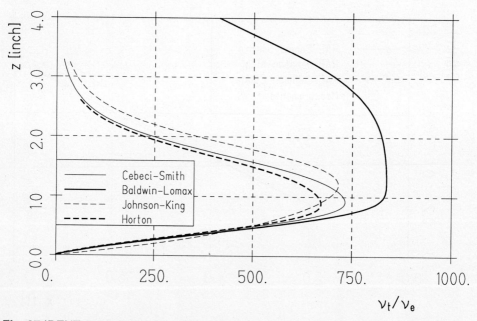

Fig. 37 IDENT-2600: Eddy-viscosity profiles at x = 3.917 ft calculated by Dornier

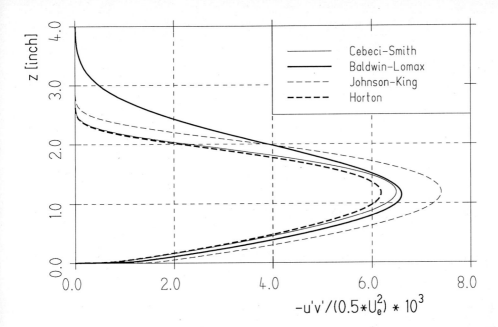

Fig. 38 IDENT-2600: Reynolds-stress profiles at x = 3.917 ft calculated by Dornier

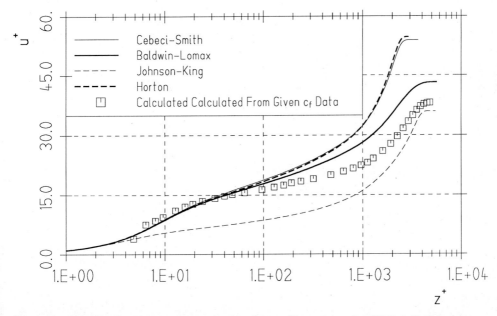

Fig. 39 IDENT-2600: Dimensionless-velocity profiles at x = 3.917 ft calculated by Dornier

THREE-DIMENSIONAL BOUNDARY LAYERS

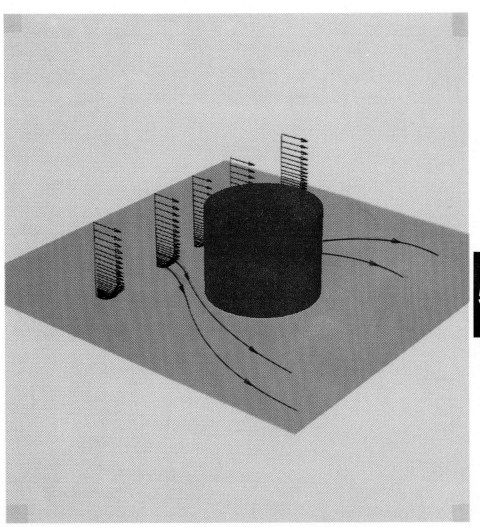

5.6

Colour figure - provided by UMIST - shows:

3D boundary layer flow for cylinder-on-flat-plate (Dechow-Felsch test case)

5.6 Three-Dimensional Boundary Layers

Author: **P.D.Smith**, RAE
Plot Coordinator: **F.Thiele**, TUB

5.6.1 Introduction

In this section the concept introduced in section 5.5, of evaluating turbulence models in the absence of numerical viscosity by the use of boundary-layer methods, is extended to the three-dimensional case. Two test cases are examined; a third, the DLR F5 wing, is described in section 5.7.

5.6.2 Partners and Methods

Three partners have contributed to this task; DA, RAE and TUB. Full details of their methods will be found in sections 4.7 (DA), 4.11 (RAE) and 4.13 (TUB) and they are briefly reviewed below.

DA: Implicit space marching method
Locally infinite swept (LISW) and 3D version (from TUB)
2nd order accurate in surface tangential directions
4th order (Hermitian) in normal direction
direct, semi-inverse and inverse modes
direct mode uses two components of edge velocity
semi inverse mode uses mainflow component of displacement thickness and spanwise component of edge velocity
inverse mode uses two components of displacement thickness

RAE: 3D time marching finite difference method
2nd order accurate in space
first order accurate in time
direct and inverse modes
direct mode uses pressure coefficient
inverse mode uses two components of skin friction

TUB: 3D implicit space marching method
2nd order accurate in surface tangential directions
4th order (Hermitian) in normal direction
direct and inverse modes
direct mode uses two components of edge velocity
inverse mode uses two components of displacement thickness

5.6.3 Turbulence Models

Three different turbulence models have been used. DA and RAE used the Cebeci-Smith (CS) and Johnson-King (JK) models whilst TUB used the CS, JK and Johnson-Coakley (JC) models.

5.6.4 Discussion of Results

5.6.4.1 Infinite Swept Wing Test Case BEEL72

This incompressible three-dimensional turbulent boundary-layer experiment was specifically designed as a test for calculation methods and was performed at the NLR Amsterdam in 1972. The experiment closely simulates the flow on an infinite swept wing (flow properties are constant along the span), making the measurement and computational tasks much simpler, whilst still providing a full check for three-dimensional turbulence models. The boundary-layer develops from zero pressure gradient conditions on a flat plate swept at 35° under the influence of an adverse pressure gradient generated by a wing-like body, also swept at 35° and suspended above the plate. Boundary layer separation was produced towards the end of the test region by adjustment of downstream blockage. The pressure distribution on the plate was measured together with boundary-layer mean velocity and Reynolds stress measurements at ten stations which covered the flow development from a nearly two-dimensional boundary-layer to a separated three-dimensional boundary-layer. Full details are given in van den Berg and Elsenaar (1972), Elsenaar and Boelsma (1974) and van den Berg (1976). This experiment has previously been used as a test case for boundary-layer calculation methods in workshops at Trondheim in 1975 (East 1975) and Berlin in 1982 (van den Berg et al 1988). Only direct calculation methods were available at the time of these workshops and they all tended to fail upstream of the separation line.

This failure is repeated in the results of the direct calculations reported here. As will be seen from Figures 1 to 5 all the predictions diverge fom the experimental results downstream of x=1.0m, well ahead of the observed separation line at approximately x=1.3m. The differences between the predictions of the three partners when using the same turbulence model seem mainly attributable to the different starting techniques used. DA and TUB assume starting values well upstream of the first measuring station and vary these in an attempt to match the values at the first station. This technique tends to confine any starting transients caused by inconsistent initial values to upstream of the region in which calculation and experiment are compared. The RAE method starts at the first station with measured values.

When inverse calculation methods are used a great improvement in the accuracy of the predictions is obtained and as, will be seen in Figures 6 to 10, in some cases the calculations are in very close agreement with the measurements. The most significant failure is the poor prediction of skin friction coefficient, c_f, by the JK model. This weakness of the JK model is well known and as will be seen in Fig. 11 is substantially overcome by the use of the JC model.

5.6.4.2 Cylinder on Flat Plate Test Case DEFE77

This experiment is described in Dechow (1977) and Dechow and Felsch (1977). An initially two-dimensional boundary-layer on a flat plate becomes three-dimensional and then separates in the pressure field of a cylinder standing on the plate. The pressure distribution on the plate was measured together with boundary-layer mean velocity and Reynolds stress measurements at two stations on the symmetry plane and ten stations on an external streamline. These ten stations covered the flow development from a two-dimensional boundary-layer to downstream of the three-dimensional separation line.

This test case exemplifies the major difficulty of fully three-dimensional boundary-layer experiments. Although making the measurements was a large task they are still insufficient for anything other than a direct calculation. For an inverse calculation (which was not a possibility at the time the measurements were made) the complete skin friction or displacement thickness fields would have to be measured; a task which is several orders of magnitude larger.

With this restriction to direct calculations the results shown in Figures 12 to 16 essentially repeat the lessons of the previous test case, BEEL72. All the methods are in reasonable agreement with the measurements initially but all fail upstream of separation. Once again the poor prediction of C_f by the JK model is significantly improved by the use of the JC modification.

5.6.5 Conclusions

It was only possible to test a limited range of turbulence models within the time and cost constraints for this task, much more remains to be done. Nevertheless the ready cooperation between the partners facilitated much faster progress than would have otherwise been possible.

Of the turbulence models tested the JC model shows the best performance for separating and separated flows whilst the CS model remains the best for flows close to equilibrium.

For the calculation of separating and separated flows it is necessary to use an inverse method of calculation. This requires extensive experimental data to be available and currently restricts comparisons to just one test case, BEEL72. Fortunately this lack of data was recognized some ten years ago, following the Berlin workshop, and led to a major European cooperative experiment, the so called GARTEUR WING (van den Berg, 1989), from which the results should become available in the next two years.

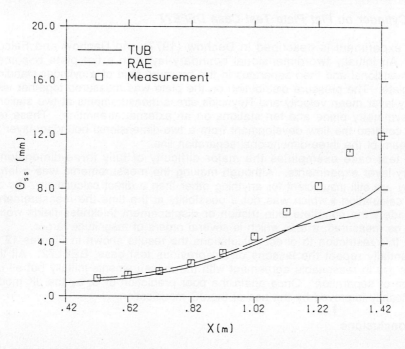

Fig.1 BEEL72. Streamwise Momentum Thickness. Direct Calculations; CS Model.

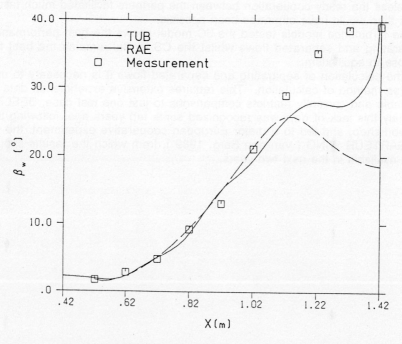

Fig.2 BEEL72. Limiting Streamline Angle. Direct Calculations; CS Model.

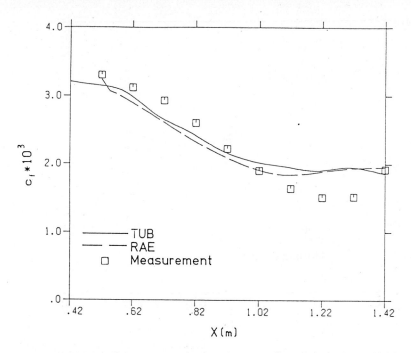

Fig.3 BEEL72. Total Skin Friction Coefficient. Direct Calculations; CS Model.

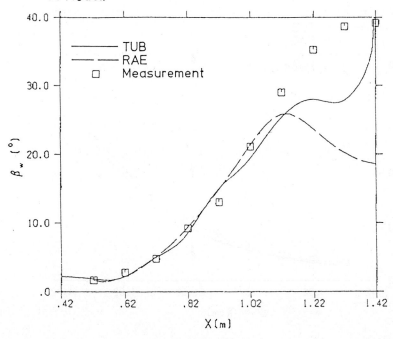

Fig.4 BEEL72. Limiting Streamline Angle. Direct Calculations, JK Model.

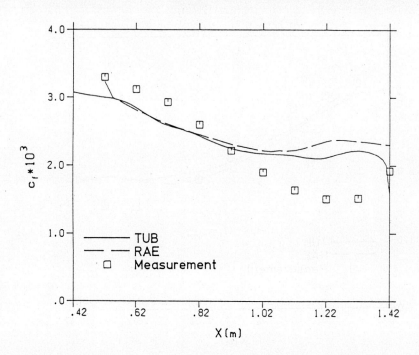

Fig.5 BEEL72. Total Skin Friction Coefficient. Direct Calculations; JK Model.

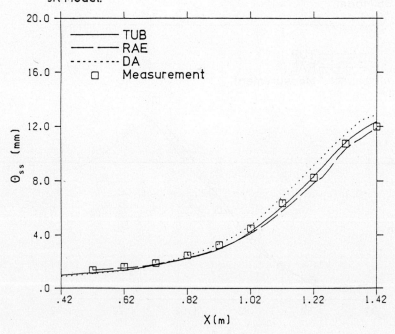

Fig.6 BEEL72. Streamwise Momentum Thickness. Inverse Calculations; CS Model.

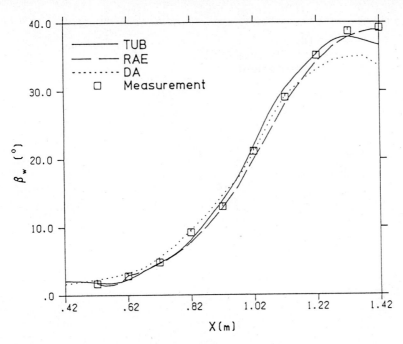

Fig.7 BEEL72. Limiting Streamline Angle. Inverse Calculations; CS Model.

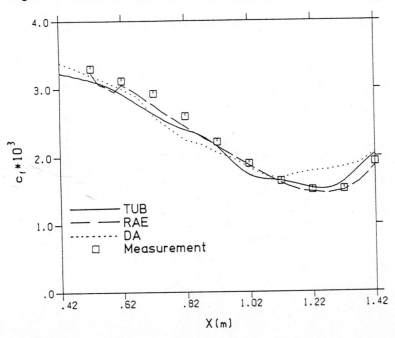

Fig.8 BEEL72. Total Skin Friction Coefficient. Inverse Calculations; CS Model.

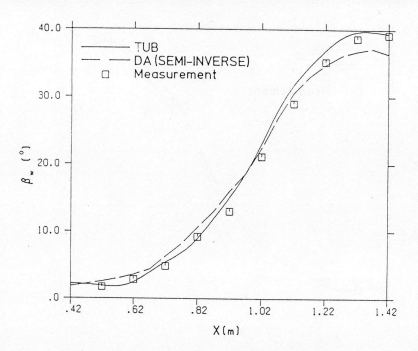

Fig.9 BEEL72. Limiting Streamline Angle. Inverse Calcultions; JK Model.

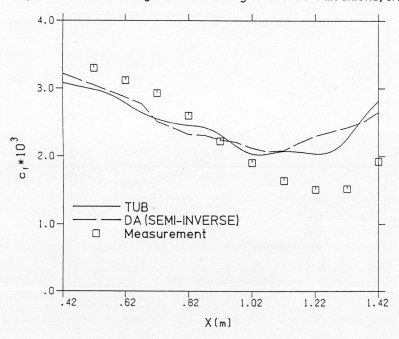

Fig.10 BEEL72. Total Skin Friction Coefficient. Inverse Calculations; JK Model.

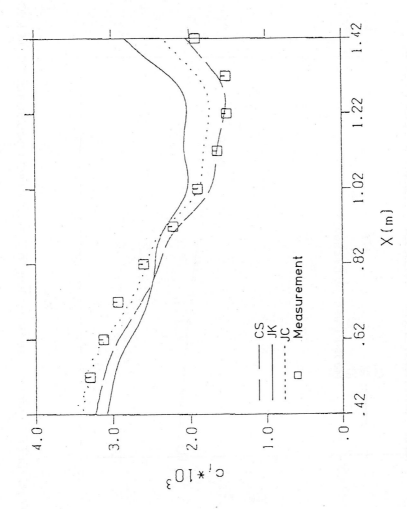

Fig.11 BEEL72. Total Skin Friction Coefficient. Inverse Calculations; CS, JK & JC Models.

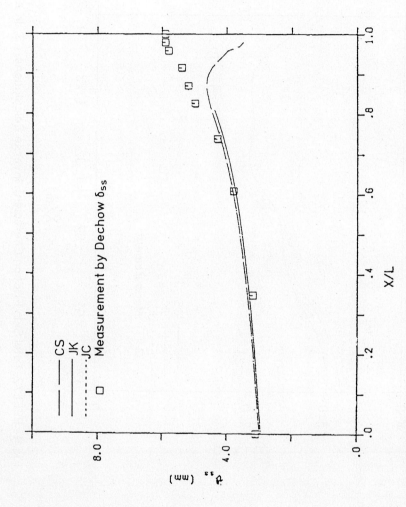

Fig.12 DEFE77. Streamwise Momentum Thickness. Direct Calculations; CS, JK & JC Models.

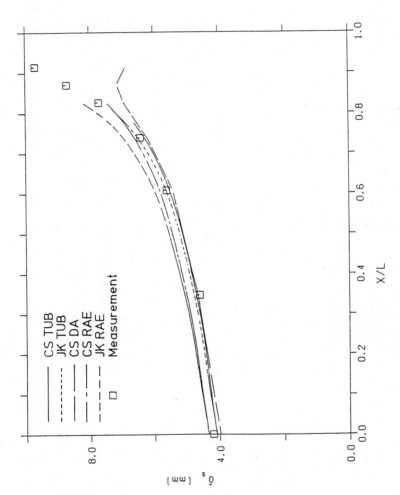

Fig.13 DEFE77. Streamwise Displacement Thickness. Direct Calculations; CS & JK Models.

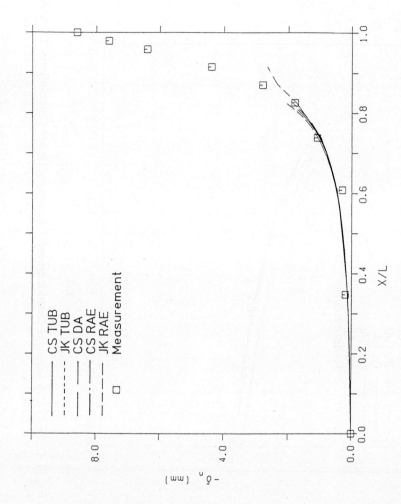

Fig.14 DEFE77. Crosswise Displacement Thickness. Direct Calculations; CS & JK Models.

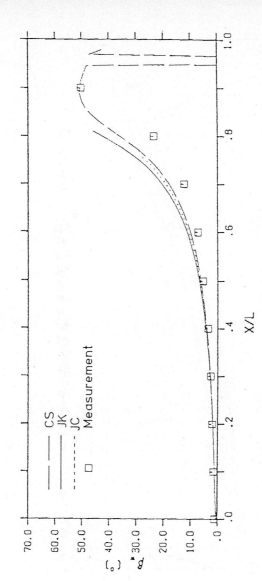

Fig.15 DEFE77. Limiting Streamline Angle. Direct Calculations, CS JK & JC Models.

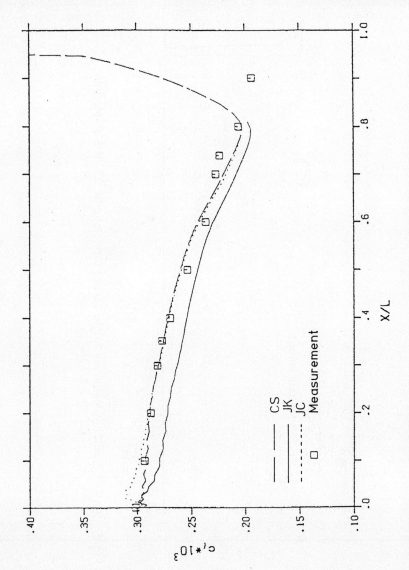

Fig.16 DEFE77. Total Skin Friction Coefficient. Direct Calculations; CS, JK & JC Models.

DLR-F5 WING TEST CASE

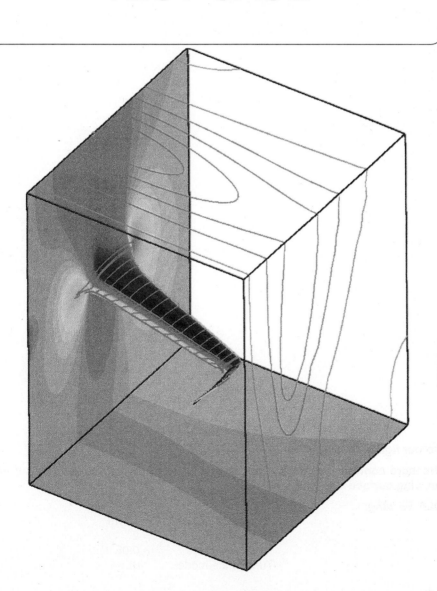

5.7

Colour figure - provided by DLR - shows:

Pressure contour plot for 3D flow over of a wing including skin friction lines on wing surface and wing-tip vortex:

DLR-F5 wing:
    Mach number $= 0.82$
    Reynolds number $= 10^7/m$
    Angle of attack $= 2°$
    Transition at x/c $= 0.05$
    Turbulence model   BL

5.7 DLR–F5 Wing Test Case

Author: **D. Schwamborn**, DLR
Plot-Coordinator: **Th. Rung**, TUB

5.7.1 Introduction

The DLR-F5 wing experiment (Sobieczky et al., 1987) was designed in an attempt to arrive at a well-posed boundary-value problem for CFD, i.e. to specify an experimentally completely defined test case with no uncertainties about boundary conditions as they exist, for example, in experiments with slotted wind tunnel walls. The wing was, therefore, mounted on a splitter plate in a closed tunnel section (Fig. 1) and pressure measurements were made not only along the wing and the splitter plate but also along all wind tunnel walls and in the inlet and exit plane forming a simply shaped control volume (Fig. 2). Together with the measured flow direction and Mach number in the inlet plane, this minimized uncertainties with respect to the boundary conditions of a computational control volume. The splitter plate with boundary-layer tripping and the extreme fairing of the wing avoided a thick boundary layer and the appearance of a horse-shoe vortex at the root of the transport-type wing allowing also for calculations with inviscid and/or boundary-layer methods. The surface of the high-aspect-ratio wing, including the fairing, is given in analytical form (Sobieczky, 1988) such that any surface discretization desired can be achieved in a numerical method.

Due to the high aspect ratio of the wing there are large regions of quasi-two-dimensional swept-wing flow where profile-design methodology can be expected to apply. The basic airfoil for the main part of the wing has been designed to be shock-free at a Mach number of .78, leading to a shock free Mach number of .82 for the wing with 20° leading-edge sweep. The design aimed to produce a pressure distribution which would maintain laminar flow over a large portion of the airfoil. A symmetric airfoil was used for experimental reasons, i.e. to allow for a dense distribution of pressure taps at ten wing sections where pressures were mainly measured along the upper surface while a few pressure taps on the lower surface were used for control purposes only. The lower surface data at finite angle of attack were obtained measuring upper surface value at negative angles of incidence.

This test case was chosen by the partners in EUROVAL to perform comparisons not only between Navier-Stokes codes but also with results of Euler and boundary-layer methods. It was decided to use the case of the wing in the wind tunnel at angle of attack (α = 2°), an inlet Mach number of .82, a Reynolds number of 10 million per meter and an exit pressure of 99 per cent of the reference pressure in the inlet as the mandatory test case. In calculations the wall of the wind tunnel was to be simulated as inviscid, since not all partners would be able to simulate no-slip conditions at all surfaces. To allow a reasonable testing of turbulence models, and since it was known from the results of the F5 workshop (Kordulla, 1988) that the original experiment with free transition near the shock-position was very difficult to simulate, it was decided to assume fixed transition at 5 per cent of local chord in the computations wherever the experimental transition location was further downstream. This decision makes a comparison between experimental and computational results questionable, but at the time of the decision it was expected that some additional data for the case of fixed transition would become available soon after. Unfortunately, this data did not materialize.

There exist, however, some experimental data on the upper surface of the wing with and without fixed transition for a slightly higher Mach number (M = .845), which give a strong indication that for free transition the pressure distribution in the accelerated part of the flow, up to the chordwise pressure minimum, is very similar to the distribution for fixed transition. This suggests that a comparison of calculated and experimental pressure distribution is still justified from the leading edge to that pressure minimum, although most of the conclusions will be based, solely on comparisons between computational results.

Taking this into account, it is of special value that there are not only results from an Euler and some Navier-Stokes codes but – as can be seen from Table 1 – also from boundary-layer methods which used the pressure distribution from a Navier Stokes solution as input. Table 1 indicates furthermore the different turbulence models used by the partners, and whether a partner also considered the influence of different grids or varying numerical dissipation, as will be discussed later.

Table 1 Partners, methods and models.
(LISW: locally infinite swept wing; CP: Chen/Patel (1987) k-ε model;
FV,CC,RK(3): Finite volume, cell-centered, Runge-Kutta (3 stage);
T.L.: Thin-Layer; MB: multi-block)
(Abbreviations of partners and models see Chapters 1 and 3).

Partner	Method	Turb-model	Remarks
DA	Boundary layer 3D and LISW	CS,JK	
RAE	3D Boundary layer (time-dependent)	CS	
TUB	3D Boundary layer	CS,JK,JC	
CASA	Euler (FV,CC,RK(4),MB)	-----	dissipation variation
CERFACS	Navier-Stokes (FV,CC,RK(3))	BL, CP	2 meshes 2 boundary conditions
CFD norway	T.L. Navier-Stokes (FV,CC,RK(3))	BL	2 meshes
DLR	Navier-Stokes (FV,CC,RK(4),MB)	BL, GR	dissipation variation 2 boundary conditions

5.7.2 Results from Euler and Navier–Stokes Calculations

In an effort to reduce uncertainties arising from the variety of meshes used by the different contributors, DLR distributed a C-O-type mesh (Fig.3) to all partners, but unfortunately the codes of many partners seem to be linked to a particular mesh topology. The best that could be achieved was that most partners used at least the wing-surface mesh from the DLR mesh data and generated their own mesh with a similar number of points normal to the surface (Table 2).

Thus, the results of most computations (even for the Euler case) can be expected to have similar resolution. Furthermore, all methods use the same approach (Table 1), i.e. all are based on cell-centered finite-volume formulations and use a flux

Table 2 Number of grid points in the Euler and Navier-Stokes calculations

Partner	Chord x span x normal	Surface	Total	Type
CASA	185 x 41 x 26	6601	197 210	C - O
CERFACS	193 x 41 x 49	7913	387 737	O - O
	97 x 21 x 25	2037	50 925	
CFD norway	149 x 36 x 38	4356	203 832	C - O
	209 x 41 x 33	6601	282 722	
DLR	209 x 41 x 33	6601	282 722	C - O

approach equivalent to central-differencing with Jameson-type artificial dissipation and explicit Runge-Kutta time-stepping with 3 or 4 stages. For a more detailed description of the methods, the reader is refered to the appropriate section in Chapter 4.

In this and the following two sections we will present the results of the partners employing the following informations:

- Lift and drag coefficients c_L and c_D for the complete wing.

- Isolines of pressure on the upper wing surface (for Euler and Navier-Stokes solutions only).

- Wall streamlines, i.e. integrated wall shear-stress lines on the upper wing surface (for Navier-Stokes only):

- Chordwise distributions of the pressure coefficient c_p at five spanwise sections (for Euler and Navier-Stokes only). The spanwise location of these sections is given in Table 3, with the section numbering as in Kordulla (1988) from which also the experimental data are taken. Note that all chordwise distributions are presented as a function of the relative local arc length from the leading edge, x_1/L, which is equal to the relative local chord for data not too close to the leading edge. (Since the differences are usually within the plotting accuracy, data are even presented in this frame of reference when they were actually delivered as a function of the local chord).

- Chordwise distributions of skin-friction coefficients (total, stream- and chordwise : c_f, c_{fs}, c_{fn}) and the angle β between the skin-friction vector and the x_1-direction as well as stream- and crosswise components of the displacement- and momentum-loss thicknesses (δ_s, δ_n, δ_{ss}, δ_{nn}) at three of the five sections, namely section 4, 6 and 8. The skin-friction coefficents are based on free-stream values of the velocity and density.

- Velocity profiles in streamwise and crosswise direction (v_s and v_n) at twelve stations in sections 4, 6 and 8, as indicated in Table 3.

Boundary-layer data are only compared for the three inner sections because the boundary-layer codes used are not capable to compute results at the root and the tip sections and because a comparison of a few NS results seemed to be not meaningful without corresponding data from experiments or from boundary-layer methods.

Table 3 Sections and stations, where computations are compared. Section numbering as in (Kordulla, 1988).
(x_2/B: relative spanwise position)

Section	x_2/B	Stations	x_1/L
1	.001.	–	–
4	.205	1,4,7,10	.2, .4, .6, .8
6	.492	2,5,8,11	.2, .4, .6, .8
8	.800	3,6,9,12	.2, .4, .6, .8
9	.952	–	–

To obtain the above-mentioned representations of components of the skin-friction vector or of integral boundary-layer properties, it is necessary to find the outer edge of the boundary layer from the Navier-Stokes data. Thus all partners who employed NS codes used the procedure outlined in Chapter 3.6.

CASA

CASA, as the only contributor of an Euler solution delivered a study of the influence of the coefficients of the second- or fourth- order dissipation and of a coefficient β that changes the influence of the mesh aspect ratio on the dissipation. From the force coefficients alone (see Table 4), no general conclusions can be drawn, although the high second-order dissipation (D2) results in the lowest lift and highest drag while the three cases with lower second-order damping yield at least about the same lift. If we consider also the c_p distribution for the 4 cases at the five wing sections of Table 3 (Fig. 4) we find that the increase in lift and the reduction in drag with reduced second-order dissipation coefficient (D2) is completely due to changes on the upper surface where a lower pressure, a higher suction peak and a steeper shock result from the lower dissipation.

An additional reduction of the fourth-order dissipation coefficient (D4) results in an additional pressure decrease on both upper and lower surfaces, a slight downstream movement of the shock and more pronounced suction peaks at the lower surface. All this increases the lift only marginally, while the drag is mainly increased due to the shock movement.

Reducing β instead of D4 has a similar effect but with a less pronounced c_p reduction on the lower surface, and also in the tip section on the upper surface, resulting in a higher lift but the same drag.

In comparison with the experimental data, which are also given in Fig. 4, we find a fair to good agreement along the lower surface (note that only the accelerated part of the flow is comparable, as mentioned above), where the result with lower numerical damping seems to be better in the outward sections. The latter is also true on the upper surface, although the discrepancies between experiment and computation are larger here. While the results in the three inner sections suggest that it is mainly the reduction of D2 to one fourth which is important, it follows from the sections near the tip that an additional reduction of the fourth-order term is advisable. As CASA reports (Abbas, 1992), the reduction of neither D2 nor β has an adverse influence on the convergence of the code, while the reduction of D4 to 1/256 results in a considerable deterioration of convergence. Taking the smaller variation in the pressure distribution

Table 4 Force coefficients predicted by the partners for various cases
(R: Riemann invariant formulation at outflow boundary instead of mandatory exit pressure; F / C: result for fine / coarse grid; (~) unsteadiness in result; CP: Chen/Patel k-ε model)

Partner	Case	C_L	C_D
	Euler		
CASA	1) D2=1; D4=1/32; β=.8	.301	.0229
	2) D2=1/4; D4=1/32; β=.8	.324	.0221
	3) D2=1/4; D4=1/256; β=.8	.327	.0216
	4) D2=1/4; D4=1/32; β=.4	.332	.0227
	Navier-Stokes		
CERFACS	1) BL; R; F	.249	.0239
	2) BL; F	.248	.0278
	3) CP; R; C	.125	.0228
	4) CP; R; F	.151	.0210
CFD norway	1) BL; own mesh	.211	.0263
	2) BL; DLR mesh	.199	.0226
DLR	1) BL; D2=1/4 D4=1/128	.212	.0288
	2) BL; D2=1/2 D4=1/64	.200	.0291
	3) BL; D2=1/4 D4=1/128; R	.227	.0266
	4) GR; D2=1/4 D4=1/128	.128 (~)	.0201 (~)

due to the latter reduction into account, it seems to be a reasonable compromise to reduce the fourth-order coefficient D4 to only 1/64 or 1/128 in order to minimize the influence of the artificial viscosity on the pressure, while increasing the stability of the method compared to the D4 = 1/256 case. Those partners solving the Navier-Stokes (NS) equations seem all to have used values for D2 and D4 which are of the suggested magnitude.

Since the variation of the pressure with the damping does not become very evident in isoline plots only one such plot for the case with low second-order damping is presented in Fig. 5 showing the qualitative behaviour of the solution with the shock for the upper wing surface.

Cerfacs

Turning attention to the NS results, we start with the results of Cerfacs who performed calculations on their own mesh which exhibits the highest resolution normal to the wing and on the wing surface (see Table 2). They employed an outlet boundary condition with the mandatory pressure (p_{exit} = .99p_∞) and made additional calculations with a Riemann-invariant formulation. For the latter they communicate a higher average exit pressure $p_{exit} \approx .994 p_\infty$. This can also be infered from the differences in the c_p distributions on the wing. Cerfacs furthermore employed two turbulence models (see Table 1) producing results for their k-ε model, however, only with the Riemann-type outflow condition. For this case they obtained a solution not only on the fine grid but also on a coarser mesh obtained by omitting every other point in all directions.

Figs. 6 and 7 show the isolines of c_p and the integrated skin-friction lines, respectively, on the upper surface of the wing for the Riemann-outflow condition and the BL model. The foot of the shock and the resulting separation can be clearly identified. Changing the outflow condition to the mandatory pressure, we find some differences which are more easily identified from the pressure distribution of Fig. 8 than from the minor changes in an isoline plot or in the skin friction lines (Fig. 9). In Fig. 8 a four-letter code is used to distinguish between the different solutions: the first two letters identify the turbulence model, the third denotes the outlet conditions (R: Riemann; M: mandatory) and the last the mesh (F: fine: C: coarse). Thus the two results under discussion are denoted BL-MF and BL-RF.

The lower exit pressure for the case of fixed p_{exit} results in higher acceleration of the flow on both surfaces and in more pronounced suction peaks, mainly on the lower surface, and yields also some downstream shift of the shock position, especially near the wing tip. Although the fixed exit pressure is taken from the experimental data, it seems that the Riemann boundary condition gives a result which is closer to the experiment. This may be due to the tunnel wall boundary layer existing in the experiment but not in the calculation which was performed with inviscid tunnel walls, i.e. the fixed pressure results in a wrong mass flow through the tunnel in the numerical simulation.

Changing from the BL to the CP model, results in a dramatic drop of the lift, especially for the coarse mesh, while the much smaller drag reduction is mainly restricted to the fine mesh. These changes are mainly caused by an upstream movement of the shock for the CP model calculations, as can be seen from the c_p-distribution (Fig. 8). Of course, the shock is more smeared on the coarse grid, which yields no suction peaks on either surface, while the fine grid result on the lower surface is very similar to the result with the BL model under mandatory conditions. This seems to indicate that the flow on the lower surface is near equilibrium and that an algebraic model is thus able to predict the turbulence reasonably well. Furthermore, it should be noted that the calculations with the CP model yield a lower trailing-edge pressure than those with the BL model, probably due to the weaker shock.

Fig. 10 exhibits the wall flow pattern for the CP model and the fine grid result indicating the more upstream position of the shock-induced separation compared to the BL model results of Figs. 7 and 9 and to the mandatory BL model result of DLR (Fig. 11) to be discuss later.

Fig. 12 shows the skin-friction distribution (modulus and components) for all four results of Cerfacs. All results on the fine grid are quantitatively similar although the shock position is shifted depending on the boundary conditions and the turbulence model. The level of the skin friction with the CP model is, however, a little lower upstream of the shock and and higher in the separated region. On the coarse mesh the skin friction is generally much lower and does not drop suddenly at the shock, which results in separation much further downstream. Thus the coarse-grid result exhibits the most downstream-located separation of all four solutions. What is astonishing, however, is that none of the results indicates the position where the turbulence model is switched on, namely at 5 per cent local chord, i.e. there is no corresponding rise in the skin friction found in other results to be shown later. (On inquiry by the author Cerfacs communicated that they seem to have a problem with the numerical dissipation very close to the wall, especially near the leading egde and upstream of the shock leading to wrong values of their c_f data delivered.) Additionally it should be noted, that Cerfacs provided the angle $\beta - \beta_e$ between the flow direction at the wall and at the boundary-layer edge instead of the angle β used otherwise. Fig. 13 presents the integral thickness data, i.e. the stream- and crosswise components of displacement and momentum loss thickness. As one would expect, the change in the boundary condition from mandatory to Riemann-type form results in a slight reduction of the displacement, while the momentum loss is almost unaffected. The CP model results in a shift of the results due to the upstream-shifted shock position and in a much steeper increase of the displacement thicknesses, while the momentum loss is decreased. The coarse grid result exhibits streamwise thicknesses which are twice as large as for the fine mesh. Downstream of the shock, however, the displacement is almost identical on both grids.

The large difference between the results with the CP model on the two grids gives certainly an indication that even the finer grid is not yet fine enough to yield a grid independent solution, at least for a calculation with a k-ε model.

Velocity profiles are presented in Fig. 14. As could already be expected from the other data the coarse grid results show a large deviation from all other results at all stations. The fine grid results, however, are all very close together up to $x_1 = .4$ (stations 1 to 6) with some small differences between the turbulence models. The discrepancies between the latter become larger at $x_1 = .6$ – especially for the crossflow component indicating the more upstream separation with the CP model – while the boundary conditions show only an influence near the tip (station 9). In the separated flow regime all result are different although still qualitatively similar.

DLR

Only the results obtained for the "mandatory" DLR mesh are discussed here. With respect to preliminary calculations employing a similar 199×39×33 grid as well as a four-block mesh (resolving the boundary layers along three tunnel walls but not on the splitter plate), the reader is referred to (Schwamborn, 1991a,b). Using the "mandatory mesh", DLR obtained two calculations with the BL model for the mandaory case with different amounts of numerical dissipation, and repeated the case with lower dissipation using a Riemann outflow condition as employed also by Cerfacs (see Table 4). Further results for the mandatory case were obtained with the GR model exhibiting some convergence deterioration due to an unsteadiness of the solution downstream of the shock.

Fig. 15 shows the comparsion of the c_p distributions for all cases. The different solutions are again labelled by a letter code where the first two letters indicate the turbulence model, while the letters M and R denote mandatory and Riemann outlet condition, respectively. Finally, the letter D marks the result obtained with increased numerical damping (see also Table 4). As can be seen, the differences between the two solutions with varying dissipation (for the mandatory case with the BL model) are only minor and mainly restricted to the steepness of the shock, such that one can assume that the influence of numerical dissipation is not a major issue in these and the following results. The resulting differences in lift and drag can be seen from the Table 4, with the main change in lift.

Changing the boundary condition at outflow under otherwise equal conditions results in an increased exit pressure which is almost the reference pressure ($p_{exit} \approx p_\infty$), while the mandatory outlet pressure is fixed to $p_{exit} = .99 p_\infty$. In the pressure distributions (Fig. 15), the higher exit pressure leads to a more accelerated flow with higher suction peaks and shock locations farther downstream. These changes result in a somewhat higher lift and a lower drag for the Riemann boundary condition.

The fourth result of DLR shows the influence of turbulence modelling introducing the GR model under the assumption that the Clauser pressure gradient parameter (equ. (GR4) in Chapter 3) can be calculated as the gradient along the chord direction x_1. This is certainly justified taking the mainly chordwise outer flow direction into account. It should be mentioned that the calculation with the GR model exhibited a convergence problem which resulted from some unsteadiness in the solution between the shock and the trailing edge over part of the wing and eventually also some shock movement.

The pressure level for the GR model is generally higher than for the BL model. There are, now, no excessive suction peaks on either surface and the shock position is always markedly upstream-shifted such that the upper-surface pressure minimum is at about the same location where the minimum experimental value is found. (Recall again, however, that the experiment has free transition and laminar separation there.) Due to the changes in the pressure distribution, the lift is much lower now (see Table 4) and the drag has dropped by 30 per cent.

The skin-friction data compared in Fig. 16 are very close together if we consider only the results obtained with the BL model, with the main, but still minor, differences being at and downstream of the shock position. The GR model shows a qualitatively similar behaviour upstream of the shock but with generally lower c_f values. Although the shock is upstream-shifted in this case, the separation occurs now farther downstream, i.e. the shock is not strong enough to lead to separation.

All solutions of DLR exhibit unusual waviness of the skin-friction coefficient and its streamwise component upstream of the shock position. The reason for this is not clear, but one possible explanation is that the wall shear reflects and amplifies a much smaller waviness in the pressure or its derivatives. This assumption is supported by the fact that the boundary-layer results obtained with the pressure distribution from DLR's Navier-Stokes solution (see next chapter) also show some – though less pronounced – waviness of the skin friction. An additional accuracy problem in the calculation of the wall shear stress can, of course, not be excluded.

Comparing the streamwise integral boundary-layer parameters in Fig. 17 (the crosswise components are not shown since they give no new information) the following is found:

> increased numerical damping leads to very little changes;

> the lower exit pressure (Riemann b.c.) results in a reduction of the integral thicknesses (taking only the modulus for the crosswise components into account) as would be expected from the more strongly accelerated flow;

> changing the turbulence model yields the largest changes, where especially the momentum-loss thickness shows a much slower increase for the GR model.

Finally, Fig. 18 presents the velocity profiles which are very similar for the stations in the front part of the wing, while the influence of the numerical dissipation or of the different outlet boundary conditions and even more that of the different turbulence model is seen at the stations near or downstream of the shock.

CFDnorway

The calculations of CFDnorway were performed using the mandatory conditions with the mesh provided by DLR as well as on their own mesh (see Table 2) which has less resolution in the surface-tangential direction but higher resolution in the wall normal one.

From the c_p distribution for both cases (Fig. 19), it can be seen that the DLR mesh (denoted as BL-D CFD) gives slightly higher suction peaks in the root section at the upper surface (note that CFDnorway delivered only data for this surface), while the CFD grid (denoted as BL-C CFD) yields higher suction for the outer half of the span and a shock position which is farther downstream. On the lower surface, however, the DLR grid leads to larger suction peaks over the main part of the wing, while the results near root and tip are almost identical. Due to this, the DLR mesh produces less lift but also less drag than the CFD mesh (see Table 4).

Since the methods of DLR and CFDnorway are similar (see Table 1) and since all results were obtained with the mandatory outlet pressure, the solutions of DLR and CFDnorway on the DLR mesh should be almost identical. The obvious discrepancies, however, suggest that, in fact, the numerical dissipation is not treated in the same way or not used with the same set of coefficients. Another possible source of the differences might be the treatment of transition, i.e. how the turbulence model is switched on at the transition location or in the usage of different coefficents in the essentially identical BL model implementations.

The skin-friction distribution in Fig. 20 indicates that the transition location for the calculation on the DLR mesh is not at 5 per cent but at 1 or 2 per cent local chord, although it is not clear that this is the reason for the above discrepancies. Otherwise, the behaviour of this solution is qualitatively similar to the result of DLR on the same mesh while we find a very important difference between the two results of CFDnorway: there is no separation in the result on their own grid, which is in contrast to all other results. For this grid, the streamwise skin friction component c_{fs} is always positive and the angle between the skin-friction lines at the wall and the flow direction at the boundary-layer edge is always less than 90°.

The integral parameters (streamwise direction only) for both solutions are shown in Fig. 21 but as can be estimated from the profile data of CFD at stations 10 to 12 (Fig. 22) these distributions seem to be faulty in stating too large values which are not mirrored by the velocity profiles. The discussion of the latter profiles is, however, postponed to the following section.

Comparison of selected results of the partners

In this section selected results of the partners are compared, which have been obtained at conditions as uniform as possible. Hence, these solutions employ

- the DLR mesh or the mesh of the partner which is closest to the DLR mesh,

- the same boundary conditions at the outlet,

- the same turbulence model (if applicable),

- and the lowest numerical dissipation.

In order to limit the number of figures the corresponding results of Cerfacs, DLR and CASA are included in Figs. 19 to 22 already shown in the discussion of CFDnorways data. Thus we come back to Fig. 19 to discuss the c_p distributions of CASA (number 3 of Table 4), Cerfacs (number 2), CFDnorway (number 2) and DLR (number 1 of Table 4). All data agree relatively well with each other and with the experiment in the accelerated part of the flow, while the main differences are in the suction peaks and the shock location, especially for the Euler result which predicts the shocks much further downstream than all the NS results. (Since the differences between the different Euler solutions of CASA are smaller than their difference to the Navier-Stokes results, all statements with respect to one Euler solution are valid also for the others.)

The Navier-Stokes results, generally, show the largest scatter at the root and the tip. At the root, DLR is closest to the experiment (which features fully turbulent flow in this section just as the computation), and Cerfacs is farthest off with the lowest pressure, while CFDnorway is in between the two. Over the remaining wing portion Cerfacs' result maintains the lowest upper surface pressure and predicts the most downstream shock position. On the lower surface, all results are very similar with some slight difference in the suction peak, except at the tip where Cerfacs obtains also a weak lower surface shock. CFDnorway and DLR (using the same mesh) are also relatively close together along most of the upper surface, except for the suction peaks which are more pronounced in DLR's data.

As discussed earlier the results of CASA indicate that a very low fourth-order damping results in more pronounced suction peaks on the lower surface while reduced second-order damping results in steeper, more downstream shocks and higher upper-surface peaks. These tendencies are in good agreement with the result of Cerfacs who should have very low damping due to their relatively fine mesh and the low dissipation coefficient used (see method description in Section 4.5). Since the results of DLR with doubled numerical damping coefficients showed only a slight influence of numerical dissipation (Fig. 15), and since even a factor of eight in CASA computations does not exhibit too much variation of the results, one could conclude that CFDnorway used a much higher fourth-order damping coefficent than DLR. But the possibility cannot be excluded that the discrepancies under consideration are at least partly due to differences in the formulation of the BL models used by the partners.

The comparison of the skin friction components in Fig. 20 shows the generally similar behaviour of the results of CFDnorway and DLR except for the already mentioned earlier transition in CFDnorways' computation and some small differences at the shock and in the separated area. One reason for the discrepancies in the data of Cerfacs is the already mentioned too high numerical damping near the wall at the leading edge and upstream of the shock which seems ,however, to influence mainly the modulus skin friction data while the pressure and displacement thicknesses and – what is very astonishing – even the direction of the skin-friction vector are not affected. Since Cerfacs delivered $(\beta - \beta_e)$ instead of β the latter can only be seen from a comparison of the skin friction lines of Cerfacs and DLR in Figs. 9 and 11, respectively, showing that major flow features in both figures are almost identical even with respect to the structure near the wing tip . The discrepancies are limited to the root - trailing edge region where DLR obtains a saddle point separation while the structure in Cerfacs solution does not change up to the root section.

Fig. 21 presents the integral thicknesses for the three partners indicating that Cerfacs and DLR obtain almost identical results, while CFDnorways results are obviously erroneous as already discussed above.

Comparing the velocity profiles in Fig.22 one finds that the results of DLR and CFDnorway on the DLR mesh are closest together, generally. This should be expected on the one hand, but is in contrast to the above results. Taking only the streamwise data at the first nine stations (with no separation) into account it is interesting that Cerfacs with the finest mesh produces always the fullest profile while CFDnorway with its own mesh, which is the coarsest in this comparison, obtains the least full profile. Comparing with the data in Table 2 it seems that it is not the number of points normal to the surface which is of importance here, since the DLR mesh has the least points in that direction.Since we have already seen from all the other results that the character of the flow is almost quasi two dimensional over a wide area of the wing, it is not likely that the small difference in the number of spanwise points in CFDnorways mesh is of any importance. Thus we have to conclude that the number of chordwise points is crucial for the above effect and indead we find that the DLR mesh employs 161 points along a chordwise section of the wing which is between the maximum of 193 for Cerfacs and the minimum of 149 in the mesh of CFDnorway. Certainly, we can not attribute all the differences found to this relatively small variance in mesh points, but it seems nevertheless worth to be noted.

Considering the results discussed so far, it is clear that it is not the influence of the numerical damping which is the critical issue for the Euler calculations but that it is the absense of the boundary layer which results in a shock position being far downstream compared to the NS solutions. The numerical damping has, however, certainly some influence in the NS calculations, although it is difficult to single this out as it may interact with other differences arising, for example, from the turbulence model. If everything else is identical and the mesh is not too coarse the influence of numerical damping can be shown to be small.

5.7.3 Results from Boundary–Layer Calculations

As can be seen from Table 1 three partners use boundary-layer codes to obtain results for the F5 wing. DA uses a LISW (= locally infinite swept wing), code and a three-dimensional code with essentially the same approach as in TUBs 3D code. All these codes are of the space-marching type, while RAE's 3D code is based on a time-dependent approach. For a more detailed description of these methods the reader is referred to the appropriate parts of Chapter 4. Here all these codes are used in the direct mode, although they are capable of inverse calculation. Since experimental data is not available for any boundary-layer quantities, all calculations are based on the pressure distribution obtained with DLR's Navier-Stokes code under mandatary conditions (DLR 1 of Table 4). RAE's code uses this pressure distribution directly, while DA and TUB need the outer-edge velocity components v_e^1, v_e^2 in the curvilinear surface-oriented x_1-x_2 coordinate system. These velocity components are computed from the pressure distribution by DA (John, 1991). The hope is that the use of a calculated pressure distribution will allow a meaningful comparison between boundary-layer data of these methods and the corresponding viscous information of the Navier Stokes codes.

In the following, the comparison of results which we start with a short discussion of DA's results, will be restricted to the upper surface and three spanwise locations at about 20, 50 and 80 per cent span (sections 4, 6 and 8 of Table 3).

DA

Here we compare solutions from four different computations based on the combination of the two methods and the two turbulence models (see Table 1). Figs. 23 and 24 show the skin-friction components and the streamwise integral parameters for the three spanwise locations. As can clearly be seen from all results, the region of strongly three-dimensional flow just in front of separation is shifted upstream with increased spanwise position. The general behaviour of all results is very similar for all span stations, which is already an indication that a quasi-three-dimensional method should be capable of yielding good results. This is obviously the case since the results of the two methods using the same turbulence model (be it CS or JK) are almost identical. Only in the skin friction some small differences are found, which are largest at the transition location and decay downstream. These differences become larger if one compares solutions of the same method with different turbulence models, although the general behaviour is still very similar. The integral data as well as the crosswise component of skin friction are, however, almost identical. Note that the 3D method with the JK model and the LISW method with the CS model break down already at the beginning of the highly three-dimensional regime while the 3D method employing the CS model seems to be more robust. The variation in the integral data would be more pronounced at a different scale, but we decided to use only one scale for all NS and boundary-layer results.

Fig. 25 presents the nine selected velocity profiles (Table 3), where we find that the variation of the profiles with the method or model is largest near the leading edge but decreases farther downstream. Note that the solution with the LISW method employing the CS model fails already at $x_1 = .6$ and $x_2 = .8$, i.e. there is no result for profile 9, while the JK model is unable to deliver results for $x_1 = .6$ anywhere, so that no data for profiles 7 to 9 are available. For all available stations, the LISW method results in the smallest crossflow, while the 3D calculation with JK yields the largest crossflow maxima at all stations $x_1 = .2$. However, all these discrepancies are still small taking the crossflow velocity scale into account.

TUB

The results of TUB who employ their 3D method (based on the same approach as DA's method) with 3 different turbulence models, namely CS, JK and JC, are presented in Figs. 26 and 27. As before the influence of the models on the results is only modest with the largest variations in the streamwise component of the skin friction. For this, the main difference between CS and JK/JC is found near the transition as was the case with DA's results, while the difference between JK and JC starts to develop further downstream. Also here, the one-half equation models are more sensitive to the three-dimensionality of the flow near separation, but it seems that some mechanism in TUB's code prevents the total breakdown of the calculation. The reason for the sharp jumps in the skin friction and to a lesser extent in the streamwise integral data, is far from clear, however, and a physical relevance of the results downstream of this jump must be doubted.

The velocity profiles are presented in Fig. 28. Again, the influence of the turbulence model is not very large but at least recognizable. With respect to the streamwise direction, JC and JK give similar results near the leading edge ($x_1 = .2$) while CS gives somewhat fuller profiles. Further downstream, however, the results of CS and JK are almost identical in the near-wall part while JC yields a little lower velocities, there. Near the boundary-layer edge the JK and JC model perform identical, but different to the CS model. In the crossflow velocity, both the JK and JC model produce the highest maxima, but again the difference to the other models is small if one takes the scale into account.

Since RAE only obtained a solution employing the CS model, their results are discussed in the next section in comparison to the CS data of DA and TUB.

Comparison of selected result of the partners

Before we start the comparison between the results of DA, RAE and TUB it seems worthwhile to make a statement about the accuracy of the methods. All methods employed are second-order accurate with respect to the wall parallel step size and fourth-order accurate normal to the wall. As can be seen from Table 5, the number of mesh points in the spanwise direction is by coincidence 41 for all cases except for those calculated with the LISW method which, by definition, needs only one chordwise plane for a calculation. Thus, the latter method uses actually three times 70 x 1 x 85 points to obtain results in the three selected wing sections.

As we have seen DA's two methods yield results of equal accuracy although the number of chordwise mesh points is quite different for the LISW and the 3D code. This seems to indicate that about 40 to 50 points from the leading edge to the separation are sufficient for a good resolution. Thus, since DA uses a much higher number of mesh points normal to the wall than the two others, one might expect that DA's results are more accurate than the others.

Table 5 Number of mesh points in boundary-layer calculation for upper surface from leading edge to separation.

	Average number of mesh points $x^1 \cdot x^2 \cdot x^3$		
Partner	CS	JK	JC
DA (LISW)	70 x 3 x 85	70 x 3 x 85	-
DA (3 D)	54 x 41 x 87	40 x 41 x 91	-
RAE	56 x 41 x 50	-	-
TUB	45 x 41 x 55	55 x 41 x 55	55 x 41 x 55

In Fig. 29 the skin-friction distributions are shown as obtained for the CS model. The main difference is with respect to transition which seems to be handled differently by TUB in the sense that they allow for a larger transition zone. All results show a more or less similar waviness, as already mentioned in the discussion of DLR's results who delivered the pressure distribution on which these boundary-layer solutions are based. Besides this the largest discrepancy between the different results is found in the crossflow component c_{fn} which is almost zero except at the very leading edge for all data from RAE. In contrast to this the distribution of the integral thicknesses is almost identical for all partners as can be seen e.g. from Figs. 24 and 27 and thus we refrain from showing another figure here.

Fig. 30 presents the velocity profiles for all CS model calculations of the partners. Again the differences are small for both components of the velocity taking the scale of the crossflow component into account. It is, however, evident that TUB's and also DA's results indicate decrease of the crossflow with increasing distance from the leading edge, while RAE's data show, generally, lower cross flow with not much variation in chordwise direction, but somewhat higher crossflow at the most inboard position ($x_2 = .2$). This latter difference to the other results is probably due to the completely different numerical approach. Furthermore it can be noted that TUB produces the fuller streamwise profiles than DA and RAE whose data are closer together. For the crosswise flow it is also TUB who obtain the highest crossflow maxima except for the (inboard) station 7 where RAE shows the highest value due to the above mentioned effect.

Fig. 31 compares the one-half equation models of DA and TUB with respect to the skin friction. All results are very close together up to the point where DA's method breaks down (compare also Fig. 23), while TUB's method gives rise to some jumps in the distributions, as already discussed above. We refrain from presenting the velocity profiles since, here again, the differences are very small.

5.7.4 Comparison Between Boundary–Layer and Navier–Stokes Results

Here we recall again some of the results already discussed either in the section about the NS solutions or about the boundary-layer ones. As the methods are now really different, the only possible common aspect is the turbulence model. Unfortunately, even this is not the case here since the boundary-layer solutions are based on the CS model or a one-half equation model, while the NS results are obtained with the BL, GR or the two-equation CP model.

Comparing the skin friction in Fig. 32 and taking the multitude of differences in methods, etc. into account, we find good agreement between the results of CFDnorway, DA, DLR and TUB with the largest scatter in the streamwise component for the NS methods. Taking also Fig. 29 into account it is the boundary layer result of RAE which yields the largest discrepancies in the crosswise component. Cerfacs is not considered here due to the uncertainities with respect to their skin-friction results.

For the streamwise integral thicknesses (Fig. 33) a similar statement as above can be made. The discrepancies are even smaller here considering all data for the CS or BL model except those of CFDnorway which are omitted due to their probably erroneous behaviour.

The streamwise velocity profiles (Fig. 34) indicate a relatively low scatter between the NS and the boundary layer results if one takes only stations 1 to 9 into account. The profiles of Cerfacs with the highest resolution normal to the wall are generally in between and quite similar to the profiles of TUB and DA and show a similar smoothness. The profiles of CFDnorway and DLR which tend to a little lower velocity in the boundary layer give an indication of the coarser mesh normal to the wall.

The crossflow profiles of Fig. 34 show a qualitatively similar behaviour for the first six stations, with the NS data being closer to DA's result. Again the influence of the mesh density is seen in the results of CFDnorway and DLR and we conclude that this produces also the smearing of their crossflow profiles. At the stations near separation (7 to 9) the results on the coarser DLR mesh indicate a much stronger turn of the crossflow direction than the result of Cerfacs or of the boundary layer codes although the latter used the pressure distribution of DLR.

Comparing also with Fig. 30 the somewhat different behaviour of the crossflow data from RAE already discussed above is not reproduced by any of the other data and seems to be specific to their boundary layer method. Taking, however, the small scale of the cross flow profiles into account the overall agreement between the results of the NS and boundary-layer methods is quite satisfying, although the comparison suggests that for NS calcutations at least a mesh of the resolution used by Cerfacs should be employed.

5.7.5 Concluding Remarks

The main objectives of the work presented were, first, the validation of 3D Navier-Stokes and boundary-layer codes taking the influence of turbulence modelling into account, second, the comparison between Euler and Navier-Stokes results, and third, investigations of the influence of numerical dissipation. This is an enormous undertaking even if the investigations are restricted to wing flows and if only a single test case is considered.

During the compilation of the results obtained by all partners, a number of questions could be answered while others remained open. Nevertheless, the work on this test case was found to be very rewarding. Due to the relatively high costs of three-dimensional computations and their evaluation, it was not feasible for the partners to achieve more during this project. Thus, the outcome which has been presented here must be seen as the very first step in the long process of validating methods and models for wing flows. Additional steps including new dedicated experiments, carefully designed for the validation of CFD codes and turbulence models, will have to follow in order to provide industry with better tools to improve and shorten the design process of aircraft.

Owing to the unpredictable unavailability of the expected experimental data, it is difficult to draw definitive conclusions with respect to the quality of the results obtained, except for the accelerated part of the flow. However, a good representation of this part of the flow field does not allow conclusions to be drawn about the validity of other parts. Thus, the mere comparison of different numerical methods allows only a very limited validation.

From the comparison of the Euler and NS computations alone it becomes clear that an inviscid solution is not suitable even for this type of flow with modest shock boundary-layer interaction. Nevertheless, the Euler results provided us with some insight into the influence of numerical damping.

The Navier-Stokes methods show considerable scatter depending on the grid and boundary conditions and especially on the turbulence model, which is mirrored in the variety of the force coefficients. Although some influence of the numerical dissipation is found also in the Navier-Stokes results, it seems not to be a major issue, at least not on the finer meshes and if the corresponding damping coefficents are small enough. The remaining uncertainties about the influence of the transition location and possible differences in the different BL variants of the partners make a final conclusion impossible even if the same mesh had been used.

The use of turbulence models which consider non-equilibrium effects (CP model) or local variations of the pressure gradient (GR model) results not only in quantitative but also qualitative changes in the flowfield compared to the BL model. That these two models (GR and CP) exhibit some similarities in their performance is due to the fact that the level of transport of turbulence is only modest.

The results obtained with the boundary-layer codes on the basis of a pressure distribution from a NS calculation are quite close together, although there are some discrepancies. The value of the results for the different turbulence models calculated with these methods is, however, limited due to the missing influence of the turbulence model on the pressure distribution. Nevertheless a certain degree of validation of those codes and the turbulence models by numerical cross-checks is possible and was shown. Although producing very similar results with a number of codes is not strictly a full-fletched validation, such a comparison is undoubtedly one aspect of it. A further validation of the boundary-layer codes and turbulence models through comparison with data from suitable experiments has to be left to other test cases found elsewhere in this book.

The most important aspect in the utilization of the boundary-layer methods with their generally high resolution was, however, the question of the numerical accuracy. These methods proved to be very useful in checking the boundary-layer related results from the NS codes which usually lack a sufficient resolution of the boundary layer. An indication of the latter could be demonstrated to some extent for some of the NS results. Despite the scatter among the NS solutions it is astonishing how well they compare with the results of the boundary-layer methods.

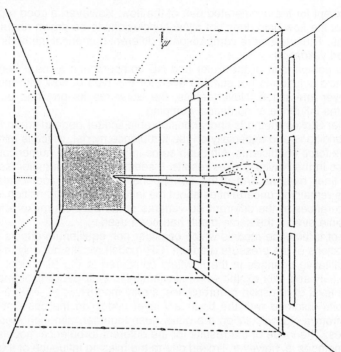

Fig. 1 Experimental set-up for the F5 experiment

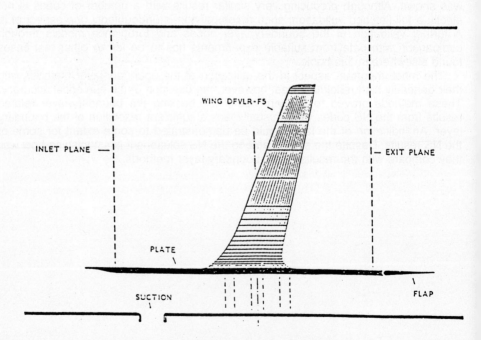

Fig. 2 Schematic of the windtunnel set-up seen from above

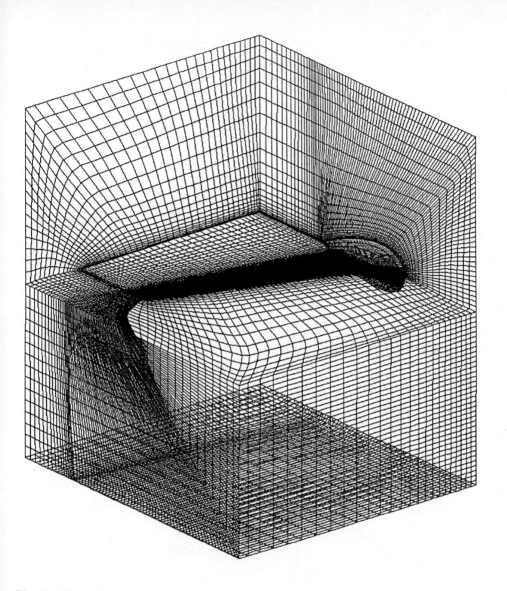

Fig. 3 View of the computational mesh distributed by DLR

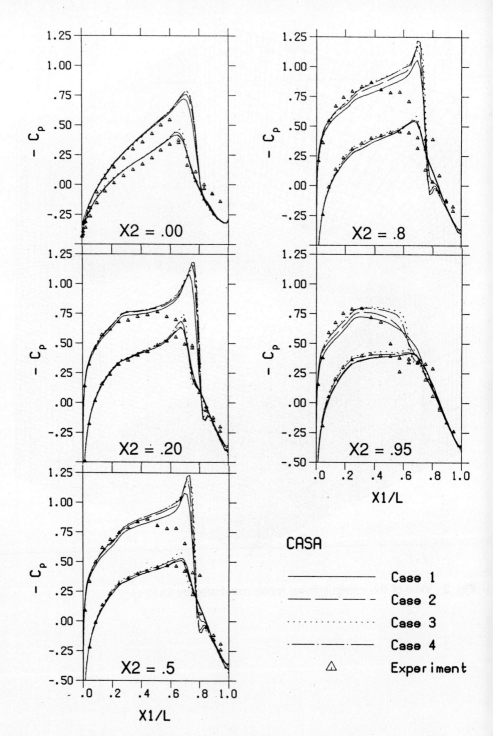

Fig. 4 C_p-distribution at five chordwise sections from four calculations of CASA (see Table 4)

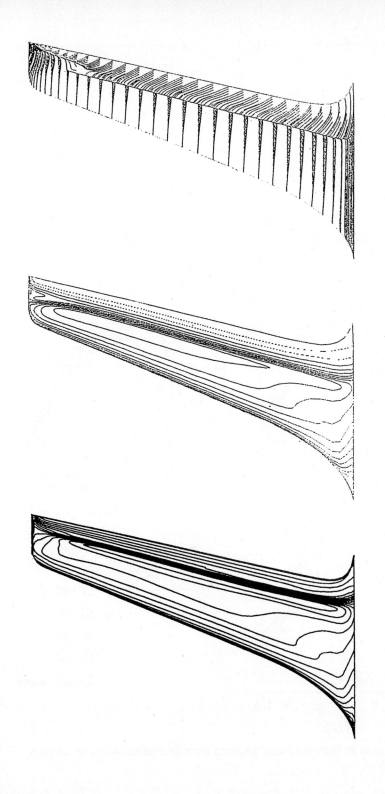

Fig. 5 Isobar distribution on upper wing surface: CASA case 3 of Table 4 ($\Delta c_p=.05$)

Fig. 6 Isobar distribution on upper wing surface: Cerfacs case 1 of Table 4 ($\Delta c_p=.05$)

Fig. 7 Skin-friction lines on upper wing surface: Cerfacs case 1 of Table 4

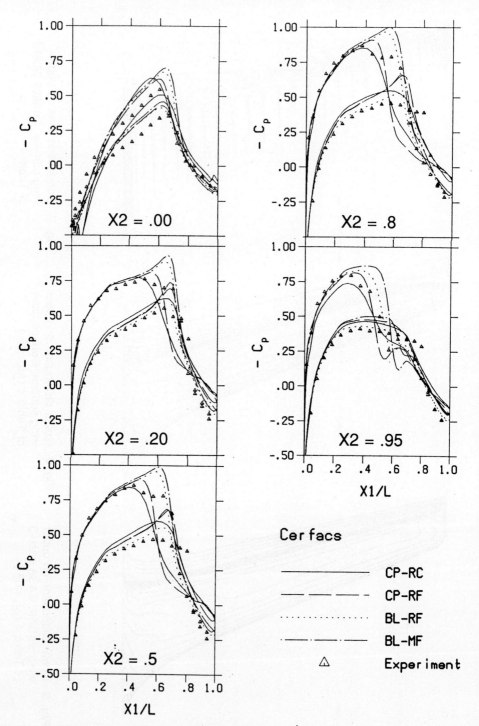

Fig. 8 C_p-distribution at five chordwise sections from four calculations of Cerfacs (see Table 4)

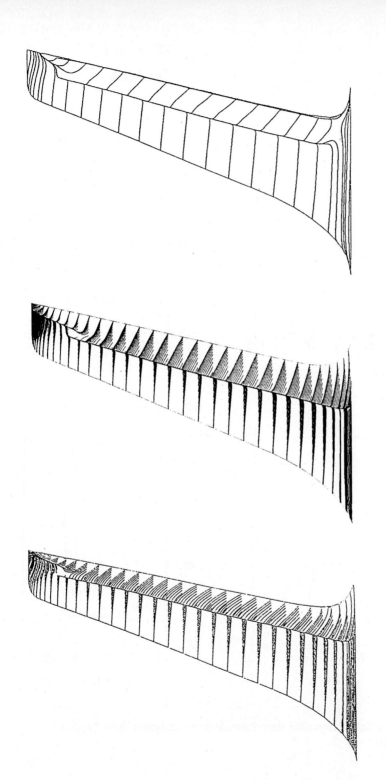

Fig. 9   Skin-friction lines on upper wing surface: Cerfacs case 2 of Table 4

Fig. 10   Skin-friction lines on upper wing surface: Cerfacs case 4 of Table 4

Fig. 11   Skin-friction lines on upper wing surface: DLR case 1 of Table 4

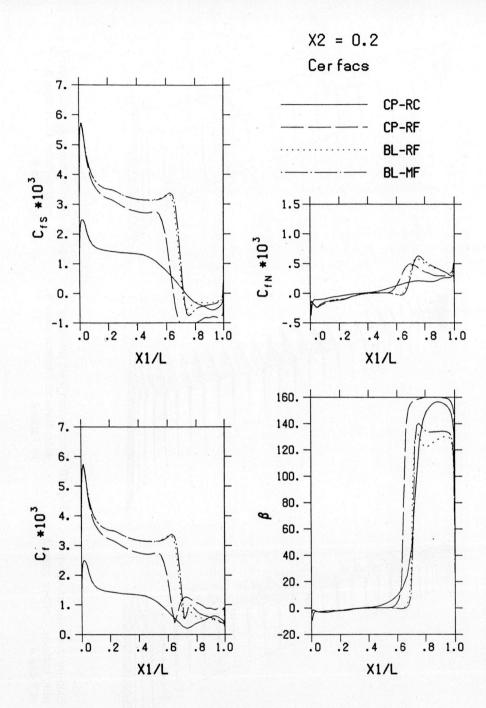

Fig. 12 Skin-friction distribution from calculations of Cerfacs (see Table 4)
a) Section x_2 = .2

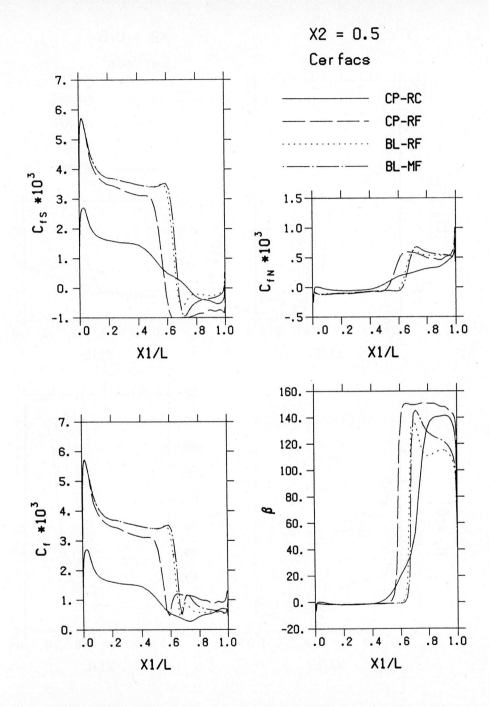

Fig. 12 Skin-friction distribution from calculations of Cerfacs (see Table 4)
b) Section $x_2 = .5$

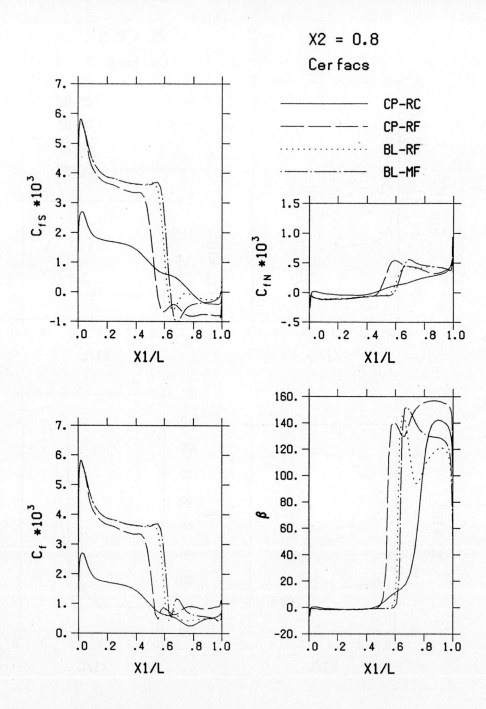

Fig. 12 Skin-friction distribution from calculations of Cerfacs (see Table 4)
c) Section $x_2 = .8$

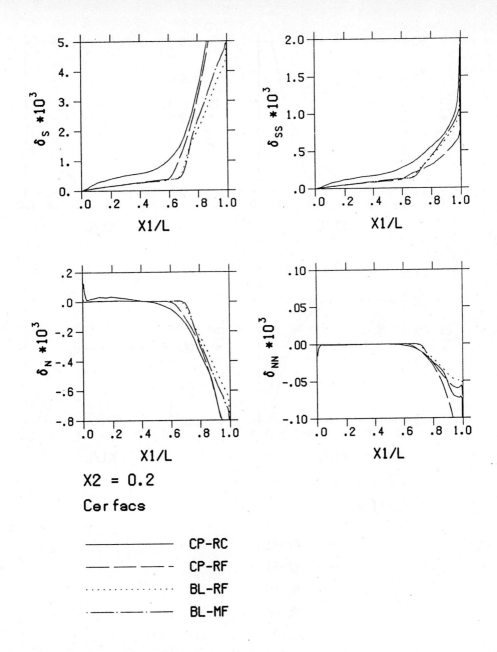

Fig. 13 Distribution of integral thicknesses from calculations of Cerfacs
a) Section $x_2 = .2$

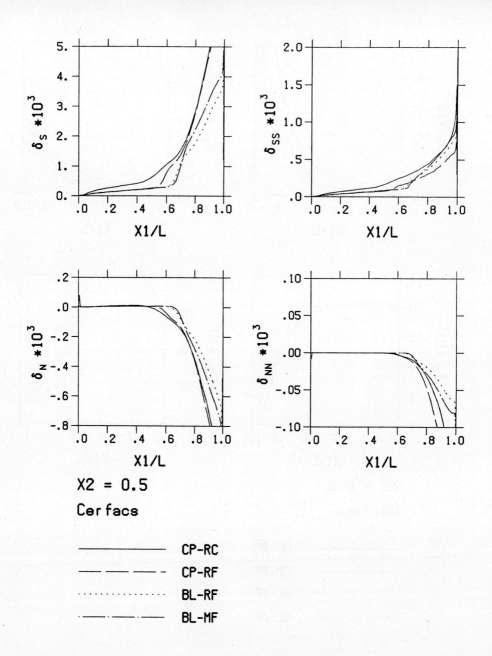

Fig. 13 Distribution of integral thicknesses from calculations of Cerfacs
b) Section $x_2 = .5$

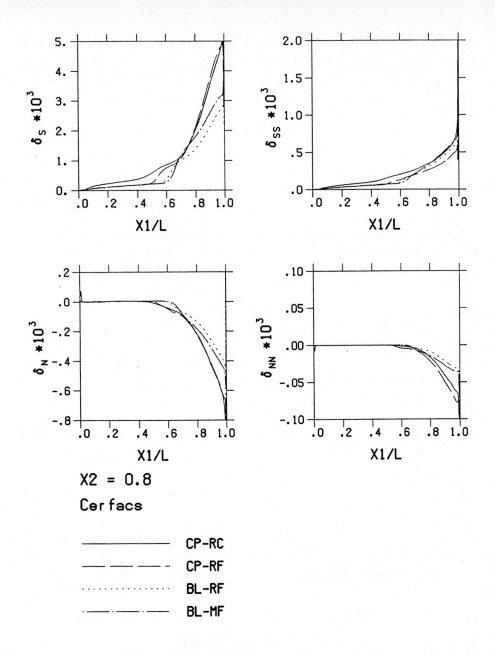

Fig. 13 Distribution of integral thicknesses from calculations of Cerfacs
c) Section x_2 = .8

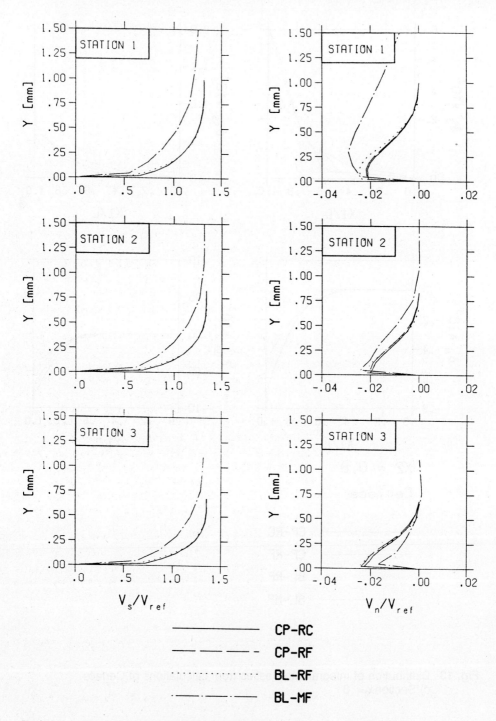

Fig. 14 Streamwise and crosswise components of the velocity: calculations of Cerfacs (see Table 4); a) stations with $x_1 = .2$

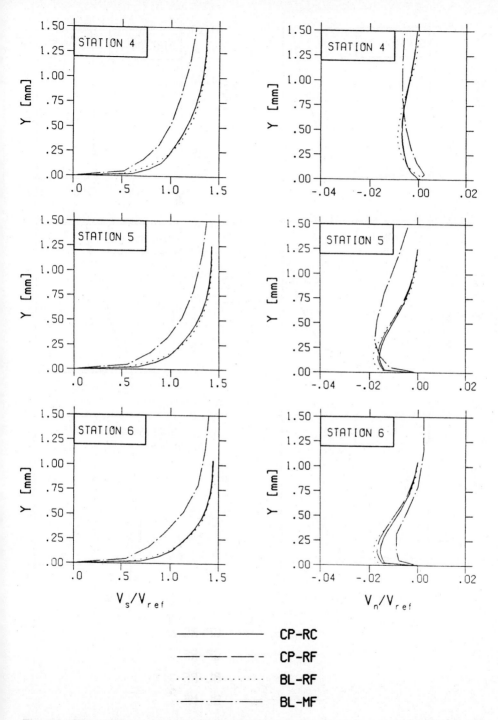

Fig. 14 Streamwise and crosswise components of the velocity: calculations of Cerfacs (see Table 4); b) stations with $x_1 = .4$

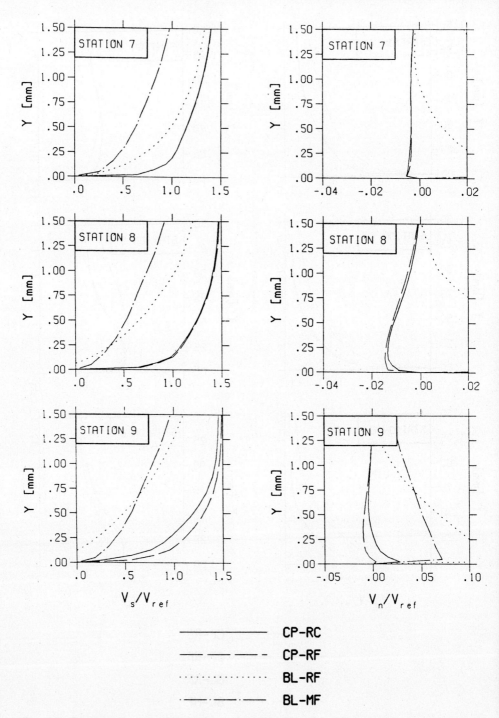

Fig. 14 Streamwise and crosswise components of the velocity: calculations of Cerfacs (see Table 4); c) stations with $x_1 = .6$

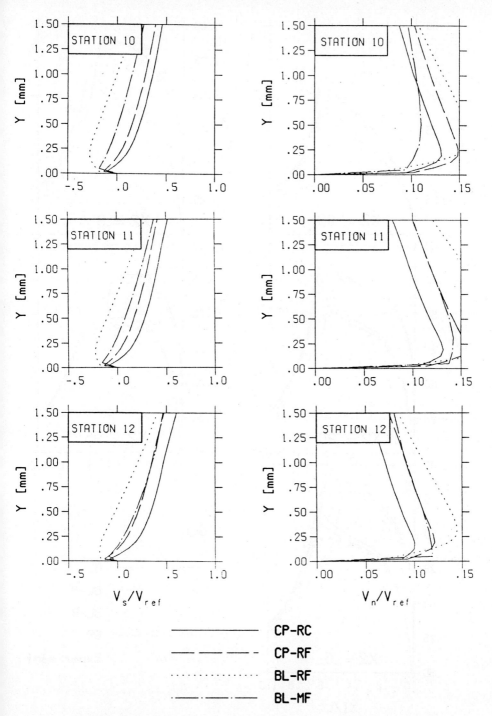

Fig. 14 Streamwise and crosswise components of the velocity: calculations of Cerfacs (see Table 4); d) stations with $x_t = .8$

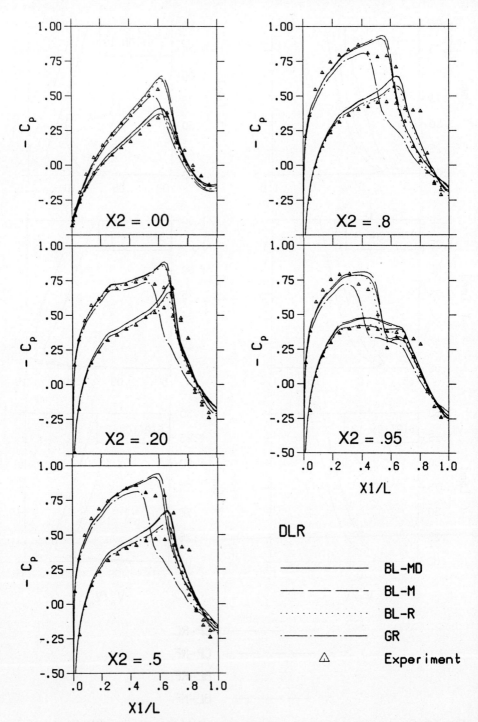

Fig. 15 C_p-distribution at five chordwise sections from four calculations of DLR (see Table 4)

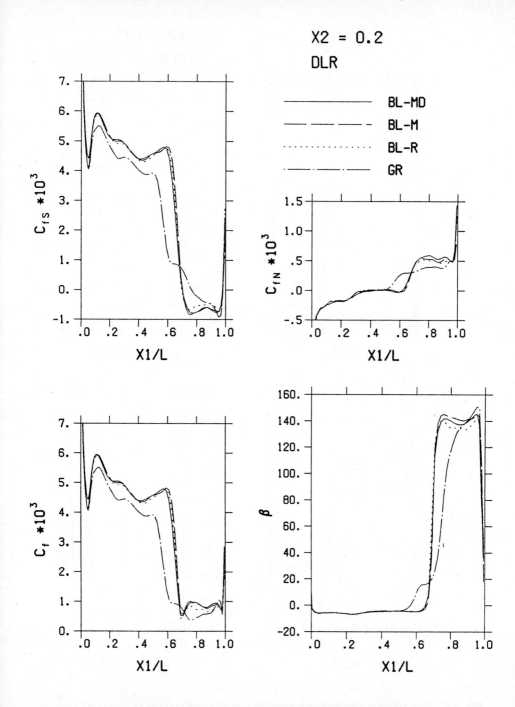

Fig. 16 Skin-friction distributions from calculations of DLR (see Table 4); a) section $x_2 = .2$

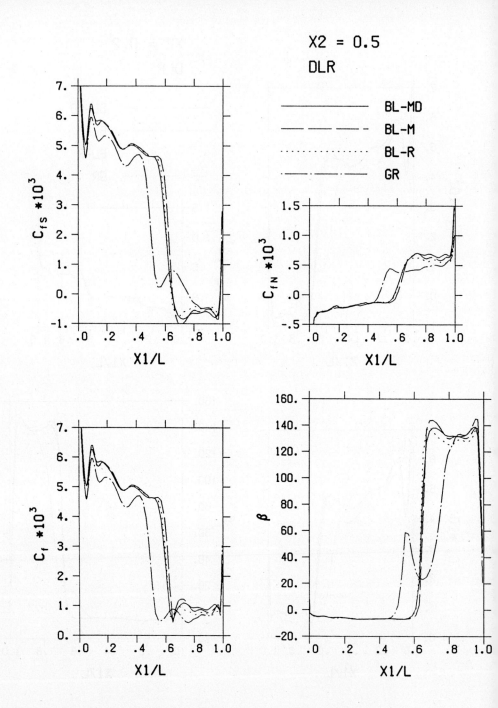

Fig. 16 Skin-friction distributions from calculations of DLR (see Table 4); b) section $x_2 = .5$

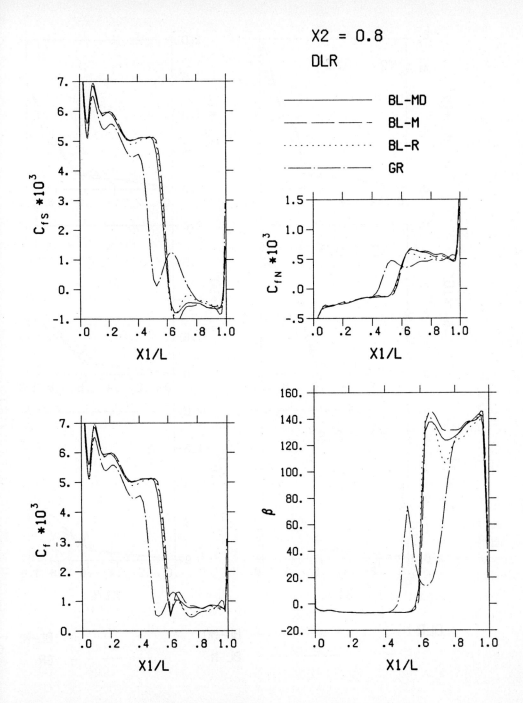

Fig. 16 Skin-friction distributions from calculations of DLR (see Table 4); c) section $x_2 = .8$

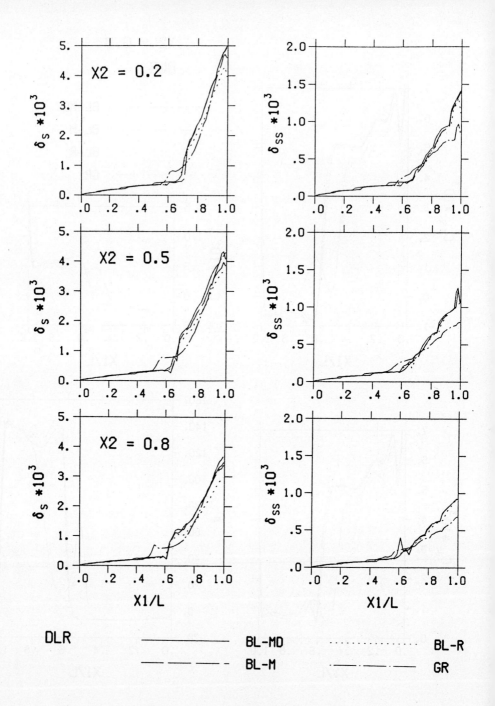

Fig. 17 Streamwise integral thickness distributions from calculations of DLR (Table 4);

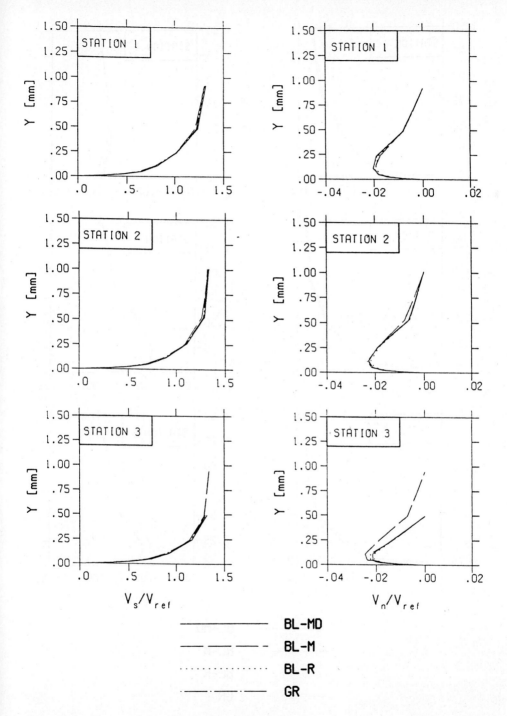

Fig. 18 Streamwise and crosswise components of the velocity from calculations of DLR (see Table 4); a) stations at $x_1 = .2$

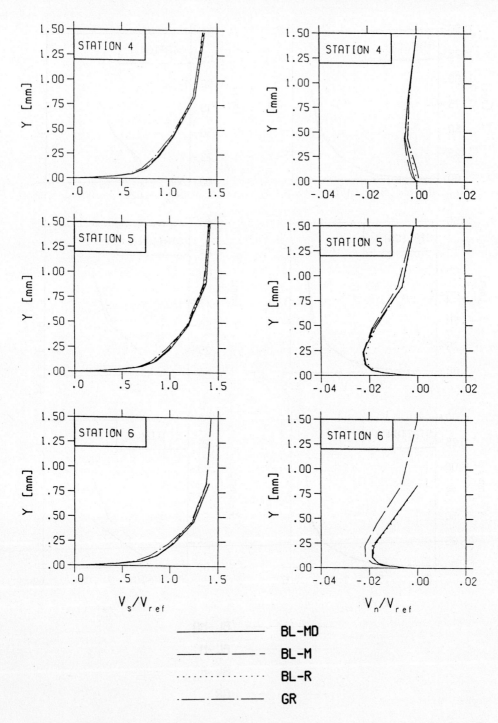

Fig. 18 Streamwise and crosswise components of the velocity from calculations of DLR (see Table 4); b) stations at $x_1 = .4$

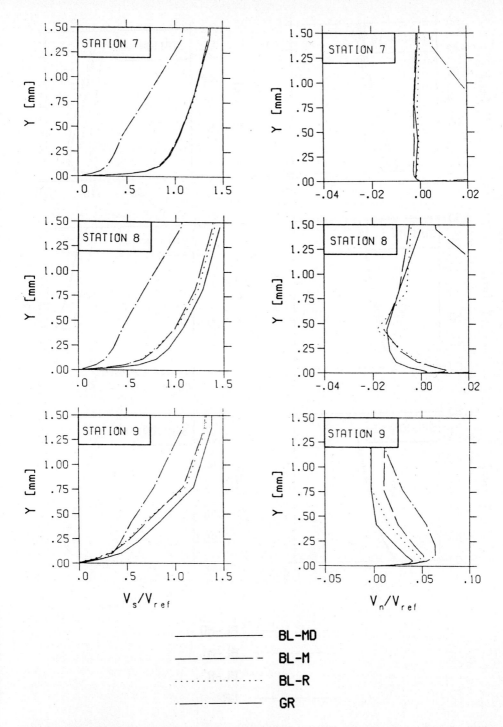

Fig. 18 Streamwise and crosswise components of the velocity from calculations of DLR (see Table 4); c) stations at $x_1 = .6$

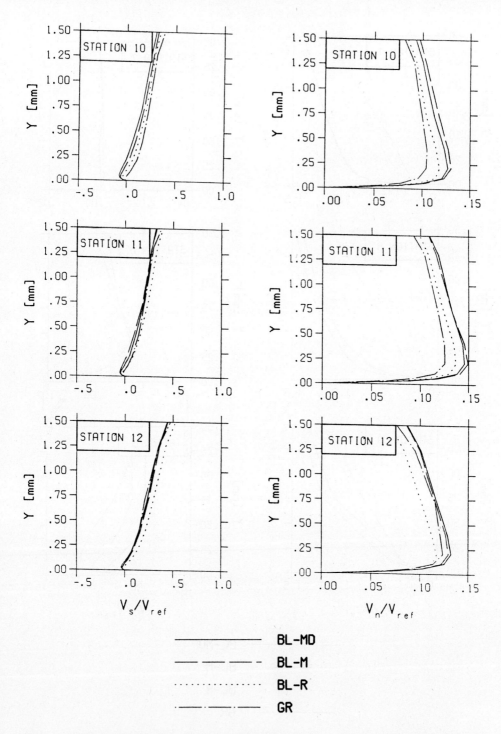

Fig. 18 Streamwise and crosswise components of the velocity from calculations of DLR (see Table 4); d) stations at $x_1 = .8$

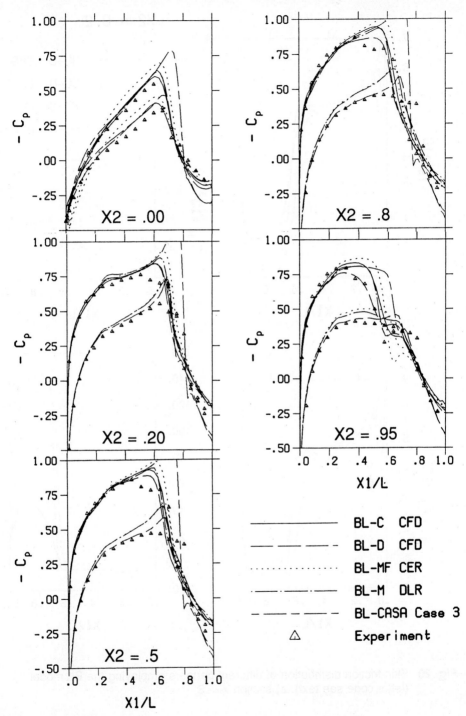

Fig. 19 C_p-distribution of different partners employing the BL model (letter code see text)

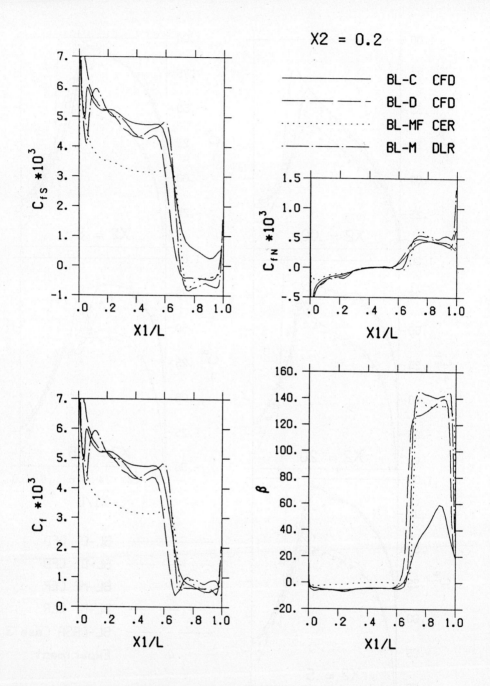

Fig. 20 Skin-friction distribution of different partners employing the BL model (letter code see text); a) section $x_2 = .2$

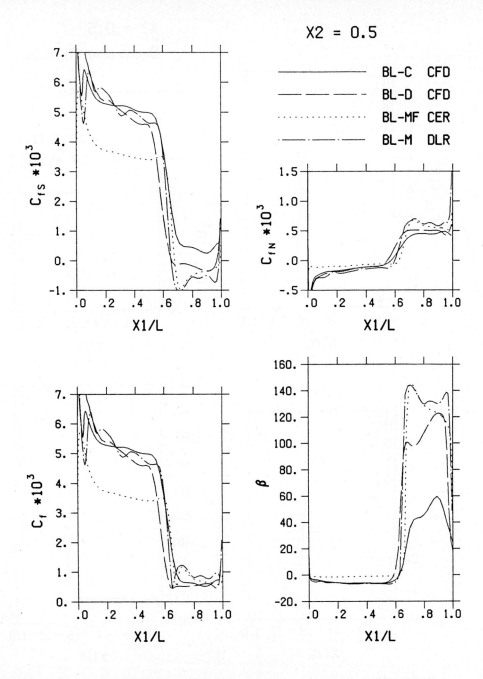

Fig. 20 Skin-friction distribution of different partners employing the BL model (letter code see text); b) section $x_2 = .5$

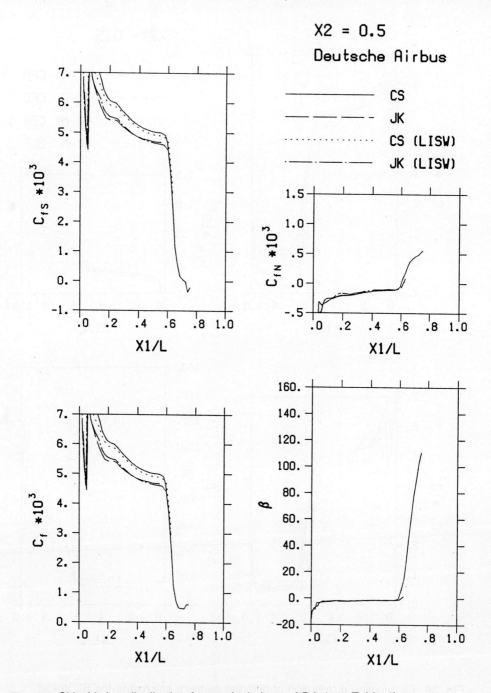

Fig. 23 Skin-friction distribution from calculations of DA (see Table 1); b) section $x_2 = .5$

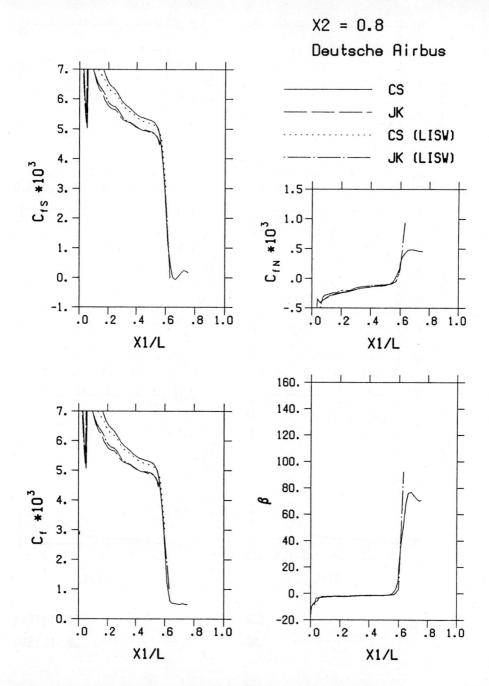

Fig. 23 Skin-friction distribution from calculations of DA (see Table 1); c) section $x_2 = .8$

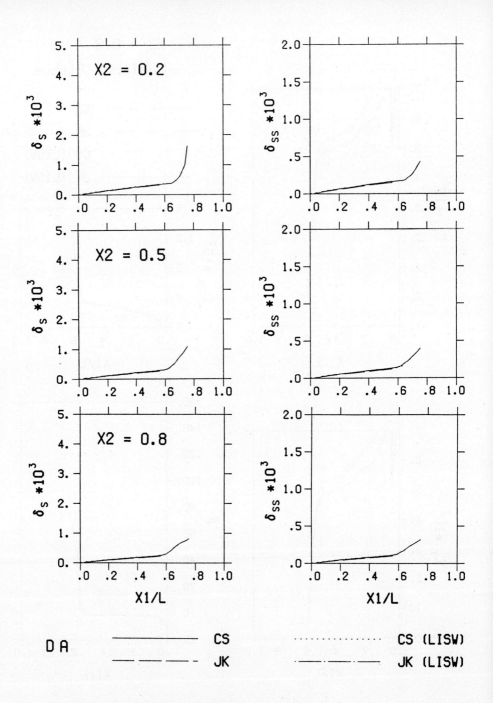

Fig. 24 Distribution of streamwise integral thicknesses at three chorwise sections from calculations of DA (see Table 1)

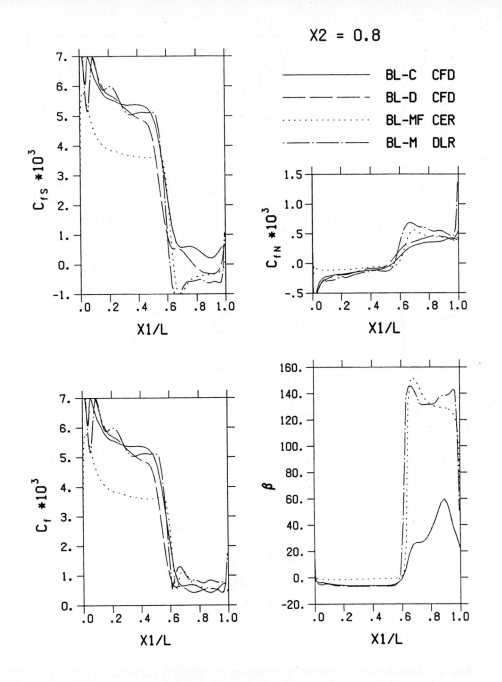

Fig. 20 Skin-friction distribution of different partners employing the BL model (letter code see text); c) section $x_2 = .8$

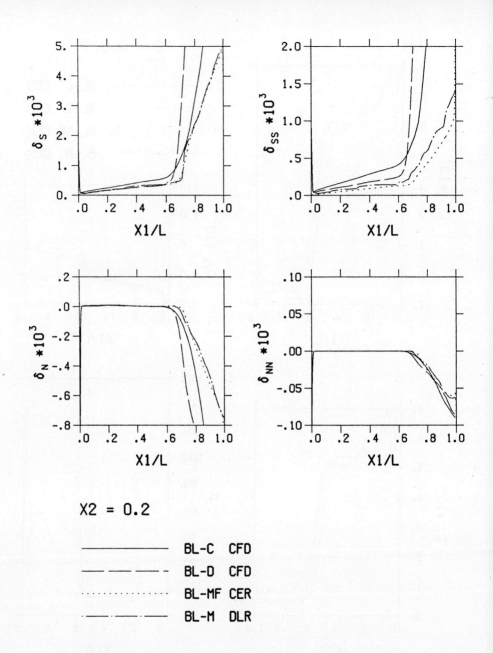

Fig. 21 Distribution of integral thicknesses of different partners employing the BL model (letter code see text); a) section $x_2 = .2$

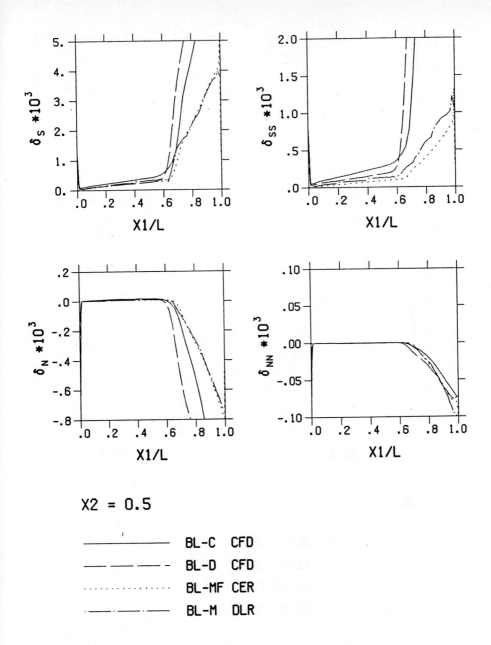

Fig. 21 Distribution of integral thicknesses of different partners employing the BL model (letter code see text); b) section $x_2 = .5$

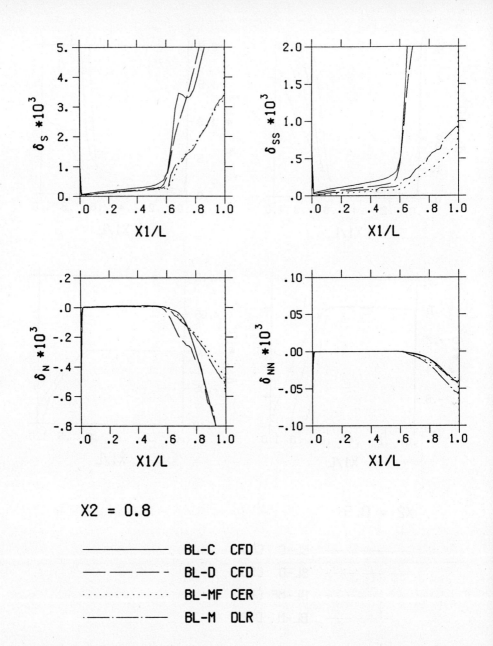

Fig. 21 Distribution of integral thicknesses of different partners employing the BL model (letter code see text); c) section $x_2 = .8$

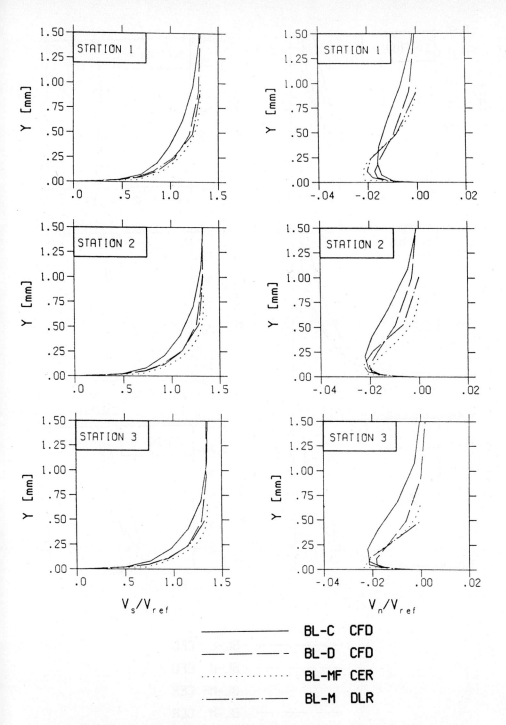

Fig. 22 Streamwise and crosswise components of the velocity of different partners employing the BL model (letter code see text); a) stations at $x_1 = .2$

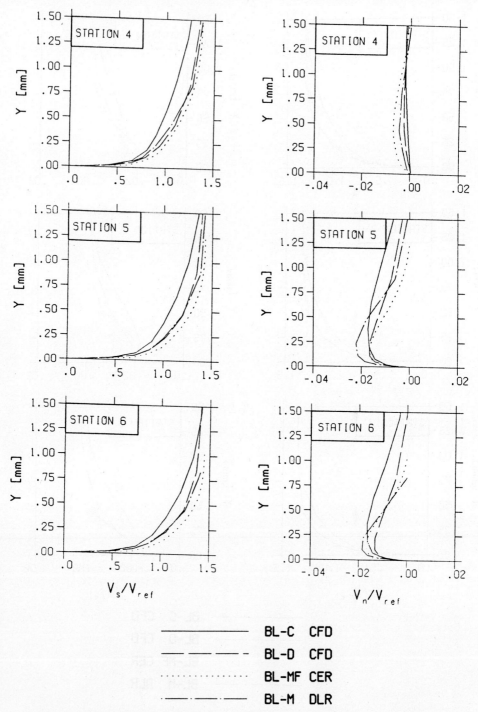

Fig. 22 Streamwise and crosswise components of the velocity of different partners employing the BL model (letter code see text); b)stations at $x_1 = .4$

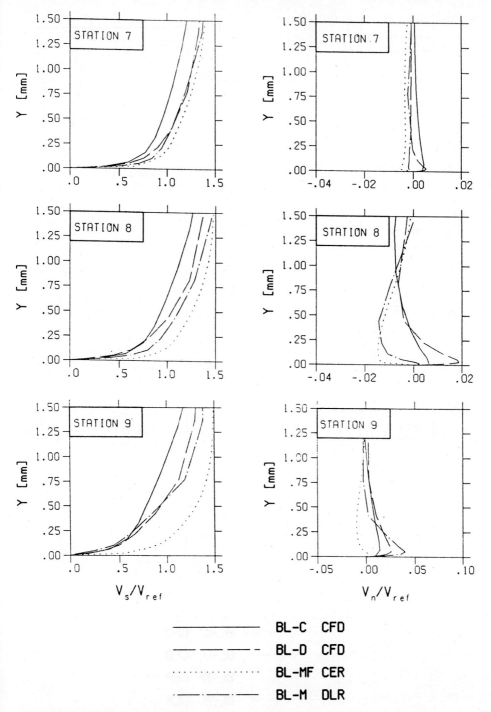

Fig. 22 Streamwise and crosswise components of the velocity of different partners employing the BL model (letter code see text); c)stations at $x_1 = .6$

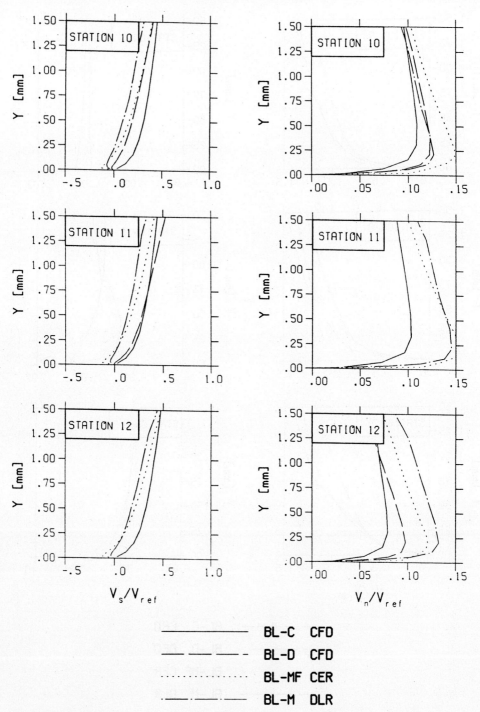

Fig. 22 Streamwise and crosswise components of the velocity of different partners employing the BL model (letter code see text); d)stations at $x_1 = .8$

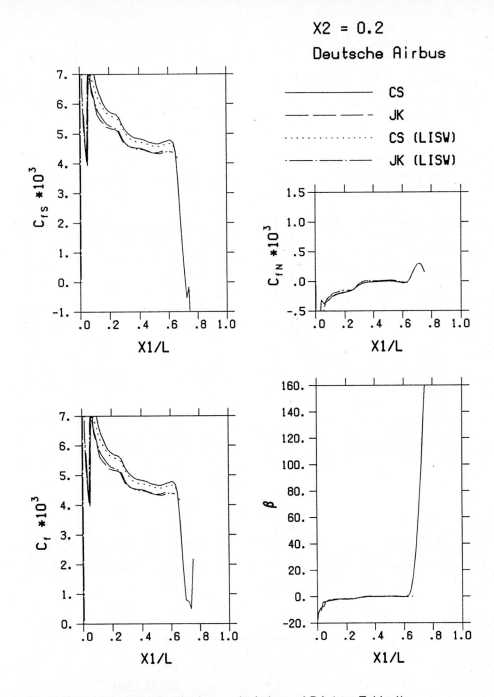

Fig. 23 Skin-friction distribution from calculations of DA (see Table 1);
a) section $x_2 = .2$

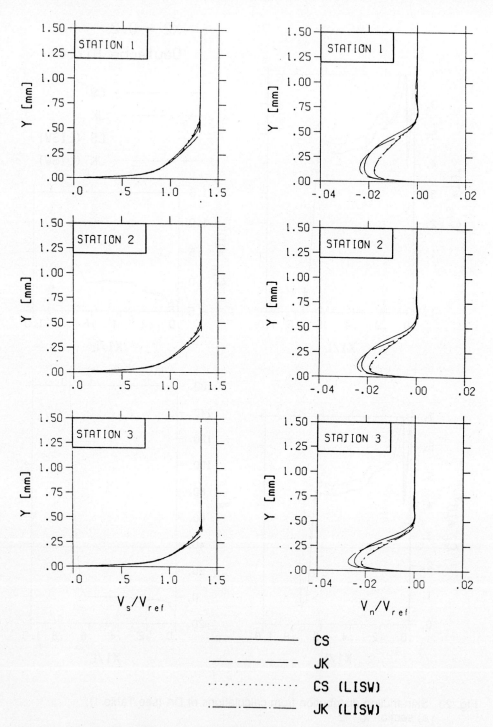

Fig. 25 Streamwise and crosswise components of the velocity from calculations of DA (see Table 1); a) stations at $x_1 = .2$

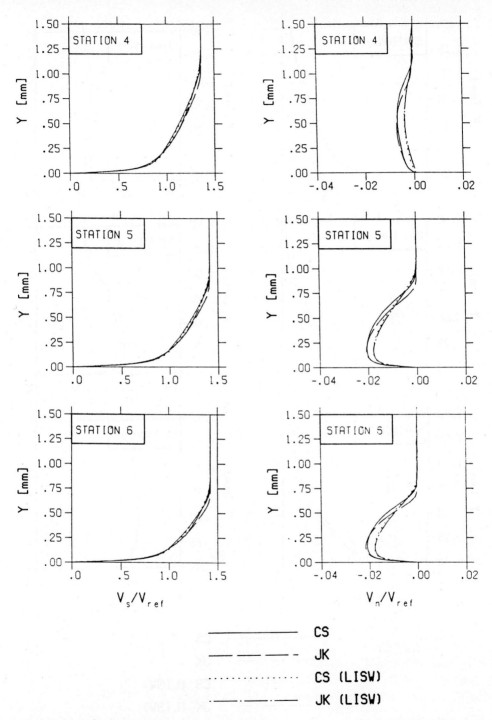

Fig. 25 Streamwise and crosswise components of the velocity from calculations of DA (see Table 1); b) stations at $x_1 = .4$

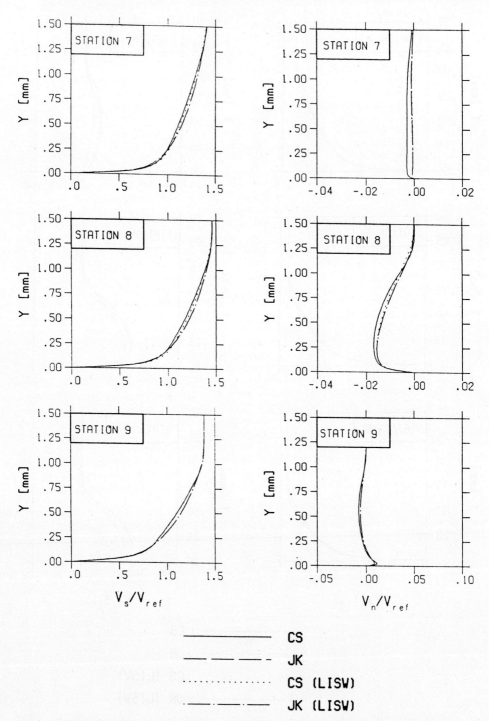

Fig. 25 Streamwise and crosswise components of the velocity from calculations of DA (see Table 1); c) stations at $x_1 = .6$

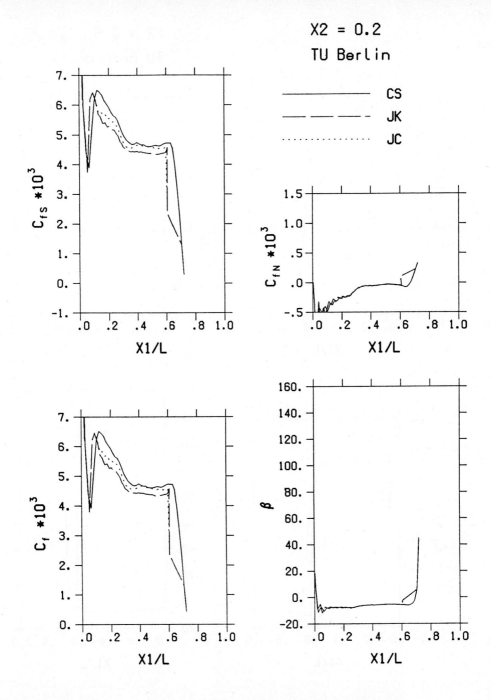

Fig. 26 Skin-friction distribution for three models of TUB (see Table 1)
a) section $x_2 = .2$

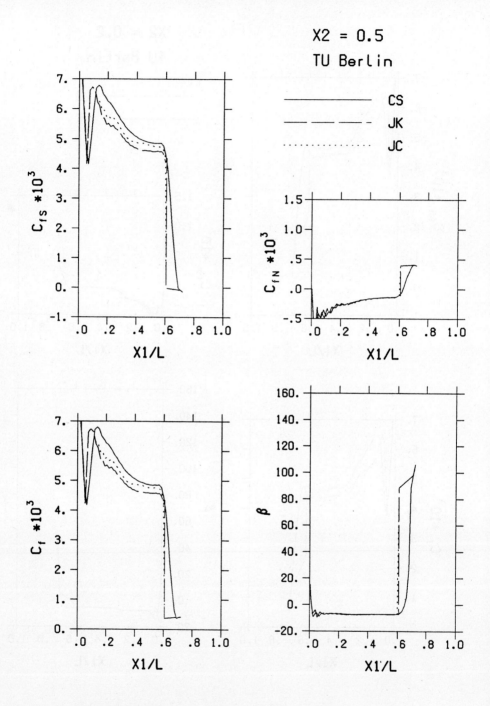

Fig. 26 Skin-friction distribution for three models of TUB (see Table 1)
b) section $x_2 = .5$

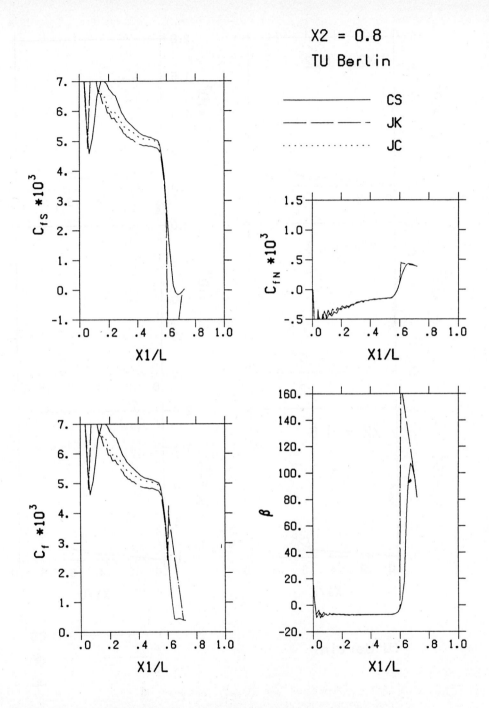

Fig. 26 Skin-friction distribution for three models of TUB (see Table 1)
c) section $x_2 = .8$

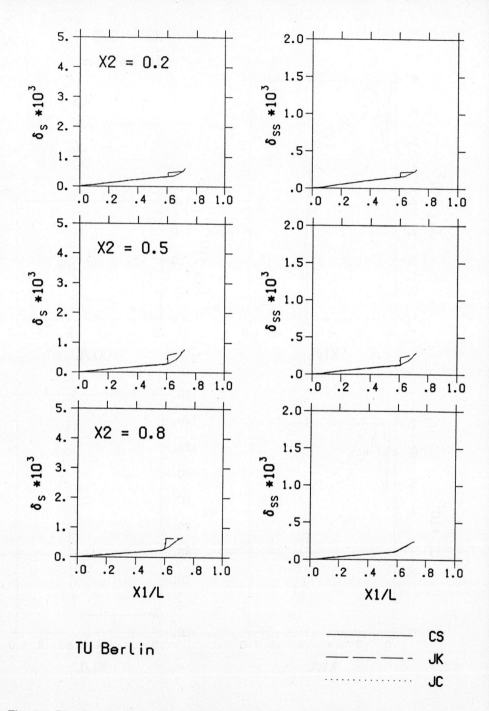

Fig. 27 Distribution of streamwise integral thicknesses at three chorwise sections for three models of TUB (see Table 1)

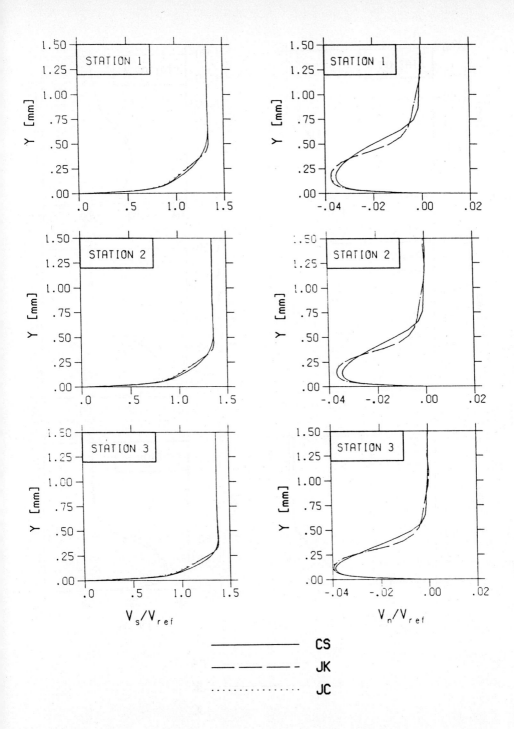

Fig. 28 Streamwise and crosswise components of the velocity for three models of TUB (see Table 1); a) stations at $x_1 = .2$

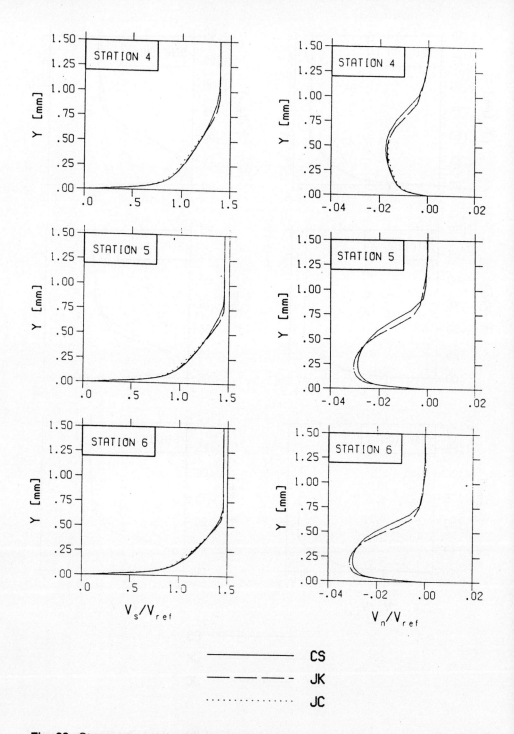

Fig. 28 Streamwise and crosswise components of the velocity for three models of TUB (see Table 1); b) stations at $x_1 = .4$

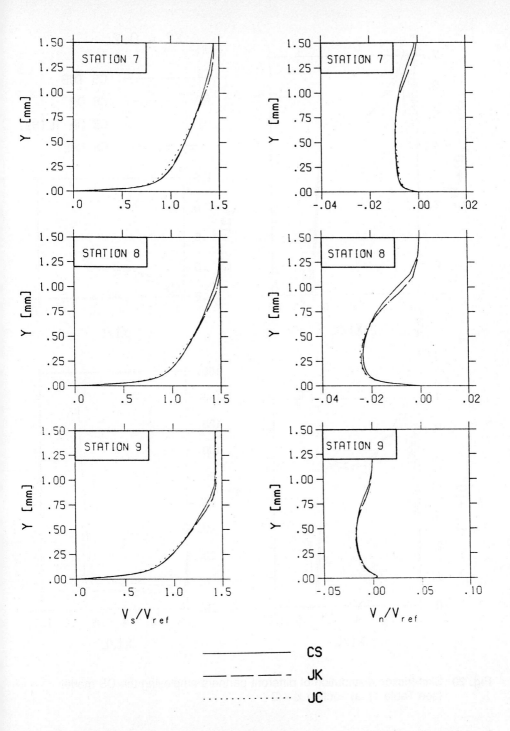

Fig. 28 Streamwise and crosswise components of the velocity for three models of TUB (see Table 1); c) stations at $x_1 = .6$

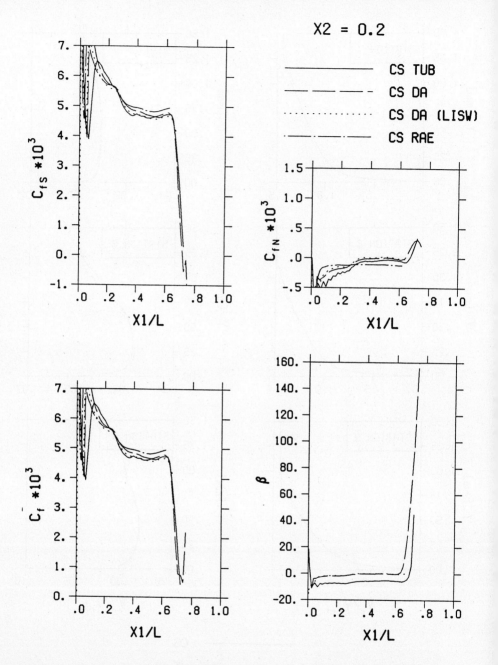

Fig. 29 Skin-friction distribution of different partners employing the CS model (see Table 1); a) section x_2 = .2

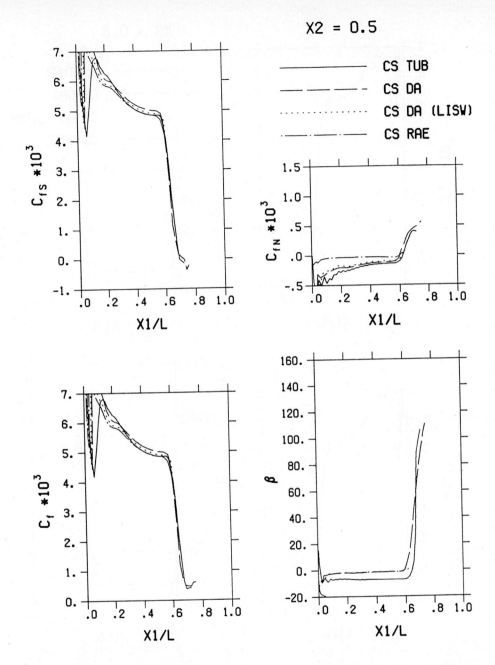

Fig. 29 Skin-friction distribution of different partners employing the CS model (see Table 1); b) section $x_2 = .5$

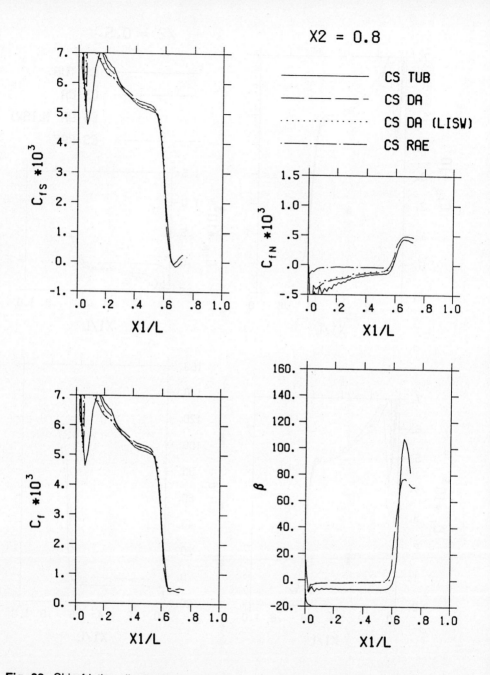

Fig. 29 Skin-friction distribution of different partners employing the CS model (see Table 1); c) section $x_2 = .8$

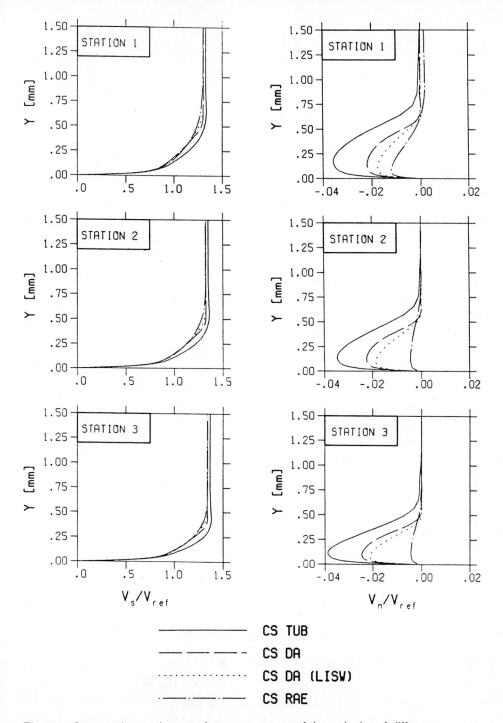

Fig. 30 Streamwise and crosswise components of the velocity of different partners employing the CS model; a) stations at $x_1 = .2$

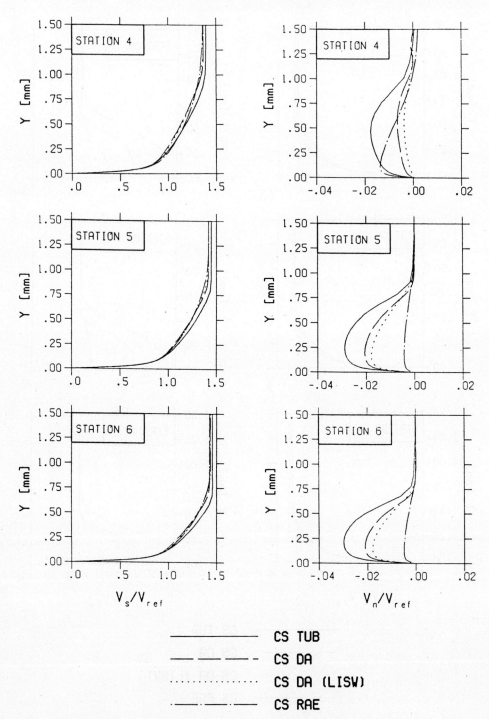

Fig. 30 Streamwise and crosswise components of the velocity of different partners employing the CS model; b) stations at $x_1 = .4$

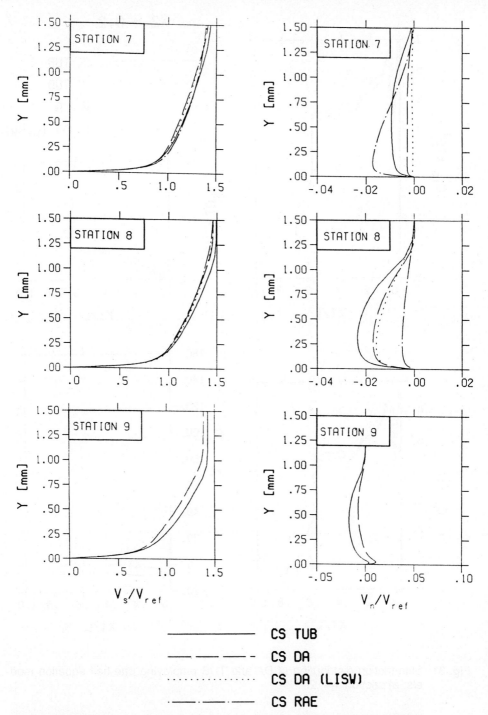

Fig. 30 Streamwise and crosswise components of the velocity of different partners employing the CS model; c) stations at $x_1 = .6$

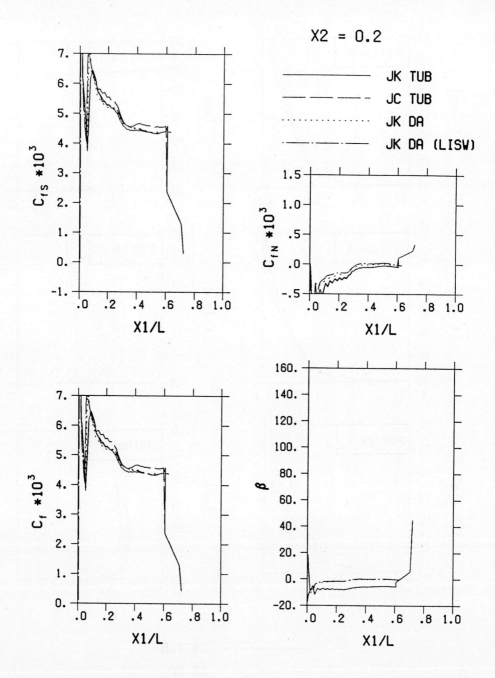

Fig. 31 Skin-friction distribution of DA and TUB employing one-half equation models; a) section $x_2 = .2$

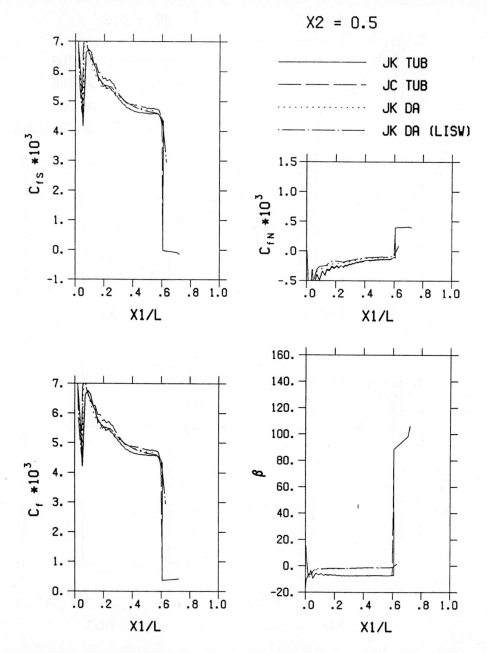

Fig. 31 Skin-friction distribution of DA and TUB employing one-half equation models; b) section $x_2 = .4$

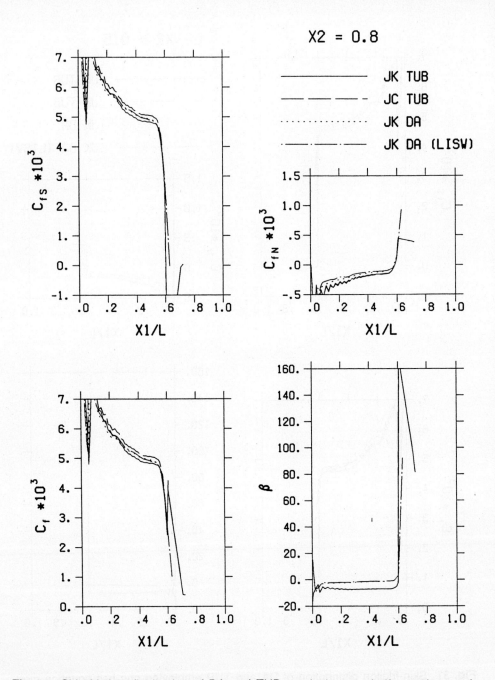

Fig. 31 Skin-friction distribution of DA and TUB employing one-half equation models; c) section $x_2 = .8$

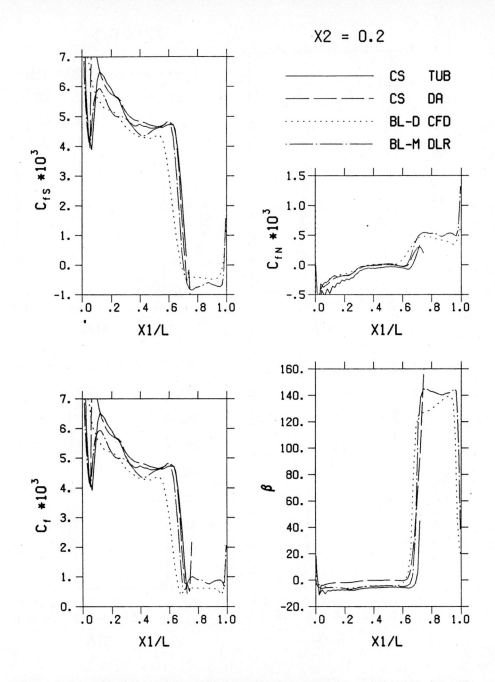

Fig. 32 Skin-friction distribution of different partners employing NS or boundary layer codes; a) section $x_2 = .2$

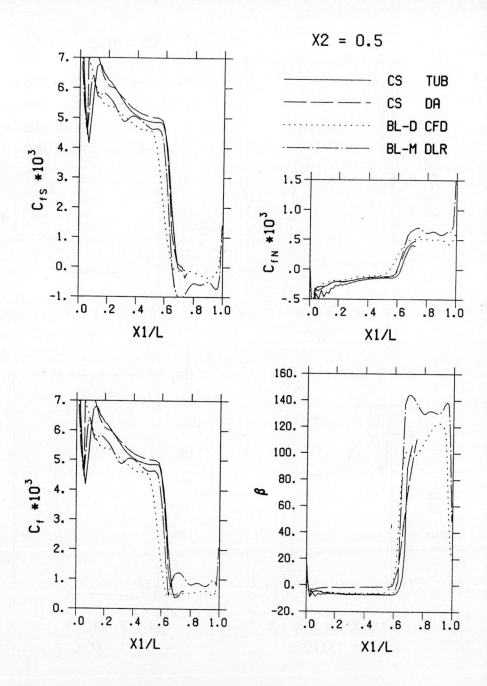

Fig. 32 Skin-friction distribution of different partners employing NS or boundary layer codes; b) section $x_2 = .5$

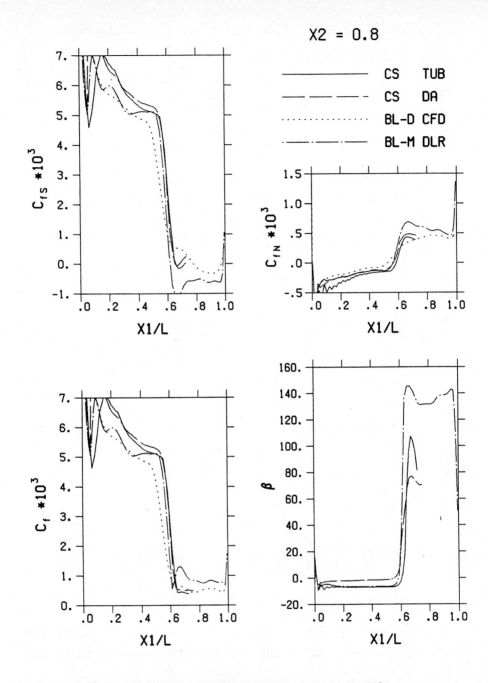

Fig. 32 Skin-friction distribution of different partners employing NS or boundary layer codes; c) section $x_2 = .8$

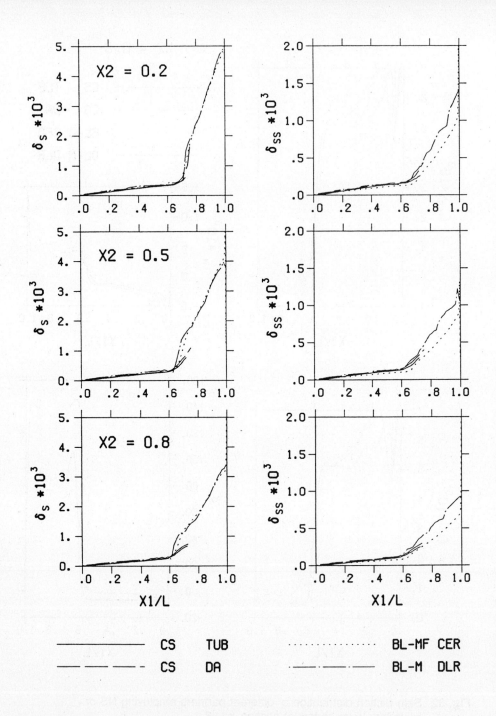

Fig. 33 Distribution of streamwise components of integral thicknesses of different partners employing NS or boundary layer codes:

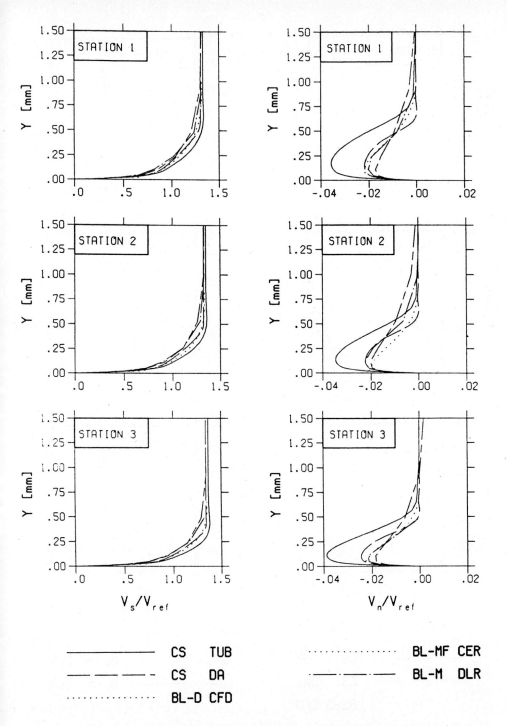

Fig. 34 Streamwise and crosswise components of the velocity of different partners employing NS or boundary layer codes; a) stations at $x_1 = .2$

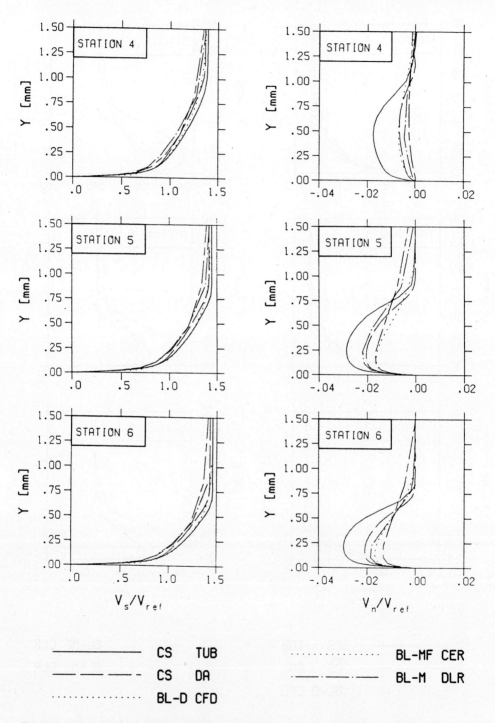

Fig. 34 Streamwise and crosswise components of the velocity of different partners employing NS or boundary layer codes; b) stations at $x_1 = .4$

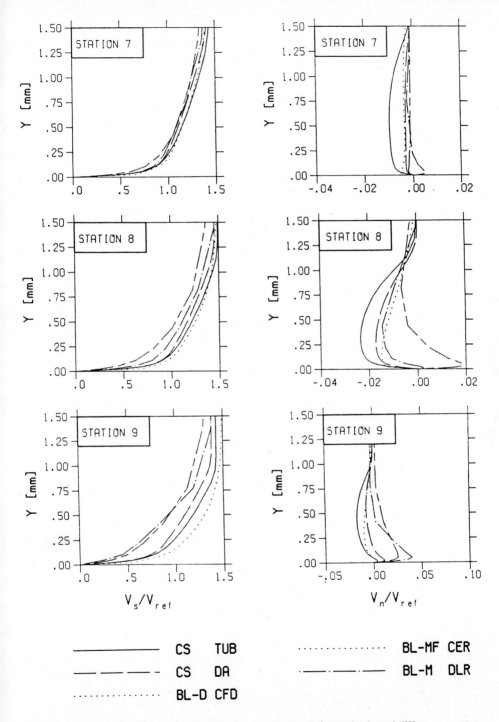

Fig. 34 Streamwise and crosswise components of the velocity of different partners employing NS or boundary layer codes; c) stations at $x_1 = .6$

2D AND 3D WIND TUNNEL SIMULATIONS

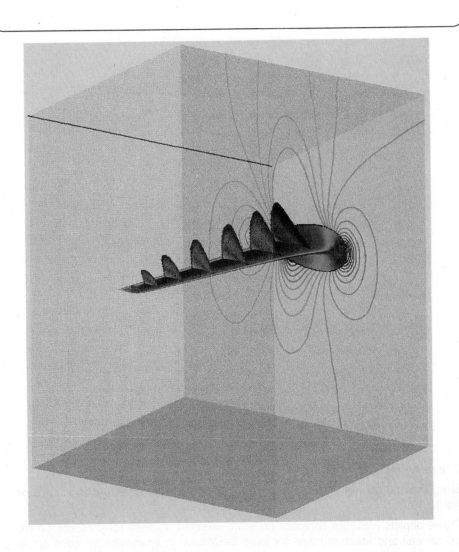

5.8

Colour figure - provided by DLR - shows:

Wing in wind tunnel with "transparent walls" (wing is attached to wind tunnel wall).

Figure depicts pressure contours on wing, selected iso-lines for pressure on wind-tunnel wall and Mach-number contours for $M \geq 1.0$ in span-wise sections on wing surface.

5.8 2D and 3D Wind–Tunnel Simulations
Author: **J.J.Benton, M.C. Fraisse, M.E. Hubbard, D.A. King**, BAe
Plot Coordinator: **J.J.Benton, M.C. Fraisse, M.E. Hubbard, D.A. King**, BAe

5.8.1 Introduction

The development of super-computers and, more recently, very powerful workstations means that CFD is playing an increasing role in the external aerodynamic design of aircraft. The accurate validation of these methods and evaluation against test data, generally obtained from model tests in wind tunnels, is thus of great importance.

Inviscid full-potential or Euler methods coupled to a boundary-layer solver have been used routinely for the calculation of flow around aerofoils and wings for many years. More recently these methods have been extended to the prediction of flow over complex configurations by the use of multiblock and unstructured techniques. Here limitations of viscous coupling are encountered. This, coupled to the need to resolve complex flow features such as 3D effects in boundary layers, shock/boundary-layer interactions, and large scale separation, has prompted the current development of Reynolds-averaged Navier-Stokes methods. The ultimate accuracy of these methods is mainly limited by the physical model of turbulence employed, with great variations being evident in the results from different models. These developments place increasing demands on test data required for their evaluation, with respect to both the range of geometry and flow types involved, and also the level of detail necessary. For example the full evaluation of a turbulence model should ultimately involve the comparison of turbulent stress levels, as well as mean velocity profiles. Clearly the value of such test data is reduced if there is a lack of equivalence between the test setup and the CFD simulation. It is this equivalence that is the subject of the present chapter.

Traditionally wind tunnel tests aimed to provide results applicable to the free-air performance of the model. The constraint on the flow introduced by the presence of the wind-tunnel walls is thus removed in the "free-air" results by applying various correction procedures to the measurements recorded in the tunnel. This procedure becomes more complex if the walls are slotted as is the case in many transonic tests. Today increasing use is being made of solid wall tunnels in order to perform model tests that are directly applicable to CFD evaluation. The intention is that the solid tunnel walls should be included in the CFD simulation so that the complete flow regime is represented in the computation; results can then be compared directly with uncorrected tunnel measurements.

The work reported here makes use of available results in solid wall tunnels, in both 2D and 3D, to assess the influence of the tunnel walls on the flow over the model, and to evaluate in 2D the effectiveness of free-air corrections by means of CFD simulations. The motivation is that the relationship between tunnel and free-air results is of significance in CFD evaluation because much of the aerofoil test data currently in use is of the free-air corrected type. In addition the relationship is critical for the use of this free-air corrected data in aircraft design.

Only transonic conditions are considered for three reasons: (i) they are most important for cruise drag and buffet boundary performance of aerofoils and wings; (ii) the location and hence strength of shock waves plays a dominant role in this performance and yet is very sensitive to perturbations in local flow conditions such as Mach number due to the tunnel walls; (iii) the correction procedures to be evaluated are based on linear theory so may be expected to break down in the fundamentally non-linear situation offered by transonic flow. This forms one of the main motivations behind the present study. The cases considered are the 2D aerofoil RAE 5225 and the DLR 3D wing F5.

The approach used in both 2D and 3D involves comparing the measurements with results of simulations of the model in the tunnel and in free-air. The free-air calculation entails placing the mesh outer boundary in the far-field. Comparison of these results with the tunnel simulation is greatly facilitated by the use of multiblock methods. Thus the free-air mesh is constructed by using the tunnel mesh as a set of inner mesh blocks and adding further blocks around them to place the new outer boundary a large distance from the model. In this way variations between the two sets of calculated results must be due to the change in outer boundary treatment from tunnel to free-air, as the inner mesh remains unchanged. In both cases an Euler flow solver is used, with and without viscous coupling.

5.8.2 2D Tunnel Simulation: RAE 5225 Aerofoil

This section presents results for the 2D aerofoil RAE 5225 to investigate the effectiveness of free-air corrections applied to solid-wall wind tunnel results. These corrections concern the effects of blockage and wall-induced velocities which are accounted for by altering the Mach number, incidence and aerofoil camber, to produce an equivalent free-air flow. The study is based purely on CFD simulations: the BAe 3D multiblock system is used for mesh generation and to produce viscous coupled Euler flow solutions, for both wind-tunnel and free-air corrected simulations, and the validity of the corrections is then assessed by comparing these two results for equivalence. Although the assessment can be performed on the basis of simulations only, it was felt appropriate to base the test case as closely as possible on an actual experiment. Thus comparisons with free-air corrected measurements are included for reference. Comparisons are also presented with free-air corrected results from the RAE developed Garabedian and Korn potential flow aerofoil code BVGK, again with viscous coupling though now with optional higher order terms in the boundary layer solution.

5.8.2.1 Aerofoil Characteristics and Wind-Tunnel Measurements

An overview of the RAE 5225 aerofoil, the wind-tunnel procedure, and some of the results are contained in Ashill et al (1987). This aerofoil is one of a family designed to have aft-loaded pressure distributions in a convex pattern resulting in a strong adverse pressure gradient over the rear of the section. The design requirement for the RAE 5225 is shock-free flow at $C_L=0.6$, M=0.734 and Re=20×10^6, with the boundary-layer approaching separation just upstream of the trailing edge. Thus significant boundary-layer displacement effects can be expected at lower Reynolds numbers.

The wind tunnel results used in the present study were obtained directly from RAE in tabular form. The measurements were performed in the RAE 8ft x 8ft solid wall transonic tunnel using a model of aspect ratio 3.84 and a model chord to tunnel height ratio c/h=0.26. Transition was fixed in the experiment by an air injection technique at 5% chord on upper and lower surfaces (Ashill et al, 1985). Measurements were made on the model centre line and included static pressure holes from which normal force was found by integration, and a wake rake of pitot-static tubes at 2 chords downstream of the trailing edge from which drag was obtained. The flow cases used for the present study have a nominal Mach number of 0.73 and Reynolds number 6×10^6.

5.8.2.2 Wind Tunnel Corrections

Details of the correction procedure to be evaluated here are contained in Ashill et al (1987). In this, the use of solid tunnel walls and the fact that the boundary layers on them are sufficiently thin means that the flow here can be approximated as inviscid. This together with measurements of the static pressure on the walls allows wall-induced velocities to be calculated using the method of Ashill and Weeks (1982). The two components of these velocities represent the effects of blockage and wall-induced upwash respectively.

A single correction for blockage is applied to Mach number, dynamic pressure, and static pressure, as the effect is uniform over the aerofoil. The experimenters report a corrected Mach number M_c of 0.734 for the case considered here. This is based on a tunnel Mach number at the model position M_m, which in turn is derived from the nominal upstream (inflow) tunnel Mach number $M_u = 0.7216$ adjusted for the change in flow between the upstream station and the model. This change was measured in the empty tunnel in terms of static pressure. Using the same subscripts:

$$\frac{p_m - p_u}{p_{ou} - p_u} = -0.00505 \quad (1)$$

The value of M_m rather than M_u is the required inflow value for CFD simulations of this tunnel flow, as these are not subject to the change in flow between inflow station and model. M_m was not given in the measured data but can be determined from the above tunnel data by assuming isentropic flow in the tunnel between upstream station and model:

$$P_o = p\left(1 + \frac{\gamma-1}{2} M^2\right) \quad (2)$$

so that equation (1) and the quoted value of M_u give $M_m = 0.7239$.

The induced upwash varies over the model chord so the correction to incidence is augmented by an adjustment of the aerofoil camber. Garner et al (1969), show that the upwash at the aerofoil can be approximated by linearised theory which gives the incidence and camber corrections as:

$$\Delta\alpha = \frac{c}{h} C_L \delta_0 + \frac{1}{\beta}\left(\frac{c}{h}\right)^2 \delta_1 \left(\frac{1}{4} C_L + C_m\right) \quad (3)$$

$$\frac{\Delta z}{c} = \frac{1}{2}\frac{\delta_1}{\beta} C_L \left(\frac{c}{h}\right)^2 \frac{x}{c}\left(1 - \frac{x}{c}\right) \qquad \beta = \sqrt{1 - M_c^2} \quad (4)$$

where M_c is Mach number after correction for blockage, x is chordwise distance from the leading edge, and C_L and C_M are sectional lift and pitching moment coefficients. Moments are about the quarter-chord, with positive nose up. The wall interference parameters δ_0 and δ_1 depend on the type of tunnel wall: for solid walls they have the values $\delta_0 = 0.0327$ and $\delta_1 = 0.1309$.

The implication of the blockage correction and the upwash correction expressed in equations (3) and (4) is that if these are applied to the in-tunnel aerofoil and the measurements are then repeated in free-air then the C_p and hence C_L and C_M

obtained would be the same as in the tunnel. This is consistent with the theoretical basis of the corrections in which the unchanged C_L and C_M represent the values that are obtained in free-air. This consideration has implications for the manner in which CFD simulations must be performed in order to assess the validity of the corrections, and is discussed in Section 5.8.2.4 below.

In view of the relatively large aspect ratio of the model no correction for sidewall boundary layer effects was considered necessary by the experimenters. However, as the model was attached at one end only, a correction to allow for elastic twist is included in all results, including in-tunnel simulations.

5.8.2.3 Mesh Generation

A multiblock approach is used not only to facilitate the generation of meshes to represent the tunnel flow, but also to allow that mesh to be extended by adding outer blocks to form a free-air mesh with the outer boundary a large distance from the aerofoil, but in which the original tunnel mesh appears unchanged as an inner region. This ensures that the results reflect changes in the type and location of boundary conditions rather than variations in mesh near the aerofoil thus minimising the effects of any mesh dependence there may be in the solution near the aerofoil.

Although the computational problem being studied is a 2-dimensional one, the multiblock system used is a 3D method, and includes both mesh generators and viscous-coupled Euler flow solver. A description of the system is given by Allwright (1988). Thus the aerofoil is represented as a constant chord wing between vertical side walls. 14 blocks are used for the tunnel mesh. After interactive positioning of boundary points (ie generation of line grids), a surface grid generator is used to produce meshes on the wing and the outer boundaries, and on any extra control surfaces that are felt to be necessary to constrain and control the mesh. This uses partial differential equation techniques due to Thompson et al (1974), with boundary point distributions propagated by means of the Thomas and Middlecoff (1979) form of the control terms. A field mesh could now be generated using a 3D Thompson method, but for this 2D problem the mesh on one of the side walls, as shown in figure 1, is simply replicated at several spanwise stations to give the 3D mesh required by the flow solver. As can be seen the mesh has a C-type structure around the aerofoil, embedded in an H-type outer region which extends to the tunnel walls located two chords above and below the aerofoil. The upstream and downstream boundaries are located far from the aerofoil. Note that the aerofoil must be rotated to the required incidence relative to the tunnel walls, so a separate mesh is needed for each incidence.

For free-air calculations the tunnel mesh is extended by adding further H-type mesh blocks above and below. These place the far-field boundary at 50 chords which was found to be adequate following an investigation of its influence on solutions that do not use a vortex boundary condition, which was the case for all multiblock solutions presented here. This showed that at least 100 chords would be necessary for full convergence of the solution.

5.8.2.4 Flow Calculation and Evaluation of Corrections

The solver used is the 3D multiblock viscous-coupled Euler code described in Section 4.2.3. As pointed out above the far-field boundary condition does not include the circulation due to the vortex model of a 2D lifting body; this is because it is a 3D scheme generally used for flow around full aircraft geometries. The turbulent boundary layer solution is a 2D integral lag-entrainment method with quasi-simultanious coupling to the Euler solution which enables small separated regions to be handled. This boundary layer solution is referred to here as being of first order since various higher order effects are not included. In tunnel simulations the tunnel walls are treated as inviscid while the inflow and outflow boundary conditions, situated far from the aerofoil, utilise 1D Riemann invariants. These are also used at the far-field boundary in free-air cases.

A Cray YMP and a Meiko parallel Transputer array are used to run the calculations. Adequate convergence is obtained in 1000 time steps, the first 100 being purely Euler cycles, with a boundary layer solution then being performed every 30 Euler cycles.

The manner in which the wall induced upwash correction, expressed in equations (3) and (4), is applied is now described. As stated in Section 5.8.2.2, the implication of the corrections is that if these are applied to the in-tunnel aerofoil and the measurements, or simulation, are then repeated in free-air then the C_p and hence C_L and C_M obtained would be the same as in the tunnel case. C_L and C_M in the correction equations are intended to represent the values that are obtained in free-air.

The purpose of the present study is to evaluate the accuracy of this equivalence between tunnel and corrected free-air simulations, and by implication the accuracy of the corrections. In applying the corrections it is not sufficient to perform the tunnel simulation, apply corrections based on the C_L and C_M obtained, and then perform the free-air simulation for the corrected conditions. This is because different C_L and C_M will inevitably be obtained from the two simulations, and the corrections will then be based incorrectly on the tunnel result.

Instead it is necessary to ensure that true free-air values of C_L and C_M are being used in the corrections by use of an iterative procedure. Thus a tunnel simulation is performed first, then a free-air simulation using corrected conditions derived from the tunnel result, then a second free-air simulation using corrections based on the C_L and C_M from the previous free-air result, and so on. In fact the process is rapidly convergent and no more than two free-air runs are used. The result may then be compared with the tunnel simulation and if the corrections were perfect then the same C_p, C_L and C_M would be seen. Note that this assessment of the performance of the corrections does not require any comparison with measured results.

Comparisons with measurements are presented for reference, as are free-air corrected solutions from the RAE developed Garabedian and Korn 2D potential flow single-block aerofoil code BVGK (Ashill et al, 1987). Viscous coupling is again used in this code, with a boundary layer solution which is based on the same lag-entrainment concept used in the multiblock Euler code, though now a number of higher order terms may be included via a user specified switch. This produces a solution that is referred to here as second order. It is expected that the first order solutions should be very similar to those obtained from the viscous coupled Euler code.

The drag prediction from both the multiblock and BVGK codes is obtained as the sum of viscous and wave components:

$$C_D = C_{Dv} + C_{Dw} \qquad (5)$$

in which C_{Dv} is calculated from the wake momentum deficit and C_{Dw} involves a shock search algorithm. Though the principle is the same for each code, the implementation, and hence algorithm details, are distinct and dedicated to each code.

All measured results from RAE were presented as free-air corrected values and this practise is continued for all results presented here. Thus Mach numbers include the blockage correction and all pressure and force coefficients are normalised by this quantity. As is the normal practise in flow solvers, pressure and force coefficients as output from the code are non-dimensionalised by the reference or "infinity" flow conditions for which the code is run and which correspond to the external flow imposed as inflow boundary conditions. The tunnel simulation results are re-normalised to use the corrected value of M_∞ as follows. With the definitions

$$C_p = \frac{p - p_\infty}{q} \qquad q = \frac{\gamma}{2} p_\infty M_\infty^2 \qquad (6)$$

we have

$$\frac{p}{P_o} = C_p \frac{q}{P_o} + \frac{p_\infty}{P_o} \qquad (7)$$

so that the re-normalised pressure coefficient is

$$C_{p(n)} = \frac{p - p_{\infty(n)}}{q_{(n)}} = C_p \frac{\dfrac{q}{P_o}}{\dfrac{q_{(n)}}{P_o}} + \frac{\dfrac{p_\infty}{P_o} - \dfrac{p_{\infty(n)}}{P_o}}{\dfrac{q_{(n)}}{P_o}} \qquad (8)$$

where subscript (n) indicates re-normalised quantities. This gives on using the definition of q and the isentropic expression (2):

$$C_{p(n)} = 0.9817734 \, C_p + 0.0245909 \, . \qquad (9)$$

Thus integrated force coefficients are re-normalised by multiplying by .9817734.

5.8.2.5 Results

Viscous coupled Euler multiblock flow simulations and the correction procedure as described in Section 5.8.2.4 are applied to the RAE 5225 aerofoil at five incidences, for the nominal Mach number 0.73 case (corrected Mach number 0.734) and Reynolds number 6×10^6 based on aerofoil chord.

The pressure distributions for nominal incidences of 0.5, 1.0 and 2.0 degrees are shown in figure 2. Four curves are overlaid for each incidence: the tunnel and uncorrected free-air calculations, and two iterations of the correction process described in section 5.8.2.4, with pressures from each stage illustrated. As can be seen the process is rapidly convergent. The second iteration should be comparable with the tunnel results if the corrections are effective. At 0.5 degrees where the flow is mostly subsonic the correction process has worked well, the corrected pressures closely matching the tunnel pressures except where the flow is locally supercritical. At 1.0 degree the correction procedure still provides a plausible solution. At 2.0 degrees extensive supersonic flow occurs together with a strong shock, and the correction process does not provide a good representation of the tunnel conditions. A large underprediction of suction levels and shock strength is seen in the uncorrected free-air results, indicating the extent of the tunnel interference in this case.

Figure 3 indicates the trend in force coefficients. The correction procedure consistently gives slightly too much lift at a given incidence. Drag is plotted against lift to remove the incidence differences. At a given lift the corrected drag is too low particularly as incidence increases beyond 1.5 degrees, which can be associated with the differences seen in the tunnel and corrected pressure distributions as the shock strength increases.

For reference, comparisons are presented in figures 4 and 5 of the viscous coupled Euler multiblock results with experimental data and with free-air corrected results from the BVGK code described in 5.8.2.4. BVGK is used with the same correction procedure as for the Euler runs, and both first and second order boundary layer solutions are presented. The viscous coupled Euler results and those from BVGK with a first order boundary layer should be very similar and this is borne out. Comparison with measured results suggest an incidence shift between the numerical and the experimental results. The second order terms in the BVGK boundary layer solution have a substantial effect and when incidence is taken out in figure 5 by plotting drag against lift a high standard of comparison with experiment is achieved. The tunnel multiblock results appear quite good with the exception of one drag point.

5.8.2.6 Conclusions

Simulations of the flow around the RAE 5225 aerofoil in a solid wall wind tunnel and in the equivalent corrected free-air conditions have been performed using an Euler solver with and without viscous coupling. Comparison of the results reveals the effectiveness of the correction procedure being studied. The work suggests that the corrections, which are based on linearised theory, work reasonably well for subcritical and mildly transonic flows but that for cases with stronger shocks they no longer provide adequate results. For the purpose of validating CFD codes against experimental data direct simulation of tunnel walls is recommended.

5.8.3 3D Tunnel Interference Simulation: DLR-F5 Wing

This section presents results for the DLR-F5 3D wing in transonic conditions as tested by DLR in a solid wall wind tunnel. As for the 2D work described above, multiblock calculations are performed for both the wing in the tunnel configuration and in free-air. Unlike the 2D study no attempt is now made to correct for the tunnel influence, the object being to evaluate the extent of this influence by comparing the results. Thus both tunnel and free-air calculation are performed with the same inflow Mach number and incidence. Euler solutions are shown for a range of incidences, with viscous-coupled solutions for 0 degrees. Experimental results are included for reference. Reference should also be made to Section 5.7 in which the F5 wing is the test case used for validation of 3D Navier-Stokes methods.

5.8.3.1 The DLR-F5 Wing and Wind-Tunnel Measurements

The wing is of symmetric section across its entire span and has a rounded tip and a large root fairing. The geometry is generated analytically by means of a computer code supplied by DLR (Sobieczky, 1987). The aerofoil section is designed for shock free flow at a free-stream Mach number of 0.78 and has sweep appropriate to a flight Mach number around 0.82. Emphasis has been placed on maintenance of laminar flow, and the symmetric section lacks the pronounced rear loading common on supercritical designs.

Descriptions of the wing and the test procedure in the 1m x 1m transonic tunnel are given by Sobieczky et al (1987), Kordulla (1988) and Kordulla et al (1988), the last two also including descriptions of extensive flow simulations. The tunnel wall slots were closed for the tests to form solid walls. The wing of 0.65m span is mounted on one of the tunnel walls which is in fact a large splitter plate to minimise the thickness of the tunnel wall boundary layer at the wing root. An impression of the geometry of the wing and the tunnel can be obtained from figure 6 which shows some of the surface meshes used in the present study. The outer surfaces correspond to the tunnel walls and to the inflow and outflow planes at which flow measurements were made to define a tunnel control volume with known boundary conditions suitable for use in CFD. The close proximity of the outer wall to the wing tip can be seen in figure 6. The nominal tunnel centreline Mach number upstream of the inflow plane is 0.82.

Tests results are available for incidences of 0 and 2 degrees at a Reynolds number of 3.6×10^6 based on the root chord at the wall. Pressure measurements are on the sections shown in figure 8. In line with the interest in laminar flow, transition was not fixed in the experiment and so occurred naturally fairly well aft on the section, being roughly at the shock location over the mid and outer parts of the wing. The approximately measured transition location is indicated in figure 8.

5.8.3.2 Mesh Generation

The principles used here follow the procedure outlined for the 2D cases as described in section 5.8.2.3. Thus the BAe multiblock system is used to generate meshes for the wing-in-tunnel configuration, and free-air meshes are then created from these by adding extra algebraically generated mesh blocks to place the outer boundaries a large distance from the wing. This ensures that the flow simulations reflect changes in the type and location of boundary conditions rather than variations in mesh near the wing thus minimising the effects of any mesh dependence there may be in the solution near the wing. Details of the mesh generation techniques are given in section 5.8.2.3.

The mesh around the wing is of C-O type, with good resolution of the tip region using the O type structure, as can be seen in figure 7 which also shows the resultant singular point on the surface. This interfaces with an outer H-mesh which has the tunnel control volume as its outer boundaries to complete the wing-in-tunnel mesh. This used 76 blocks containing 209664 cells. Note that a different mesh is required for each incidence. The free-air meshes have outer boundaries extended to around 9 semi-spans fore and aft of the wing, 6 semi-spans above and below, and 4 semi-spans outboard. They use 232 blocks and 475776 cells. The number of blocks is increased beyond the minimum to allow use of the meshes on a Meiko parallel Transputer array.

5.8.3.3 Flow Calculation

The solver used is again the 3D multiblock code described in Section 4.2.3, used in both inviscid (Euler) and viscous coupled mode. The turbulent boundary layer solution is a 2D integral lag-entrainment method with quasi-simultanious coupling to the Euler solution which enables small separated regions to be handled. Transition is set at 10% chord, a compromise which does not match the location in the experiment. However it should be remembered that the important comparison to be made in this study is between tunnel and free-air predictions, rather than with experiment. The viscous coupling does not extend to the wing tip due to the occurrence of the singular point in the surface grid which could not be handled in the version of the code then in use.

In tunnel simulations the tunnel walls are treated as inviscid while the inflow and outflow boundary conditions utilise 1D Riemann invariants. These are also used at the far-field boundary in free-air cases. The inflow condition uses the nominal tunnel centreline Mach number of 0.82 rather than the measured inflow plane conditions. Incidences of 0, 2, and 4 degrees are considered but viscous results are presented for 0 degrees only, due to unrealistic shock induced separations in modelling the strong shock in the lifting cases.

5.8.3.4 Results

Inviscid multiblock Euler results are presented for incidences of 0, 1 and 2 degrees, with in addition viscous coupled results at 0 degrees. The Reynolds number is the tunnel value of 3.6×10^6 based on root chord and inflow Mach number is the nominal upstream tunnel figure of 0.82. Tunnel and free-air simulations use the same inflow conditions.

Although comparisons can be drawn between measurements and the inviscid and viscous results obtained here, and the flow around the wing could be examined in great detail, particularly where 3D effects predominate near the gross fairing at the root and at the tip, these aspects are of secondary importance to the study of the tunnel interference effects based on comparison of calculated tunnel and free-air results.

Figure 9 shows pressure distributions from inviscid and viscous wing-in-tunnel simulations at 0 degrees incidence, at four of the sectional measurement stations across the span. A shock at about 70% chord is observed over the central part of the span. At the root this shock is weakened and moves back slightly, and at the tip it moves forward and weakens until it disappears. The viscous calculations show the expected significant reduction in pressure around the trailing edge, exceeding that seen in the experiment except at the root.

The experimental results at $\eta = .205$ and $.492$ seem to indicate, by the cutting off of the suction peaks in the central sections, that a separation bubble may be present immediately ahead of the shock, suggesting the presence of laminar flow here and hence that transition occurs at the shock, causing large discrepancies here in all of the comparisons made with calculated results. This is consistent with the approximately measured transition line shown in figure 8.

The inviscid and viscous tunnel results at 0 degrees incidence are repeated with the corresponding free-air simulations in figures 10 and 11 respectively. The main observation to be made here is that for both set of results at all spanwise locations the tunnel simulation shows a more prominent suction peak ahead of the shock than is seen in the free-air calculation, with an associated increase in shock strength. This suggests that the tunnel flow is strongly influenced by the blockage effect causing a higher effective Mach number at the wing. Note that the absence of lift and the symmetry of the tunnel configuration means there are no wall induced upwash effects in this case.

Figure 12 shows inviscid results for tunnel and free-air simulations at an incidence of 2 degrees. Again the tunnel calculation shows the stronger suction peak and shock, but now only on the upper surface. The effects of blockage may here be augmented by wall-induced suppression of the wing-induced downwash field. The tunnel and free-air results are closely matched on the lower surface. The measurements again show evidence of laminar separation ahead of the shock, now extending to the tip on the upper surface but largely suppressed on the lower surface.

Figure 13 shows inviscid results for tunnel and free-air simulations at an incidence of 4 degrees. Observations to be made here are exactly as for the 2 degree calculated results.

Lift coefficients for the inviscid results presented are given in the table below. The probable reason for the value not being zero at zero incidence, even though the wing is symmetric, is that the wing is not set exactly symmetrically within the tunnel. It can also be seen here that, considering the inviscid results which are the most reliable, the lift is about 10 to 15% less in free-air conditions than in the tunnel:

	$0°$	$2°$	$4°$
Tunnel	0.00163	0.36746	0.67380
Free-air	0.00152	0.31360	0.59293

5.8.3.5 Conclusions

Results of Euler and viscous coupled Euler simulations of transonic flow over a wing in a solid wall wind tunnel and in free-air using the same inflow conditions have been presented. Differences in the results indicate the level of tunnel interference. Calculations done with the tunnel boundary conditions give increased suction levels ahead of the shock and an increase in shock strength compared with the free-air results. This can largely be associated with the effects of blockage and is consistent with the uncorrected free-air results seen in the 2D aerofoil study in section 5.8.2. It is recommended that CFD validation should include the tunnel walls in the simulation, and that the walls should be solid to facilitate this. While the use of slotted wall tunnels can be effective in minimising blockage effects, other constraints on the flow remain which are difficult to quantify and difficult or impractical to simulate in the CFD model.

5.8.4 Overall Conclusion

Flow simulations for the 2D aerofoil RAE 5225 and the 3D wing DLR-F5 have been performed for both solid wall tunnel and free-air configurations. In the 2D study the free-air results include the use of corrections for tunnel interference. These show that the corrections are effective for subcritical and mildly transonic cases but break down at higher incidence for which a stronger shock occurred. The 3D results show the effect of the tunnel blockage constraint as increased suction levels and shock strength. It is recommended that for accurate validation of CFD methods against wind tunnel results, the tunnel walls should be included in the simulation and tests using solid walls are preferred to facilitate this.

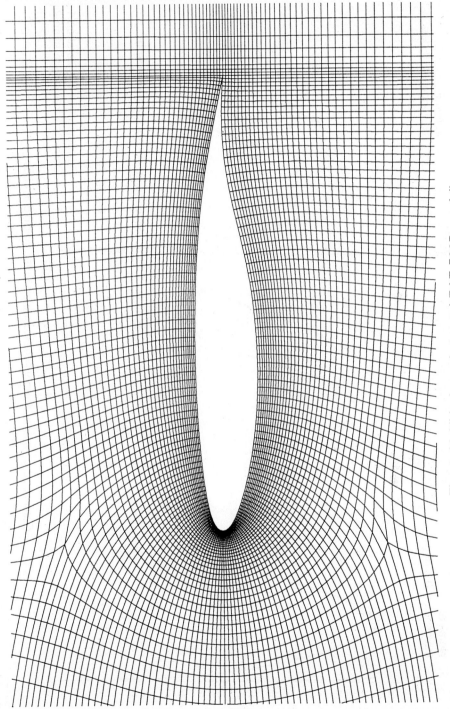

Fig. 1 Multiblock mesh around RAE 5225 aerofoil

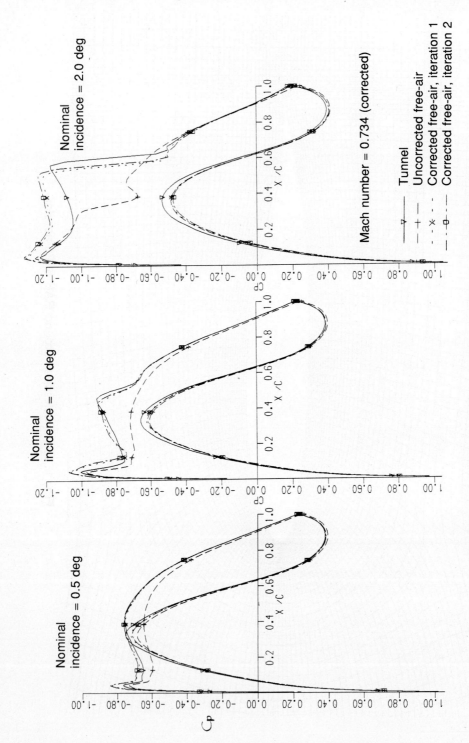

Fig. 2 Calculated pressures from viscous-coupled Euler multiblock code

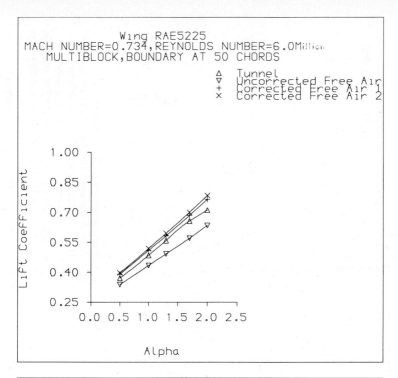

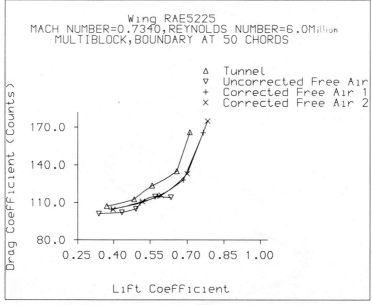

Fig. 3 Calculated forces from viscous-coupled Euler multiblock code

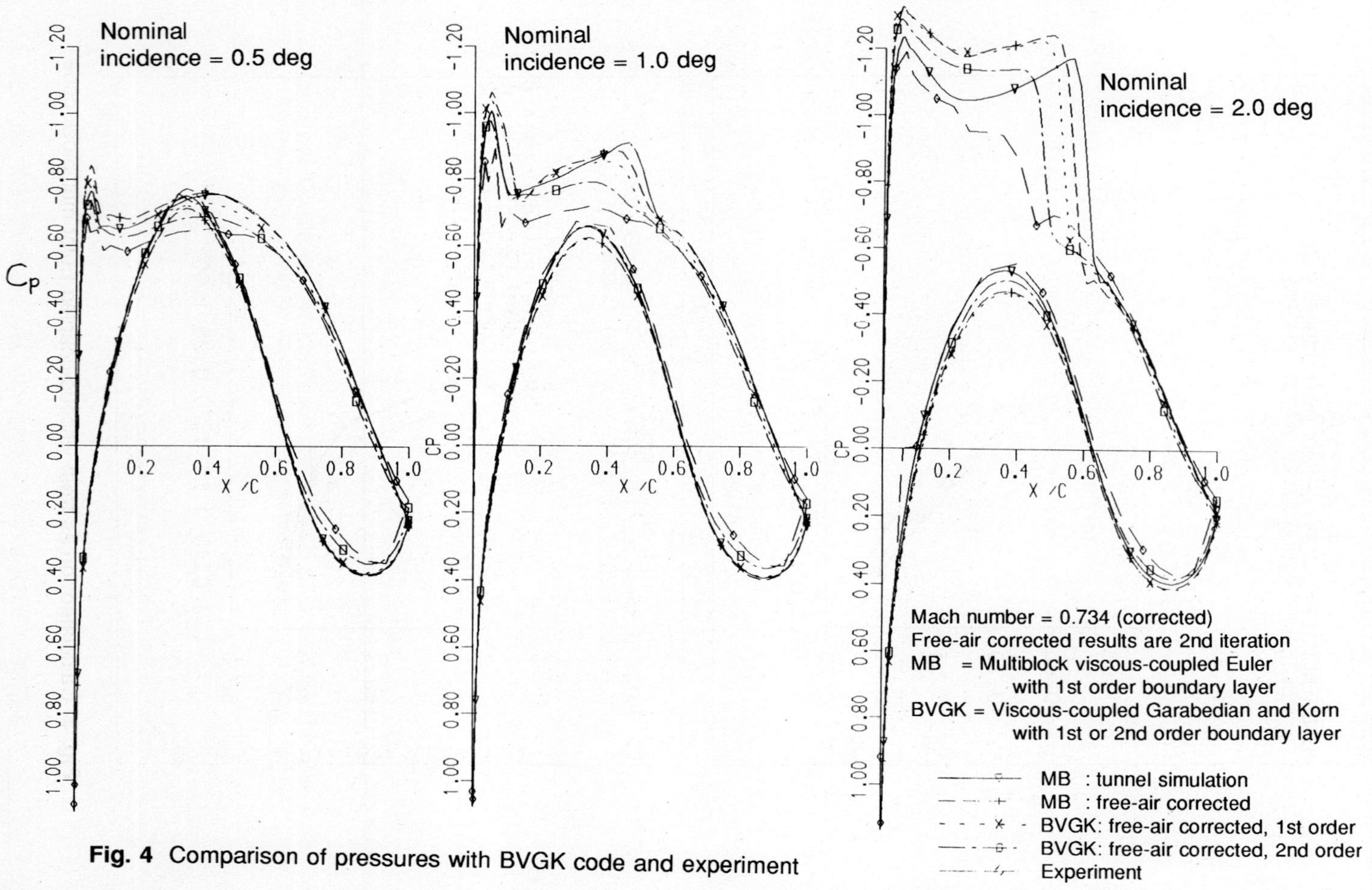

Fig. 4 Comparison of pressures with BVGK code and experiment

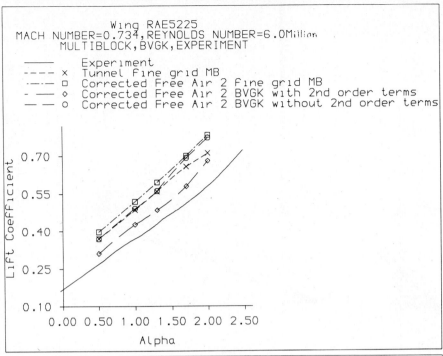

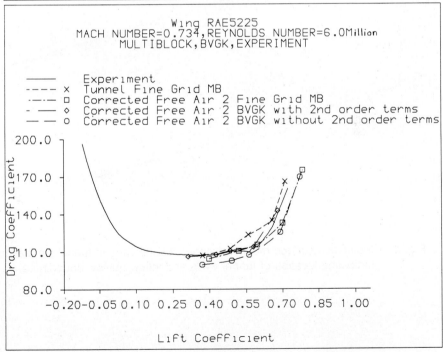

Fig. 5 Comparison of forces with BVGK code and experiment

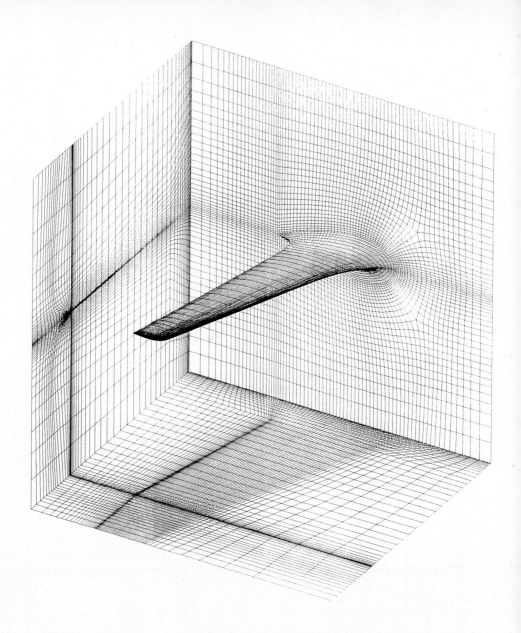

Fig. 6 Multiblock surface meshes for the DLR F5 wing in tunnel simulation, indicating location of tunnel walls and inflow/outflow boundaries.

Fig. 7   Multiblock surface mesh on the DLR F5 wing: tip region.

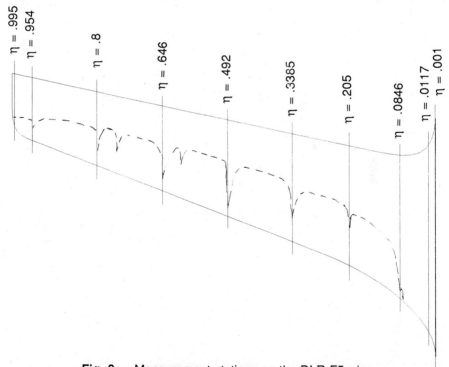

Fig. 8   Measurement stations on the DLR F5 wing.
Dashed line shows approximate measured location of transition.

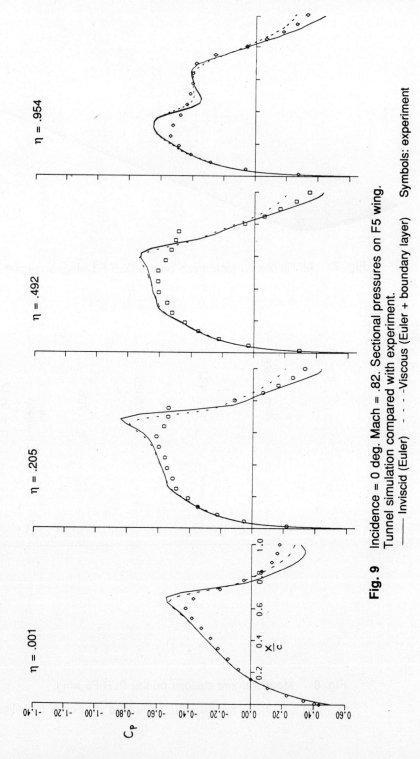

Fig. 9 Incidence = 0 deg. Mach = .82. Sectional pressures on F5 wing. Tunnel simulation compared with experiment.
——— Inviscid (Euler)   - - - Viscous (Euler + boundary layer)   Symbols: experiment

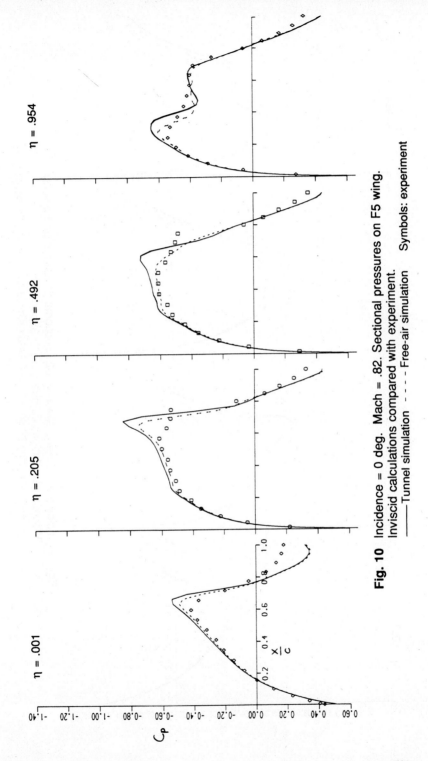

Fig. 10 Incidence = 0 deg. Mach = .82. Sectional pressures on F5 wing. Inviscid calculations compared with experiment. Symbols: experiment ——— Tunnel simulation – – – Free-air simulation

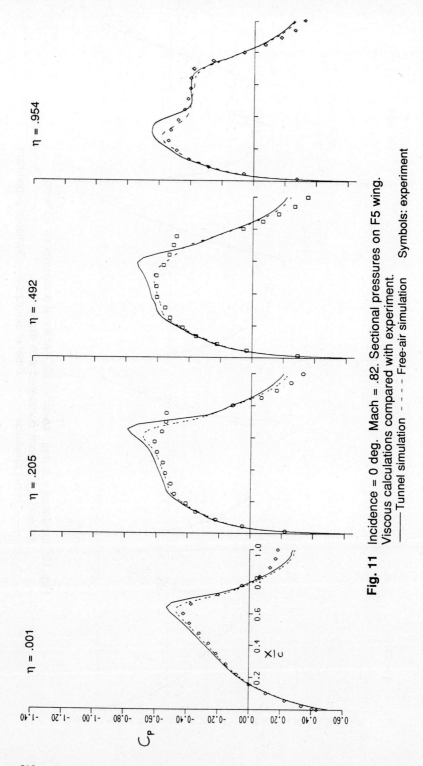

Fig. 11 Incidence = 0 deg. Mach = .82. Sectional pressures on F5 wing. Viscous calculations compared with experiment. Symbols: experiment
——— Tunnel simulation - - - Free-air simulation

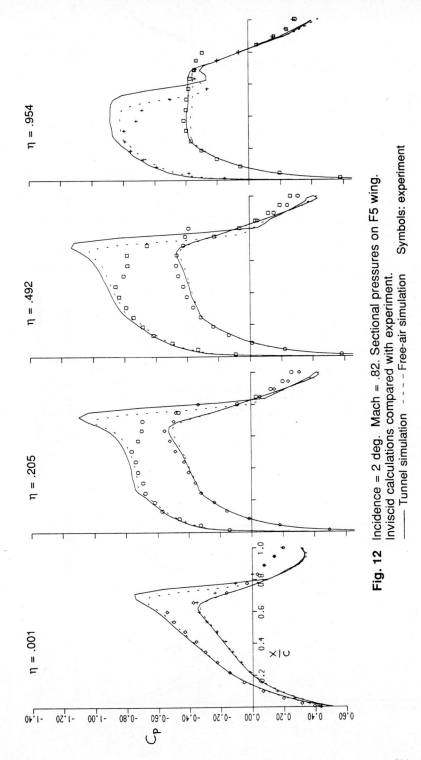

Fig. 12 Incidence = 2 deg. Mach = .82. Sectional pressures on F5 wing. Inviscid calculations compared with experiment. Symbols: experiment
——— Tunnel simulation - - - - Free-air simulation

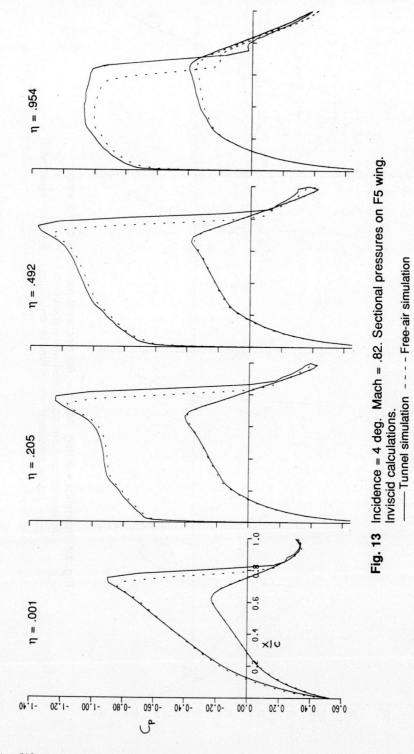

Fig. 13 Incidence = 4 deg. Mach = .82. Sectional pressures on F5 wing. Inviscid calculations.
——— Tunnel simulation   - - - - Free-air simulation

VORTEX BREAKDOWN

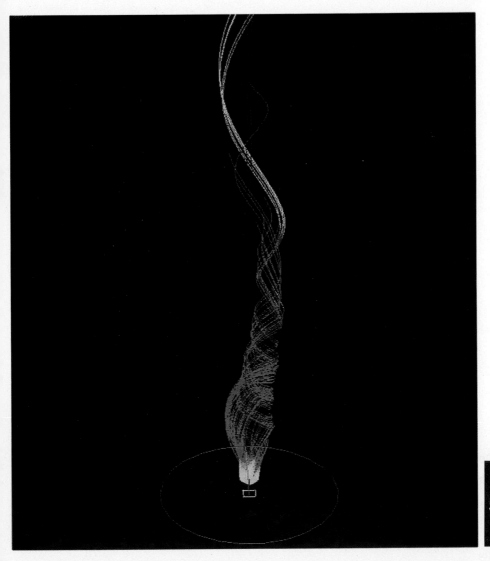

5.9

Colour graphics shows:

Direct simulation of vortex flow in non-axisymmetric shear flow:

Reynolds number = 200 (based on core radius (δ) and mean axial velocity)

Computational domain: L = 20δ and R = 3δ

5.9 Vortex Breakdown

Authors: **J.N. Sørensen, M.O.L.Hansen**, TU Denmark
Plot Coodinators: **J.N. Sørensen, M.O.L. Hansen**, TU Denmark

5.9.1 Introduction

In highly rotating flows a sudden appearence of a stagnation point followed by reversed axial flow may sometimes be seen. This phenomenon, termed vortex breakdown, was first observed in vortices shedded from delta wings. Later investigations demonstrated that vortex breakdown could also appear in confined flows of industrial interest, as e. g. in combustion chambers, cyclone separators, and in swirl flow in diverging ducts.

Several theories have attempted to define the conditions under which vortex breakdown is likely to occur. A detailed description of the internal structure of the breakdown region, however, demands a complete three-dimensional simulation. Such a description is useful for many industrial applications, such as in certain combustion processes.

To validate the ability of a Navier-Stokes algorithm to simulate vortex breakdown, the dynamical behaviour of an unconfined vortex is studied by direct simulation of the equations, i.e. without introducing any kind of artificial viscosity or turbulence model.

5.9.2 Test Cases

Consider a vortex which is defined in a cylindrical flow domain (r,Θ,z) with velocity components (u,v,w) and embedded in an axial shear flow. Let the shear flow be characteristic by a mean axial velocity W_o and let the core of the vortex be defined by a characteristic radius δ. Employing δ to non-dimensionalise the variables the Reynolds number is given as $Re = W_o\delta/v$, where v is the kinematic viscosity. In non-dimensionalised coordinates the lateral boundary is located at $r = R/\delta$ and the lenght of the flow domain is given by $z = L/\delta$, thus $r \in [0, R/\delta]$, $\Theta \in [0, 2\pi]$, and $z \in [0, L/\delta]$.

The vortex is defined by the tangential velocity profile (see Grabowski and Berger, 1976)

$$v_0 = \begin{cases} r \cdot (2-r^2) & \text{for } 0 \le r \le 1 \\ 1/r & \text{for } 1 \le r \le R/\delta \end{cases},$$

and the shear flow is given by the expression

$$w_o(r,\theta) = 1 + (w_{max} - 1)(\frac{\delta}{R})r\sin\theta ,$$

where w_{max} denotes the maximum axial velocity.

Two test cases are studied. In the first one, the base flow consists of a vortex embedded in a constant axial flow field, i.e. $w_{max} = 1$. This case is particularly interesting for analysing the code's ability of maintaining axial symmetry. In the second test case, the axial base flow is a non-axisymmetric shear flow with w_{max} = 1.25.

This test case serves to analyse non-axisymmetric behaviour, such as vortex tilting and streching, and to validate the codes ability of treating the polar singularity at the center line of the cylindrical coordinate system. In both cases, Re = 200 and a domain of L = 20 δ and R = 3δ is employed. Employing the three-dimensional Navier-Stokes algorithm described in section 4.14.2, only TUD contributed to this task.

5.9.3 Discussion and Results

5.9.3.1 Vortex Embedded in a Constant Axial Flow Field

Setting w_{max} = 1, the axial velocity becomes constant, equal to unity, and the base flow is axisymmetric. From axisymmetric calculations of Grabowski & Berger (1976) this case is known to exhibit vortex breakdown. In Fig. 1 the resulting flow field is shown after 300 timesteps corresponding to t = 25. A vortex breakdown is here seen to appear with a stagnation point at about z = 1.5. As seen from the velocity-vector plot the flow exhibits a high degree of axisymmetry. A closer view of the breakdown bubble (Fig. 2) shows that the breakdown consists of a closed bubble of recirculating fluid. In Fig. 3 velocity vectors are shown along a radius at various z-locations. The first (Fig.3a) shows the inlet velocity profile which corresponds to the basic tangential velocity distribution. At z = 1 (Fig. 3b), owing to retardation of axial velocity, radial outflow occurs. This results in a decrease of the swirl velocity, which is to be expected, since the angular momentum of a fluid particle moving radially outwards is conserved when viscous effects are negligible. At z = 10 (Fig. 3c) the maximum swirl velocity has decreased to about half the value of that at the inlet and the viscous core, defined as the radial distance from the vortex center to the point where the tangential velocity obtains its maximum value, has increased about 50%. Further downstream the velocity field does not change noticeable, thus at the outflow plane (Fig. 3d) any difference from the velocity distribution at z = 10 is hardly seen.

5.9.3.2 Vortex Embedded in Plane Shear Flow

To study the code's ability of calculating non-axisymmetric flows we simulate the dynamics of a vortex embedded in an axial shear flow with w_{max} = 1.25. The simulation is carried out by employing the former flow field shown in Fig. 3 as initial condition. In Fig. 4 a vector plot of the velocity field is shown in a plane along the center axis after 100 timesteps. The symmetry has disappeared and the vortex exhibits a spiralling form. There is now a strong interaction between the vorticity components resulting in a significant retardation of the axial flow component. A closer view of the rear part of the breakdown zone (Figs. 5 and 6) shows a rather complicated topology with a single recirculating bubble about which streamsurfaces are folded. In Figs. 7a and 7b velocity vectors along a radius are shown at z = 10 and z = 20 (outflow), respectively. At both positions the center of the vortex, defined as where v = 0, does not coincide with the polar center axis, and in particular at z = 10 strong radial velocities are seen to appear near the vortex center.

5.9.4 Conclusion

A three-dimensional Navier-Stokes code formulated in vorticity-velocity variables and given in cylindrical coordinates has been developed and validated against vortical flows exhibiting vortex breakdown. The results show that the code is able to preserve axisymmetric conditions, when they exist, and to simulate non-axisymmetric behaviour with strong vortical interaction.

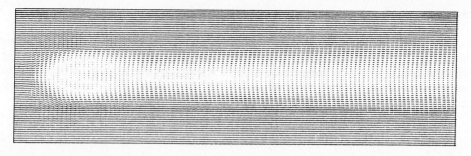

Fig. 1 Axisymmetric vortex breakdown; Velocity vectors at Θ = constant plane.

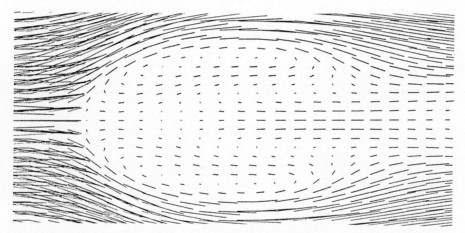

Fig. 2 Axisymmetric vortex breakdown; Magnification of bubble region.

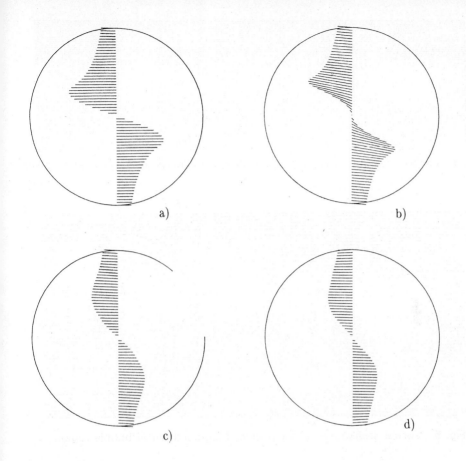

Fig. 3 Axisymmetric vortex breakdown; Velocity vectors at various z = constant planes; a) z = 0, b) z = 1.5; c) z = 10, d) z = 20.

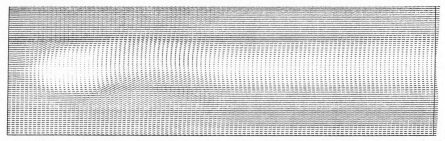

Fig. 4 Vortex breakdown in shear flow; Velocity vectors at Θ = constant plane.

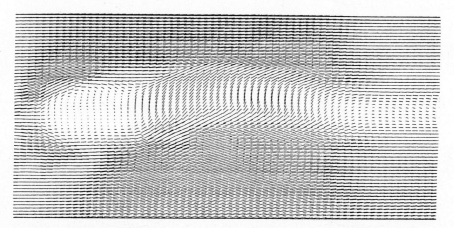

Fig. 5 Vortex breakdown in shear flow; Magnification of bubble region.

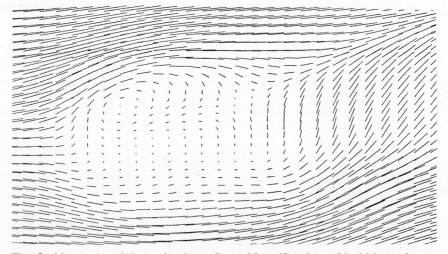

Fig. 6 Vortex breakdown in shear flow; Magnification of bubble region.

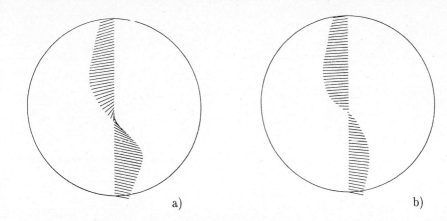

Fig. 7 Vortex breakdown in shear flow; Velocity vectors at various z = constant planes; a) z = 10; b) z = 20.

6 Reference List

Abbas, A. (1992): Private Communication

Abid, R. (1988): Extension of the Johnson-King turbulence model to 3D Flows. AIAA-Paper 88-0223.

Abid, R., Vatsa, V.N., Johnson, D.A., Wedan, B.W. (1989): Prediction of separated transonic wing flows with a non-equilibrium algebraic model. AIAA-Paper 89-0558.

Allwright, S.E. (1988): Techniques in multiblock domain decomposition and surface grid generation. Numerical Grid Generation in Computational Fluid Mechanics, Pineridge Press.

Ashill, P.R., Fulker, J.L., Weeks, D.J. (1985): The air-injection method of fixing boundary layer transition and investigating scale effects. RAE Technical Report 85025.

Ashill, P.R., Weeks, D.J. (1982): A method for determining wall-interference corrections in solid-wall tunnels from measurements of static pressures at the walls. AGARD CP-335, Paper 1.

Ashill, P.R., Wood, R.F., Weeks, D.J. (1987): An improved semi-inverse version of the Garabedian and Korn method (VGK). RAE TR 87002.

Baldwin, B.S. and Lomax, H. (1978): Thin layer approximation and algebraic model for separated turbulent flows. AIAA Paper 78-257.

Berg, B. van den, Elsenaar, A. (1972): Measurements in a three-dimensional incompressible turbulent boundary layer under infinite swept wing conditions. NLR TR 72092 U.

Berg, B. van den (1976): Investigations of three-dimensional incompressible turbulent boundary layers. NLR TR 76001 U.

Berg, B. van den (1979): Boundary layer measurements on a two-dimensional wing with flap. NLR TR 79009 U.

Berg, B. van den (1983): Comparison of theory and experiment for a simple two-dimensional airfoil with flap. GARTEUR Working Paper No. AD(AG03)/WP-3, NLR TR 83034 L.

Berg, B. van den, et al (1988): Three-dimensional turbulent boundary layers - calculations and experiments. In: Notes on Numerical Fluid Mechanics, Vol. 19, Vieweg Verlag.

Berg, B. van den (1989): A European collaborative investigation of the three-dimensional turbulent shear layers of a swept wing. AGARD Conf. Proc. No. 438.

Bergman, C.M., Vos, J.B. (1991): A multi block flow solver for viscous compressible flows. 9th GAMM Conference on Numerical Methods in Fluid Mechanics. Lausanne, Switzerland. NNFM, Vol. 35, pp 151-159, Vieweg.

Blottner, F.G. (1989): Accurate Navier-Stokes results for the hypersonic flow over a spherical nosetip. AIAA-Paper 89-1671.

Brandsma, F.J., Kuerten, J.G.M. (1990): The ISNaS compressible Navier-Stokes solver; first results for single airfoils. Proc. of the 12th Int. Conf. on Numerical Methods in Fluid Dynamics, July 1990. Ed.: K.W. Morton, Springer Verlag, pp. 152-156.

Cebeci, T. (1984): Problems and opportunities with three-dimensional boundary layers. AGARD Report No. 719.

Cebeci, T., Kaups, K., Ramsey, J. (1977): A general method for calculating three-dimensional compressible laminar and turbulent boundary layers on arbitrary wings. NASA CR-2777.

Cebeci, T. and Smith, A.M.O. (1974): Analysis of turbulent boundary layers. Academic Press, New York.

Cebeci, T., Smith, A.M.O., Mosinski, G. (1970): Calculation of compressible adiabatic turbulent boundary layers. AiAA J., Vol. 8, No. 11, pp. 1974-1982.

Chen, H.C., Patel, V.C. (1987): Practical near-wall turbulence models for complex flows including separation. AIAA-Paper 87-1300.

Chien, K.Y. (1982): Predictions of channel- and boundary layer flows with a low-Reynolds number turbulence model. AIAA J., 20, pp. 33-38.

Cook, P.H., MacDonald, M.A., Firmin, M.C.P. (1979): Aerofoil 2822 - Pressure distributions, boundary layer and wake measurements. AGARD AR 138.

Dalsem, W.R. van, Steger, J.L. (1985): The efficient simulation of separated three-dimensional viscous flows using the boundary-layer equations. AIAA-Paper 85-4064.

Davidson, L. (1990): Implementation of a semi-implicit k-ε turbulence model into an explicit Runge-Kutta solver. CERFACS Report TR/RF/90/25.

Dechow, R. (1977): Mittlere Geschwindigkeiten und Reynoldsscher Spannungstensor in der freidimensionalen Wandgrenzschicht vor einem stehenden Zylinder. Diss. Univ. Karlsruhe, Strömungsmechanik und Strömungsmaschinen, Heft 21.

Dechow, R., Felsch, K.O. (1977): Measurements of the mean velocity and of the Reynolds stress tensor in a three-dimensional turbulent boundary layer induced by a cylinder standing on a flat wall. First Symp. on Turbulent Shear Flows, Univ. Park, Pennsylvania.

Deiwert, G.S. (1980): Numerical simulation of high Reynolds number transonic flows. AIAA J., Vol. 13, pp. 1354-1359.

Delery, J., Copy, C. and Reisz, J. (1980): ONERA, Rapport Technique No. 37/078 AY 014, Chatillon, August 1980.

Delery, J., Srieix, M. and Capelier, C. (1980): ONERA, Rapport Technique No. 42/078 AY 014, Chatillon, December 1980.

Delery, J. (1981): Investigation of strong shock turbulent boundary layer interaction in 2D flows with emphasis on turbulence phenomena. AIAA Paper 81-1245.

Dimitriadis, K.P., Leschziner, M.A. (1989a): Computation of shock/boundary-layer interaction with transport models of turbulence. Proc. Royal Aeronautical Society Conference on the Prediction and Exploitation of Separated Flow, London, pp. 10.1-10.15.

Dimitriadis, K.P., Leschziner, M.A. (1989b): Approximation of viscous and turbulent transport in transonic-flow cell-vertex algorithm. 6th Int. Conf. on Numerical Methods in Laminar and Turbulent Flows, Swansea, pp. 861-881

Dimitriadis, K.P., Leschziner, M.A. (1990): Computation of turbulent transonic and supersonic flow with a cell-vertex algorithm and second-moment closure. Proc. 12th Int. Conf. on Numerical Methods in Fluid Mechanics, Lecture Notes in Physics, 371, K.W. Morton, Ed.), Springer Verlag.

Dimitriadis, K.P., Leschziner, M.A. (1991): A cell-vertex TVD scheme for transonic viscous flows. Proc. 7th Int. Conf. on Numerical Methods in Laminar and Turbulent Flows, Stanford.

Douglas, I.Jr. (1962): Alternating direction methods for three space variables. Numerische Mathematik, Vol. 4, pp. 41-63.

East, L.F. (1975): Computation of three-dimensional turbulent boundary layers. Euromech 60, Trondheim. FFA TN AE-1211.

Elsenaar, A., Boelsma, S.H. (1974): Measurements of the Reynolds stress tensor in a three-dimensional turbulent boundary layer under infinite swept wing conditions. NLR TR 74095 U.

Elsholz, E. (1988): Ein inverses LISW-Grenzschicht-Differenzenverfahren. MBB-Report TE 2-1681.

Eriksson, L.E. (1982): Generation of body-conforming grids around wing-body configurations using transfinite interpolation. AIAA Journal, Vol. 20, No. 10, pp. 1313-1320.

Eriksson, L.E. (1985): Simulation of inviscid flow around airfoils and cascades based on the Euler Equations. FFA TN 1985-20, Stockholm.

Eriksson, L.E. (1987): A finite volume solution technique for the Navier-Stokes equations governing viscous compressible flow. HOG Report 1987:102(A). Division of Hydro- & Gas Dynamics, The Norwegian Institute of Technology (NTH), Trondheim.

Erisson, L.E. (1990): Algebraic block-structured grid generation based on a macro-block concept. AGARD-CP-464.

Fritz, W. (1987): Two-dimensional and three-dimensional block structured grid generation techniques. Conference on Automated Mesh Generation and Adaptation, Grenobel, France.

Garner, H.C., Rogers, E.W.E., Acum, W.E.A., Maskell, E.C. (1969): Subsonic wind tunnel wall corrections. AGARDograph 109.

Gibson, M.M. and Launder, B.E. (1978): Ground effects on pressure fluctuations in the atmospheric boundary layer. J. Fluid Mech. 86, pp. 491-511.

Goldberg, U.C. (1986): Separated flow treatment with a new turbulence model. AIAA Journal, Vol. 24, pp. 1711-1713.

Goldberg, U.C., Chakravarthy, S.R. (1987): Prediction of separated flows with a new turbulence model. Proc. of the 5th Int. Conference on Numerical Methods in Laminar and Turbulent Flows, Pt. 1, Taylor, Habashi, Hafez (Eds.), pp. 560-571.

Gooden, J.H.M., Lent, M. van (1989): Measurements in the two-dimensional turbulent wing wake above a trailing edge flap (data report). NLR CR 89274 C.

Gooden, J.H.M., Lent, M. van (1991): Measurements in the two-dimensional turbulent wing wake above a trailing edge flap (data report, BRITE/EURAM - EUROVAL version). NLR CR 91038 C.

Grabowski, W.J., Berger, S.A. (1976): Solution of the Navier-Stokes equations for vortex breakdown. J. Fluid Mechanics, Vol. 75, pp. 525-544.

Granville, P.S. (1976): A modified law of the wake for turbulent Shear flows. ASME J. of Fluids Engineering, 98, pp. 578-580.

Granville, P.S. (1987): Baldwin-Lomax factors for turbulent boundary layers in pressure gradients. AIAA J., 25, No.12, pp. 1624-1627.

Green, J.F., Weeks D.J., Brooman, J.W.F. (1973): Prediction of turbulent boundary layers and wakes in compressible flow by a lag entrainment method. RAE TN 72231

Haase, W., Wagner, B., Jameson, A. (1983): Development of a Navier-Stokes method based on a finite volume technique for the unsteady Euler equations. In: Notes on Numerical Fluid Mechanics. Vieweg Verlag, Vol. 7, pp. 99-107.

Hall, M.G. (1086): Cell vertex multigrid schemes for solution of the Euler equations. In: Numerical Methods for Fluid Dynamics II, Clarendon Press, pp. 303, W. Marton and M.J. Baines (Eds.).

Hansen, M.O.L., Sørensen, J.N., Barker, V.A. (1992): Iterative solution of the 3D Cauchy-Riemann Type Equations for the vorticity-velocity formulation of the Navier-Stokes equations. Proc. of the 10th Int. Conf. on Computing Methods in Applied Sciences and Engineering, Paris, pp. 633-646.

Hassid, S. and Poreh, M. (1975): A turbulence energy model for flows with drag reduction. ASME J. of Fluids Engineering, 97, pp. 234-241.

Hedberg, P.K.M. (1989): NONDIF: A method to avoid numerical diffusion and over- and unders-shoots. Proc. 6th Int. Conf. on Numerical Methods for Laminar and Turbulent Flows, pp. 193-202.

Hirsch, Ch. (1990): Numerical computation of internal and external flows. Volume 2, John Wiley & Sons.

Hirsch, Ch. et al (1991): A multiblock/multigrid code for the efficient solution of complex 3D Navier-Stokes flows. First Europ. Symposium on aerothermodynamics for space vehicles, ESTEC, Noordwijk, The Netherlands.

Horton, H.P. (1969): Entrainment in equilibrium and non-equilibrium turbulent boundary layers. Hawker Siddeley Aviation, Hatfield, Research Report 1094.

Horton, H.P. (1990): A non-equilibrium model for turbulent boundary layers. Queen Mary & Westfield College, University of London

Iacovides, H., Launder, B.E. (1990): Parametric and numerical study of fully developed flow and heat transfer in rotating rectangular ducts. ASME Paper 90/GT/24.

Jameson, A. (1983): The evolution of computational methods in Aerodynamics. J. Appl. Mech., Vol. 50.

Jameson, A. (1985): Multigrid algorithms for compressible flow calculations. MAE Report 1743, Princeton University.

Jameson, A. (1985): Numerical solution of the Euler equations for compressible inviscid fluids. Numerical Methods for the Euler Equations of Fluid Dynamics. Ed. Angrand, F., et al, SIAM.

Jameson, A., Baker, T.J. (1983): Solution of the Euler equations for complex configurations. AIAA-Paper 83-1929.

Jameson, A., Baker, T.J. (1984): Multi-grid solution of the Euler equations for aircraft configurations. AIAA-Paper 84-0093.

Jameson, A., Baker, T.J., Weatherill, N.P. (1986): Calculation of inviscid transonic flow over a complete aircraft. AIAA-Paper 86-0103.

Jameson, A., Schmidt, W. (1985): Some recent developments in numerical methods for transonic flows. Computer Methods in Applied Mechanics and Engineering, Vol. 51, pp. 515-528.

Jameson, A., Schmidt, W., Turkel, E. (1981): Numerical solutions of the Euler equations by finite volume methods using Runge-Kutta time-stepping schemes. AIAA-Paper 81-1259.

John, D. (1991): Numerisches Berechnung der 3D Geschwindigkeitskomponenten am Grenzschichtrand. DA Report EF1850.

Johnson, D.A. (1987): Transonic separated flow predictions with an eddy viscosity/Reynolds stress closure model. AIAA J., Vol. 25, pp. 252-259.

Johnson, D.A., Coakley, T.J. (1989): Improvements to a non-equilibrium algebraic turbulence model. AIAA J., Vol. 28, No. 11, pp. 2000-2003.

Johnson, D.A., King, L.S. (1984): A mathematical simple turbulence closure model for attached and separated turbulent boundary layers. AIAA J., Vol. 23, No. 11, pp. 1684-1692.

Johnson, D.A., King, L.S. (1985): A mathematical simple turbulence closure model for attached and separated turbulent boundary layers. AIAA-Paper 84-0175

Jones, W.P. and Launder, B.E. (1972): The prediction of laminarisation with a two-equation model of turbulence. Int. J. Heat Mass Transfer, 15, pp. 301-314.

King, D.A., Williams, B.R. (1988): Developments in computational methods for high-lift aerodynamics. Aero. J., Vol 92, pp. 265.

King, L.S. (1987): A comparison of turbulence closure models for transonic flows about airfoils. AIAA Paper 87-0418.

Kline, S.J., Morkovin, M.V., Sovran, G., Cockrell, D.J. (1968): Proceedings: Computation of turbulent boundary layers - 1968, AFOSR-IFP Stanford Conference. Stanford Univ., California, USA, 94305: L.C. 68-58871.

Kordulla, W. (Ed.) (1988): Numerical simulation of the transonic DLR-F5 wing experiment. Notes on Numerical Fluid Mechanics. Vol. 22, Vieweg, Braunschweig/Wiesbaden.

Kordulla, W., Schwamborn, D., Sobieczky, H. (1988): The DFVLR-F5 wing experiment - towards the validation of the numerical simulation of transonic viscous wing flows. AGARD CP-437, Validation of CFD, Lisbon.

Lam, C.K.G. and Bremhorst, K.A. (1981): Modified form of the k-ε model for predicting wall turbulence. ASME J. Fluids Engineering, 103, p. 456.

Launder, B.E and Sharma, B.I. (1974): Application of the energy dissipation model of turbulence to the calculation of flow near a spinning disk. Lett. Heat Mass Transfer, 2, p. 1.

Lindeberg, T. (1987): The construction of a three-dimensional finite volume grid generator for a wing in wind tunnel with application to Navier-Stokes flow solvers. FFA TN 87-62, Stockholm.

Lock, R.C., Williams, B.R. (1987): Viscous-inviscid interactions in external aerodynamics. Prog. Aerospace Sci, Vol. 24, pp.51.

Martinelli, L. (1987): Calculations of viscous flows with a multigrid method. Ph.D. Thesis, MAE Department, Princeton University.

Mavriplis, D.J. (1988): Multigrid solution of the two-dimensional Euler equations on unstructured triangular meshes. AIAA Journal, Vol. 26, pp. 824-831.

Middlecoff, J.E., Thomas, P.D. (1979): Direct control of the grid point distribution in meshes generated by elliptic equations. AIAA Paper 79-1462.

Müller, B., Rizzi, A. (1987): Navier-Stokes computations of transonic vortices over a round leading edge delta wing. AIAA-Paper 87-1227.

Müller, B., Rizzi, A. (1988): Navier-Stokes solutions for transonic flows over wings. Proc. 7th GAMM-Conf. on Numerical Methods in Fluid Mechanics, M.Deville (Ed.). Notes on Numerical Fluid Mechanics, Vol. 20, Vieweg, Braunschweig/Wiesbaden, pp. 247-255.

Pantakar, S.V. (1980): Numerical Heat transfer and fluid flow. McGraw-Hill, New York.

Radespiel, R. (1989): A cell-vertex multigrid method for the Navier-Stokes Equations. NASA TM 101557.

Radespiel, R., Rossow, C., Swanson, R.C. (1989): An efficient cell-vertex multigrid scheme for the three-dimensional Navier-Stokes equations. AIAA-Paper 89-1953.

Rhie, C.M. (1981): A numerical study of the flow past an isolated airfoil with separation. Ph.D. Thesis, Univ. Illinois, Urbane-Champaign.

Rizzi, A., Erikson, L.E. (1984): Computation of flow around wings based on the Euler equations. J. Fluid Mechanics, Vol. 148, pp. 45-71.

Rodi, W. (1976): A new algebraic relation for calculation of the Reynolds stresses. ZAMM, 56, p. 219.

Rodi, W. (1991): Experience with two-layer models combining the k-ε model with a one-equation model near the wall. AIAA-Paper 91-0216.

Roe, P.L. (1981): Approximate Riemann solver, parameter vectors and difference schemes. Journal of Computational Physics, 43, pp. 357-372

Roe, P.L. (1984): Generalized formulation of TVD Lax-Wendroff Schemes. ICASE Report no. 84-53.

Rudy, D.M., Strickwerda, J.C. (1988): A non-reflecting outflow boundary condition for subsonic Navier-Stokes calculations. Comp. Physics, Vol. 36, pp. 50-70. AGARD Paper No. 437.

Runchal A.K. (1986): A modified central-difference scheme with unconditional stability and very low numerical diffusion. Proc. 8th Int. Heat Transfer Conference, San Francisco.

Schwamborn, D. (1988): Simulation of the DLR-F5 wing experiment using a block structured explicit Navier-Stokes method. In: Notes on Numerical Fluid Mechanics. Vieweg, Vol. 22.

Schwamborn, D. (1991a): Validation of 3D Navier-Stokes codes for flows about wings. Progress Report for the First Year of the EUROVAL Project, DLR-IB 221-91 A 25.

Schwamborn, D. (1991b): Validation of 3D Navier-Stokes codes for flows about wings. Progress Report for the Third Half-Year of the EUROVAL Project, DLR-IB 221-91 A 28.

Schwamborn, D. (1991): Private Communication.

Schlichting, H. (1955): Boundary layer theory. McGraw-Hill.

Siikonen, T., Hoffren, J., Laine, S. (1990): A multigrid LU factorization scheme for the thin-layer-Navier-Stokes equations. ICAS-90-6.10.3, 17th ICAS Congress, Stockholm.

Sobieczky, H. (1985): Geometry generation for transonic design. In: Advances in Computational Transonics. Series on Recent Advances in Numerical Methods in Fluids. Vol. 4, W.G. Habashi (Ed.), Pineridge Press.

Sobieczky, H. (1987): DFVLR-F5 test wing configuration for computational and experimental aerodynamics: Wing surface generator code, control surface and boundary conditions. DFVLR Report IB 221-87.

Sobieczky, H. (1988): DFVLR-F5 test wing configuration. - The boundary value problem. In: Notes on Numerical Fluid Mechanics. Vol. 22, Vieweg.

Sobieczky, H., Hefer, G., Tusche, S. (1987): DFVLR-F5 wing experiment for computational aerodynamics. AIAA-Paper 87-2485. Also in: Kordulla (1988).

Stock, H.W. and Haase, W. (1989): Determination of length scales in algebraic turbulence models for Navier-Stokes methods. AIAA J., Vol. 27, No. 1, pp. 5-14.

Swanson, R.C., Turkel, E. (1987): Artificial dissipation and central difference schemes for the Euler and Navier-Stokes equations. AIAA-Paper 87-1107.

Thomas, J.L., Salas, M.D. (1985): Far-field boundary conditions for transonic lifting solutions to the Euler Equations. AIAA-Paper 85-0020.

Thompson, J.E., Thames, J.P., Martin, C.W. (1974): Automatic numerical generation of a body fitted curvilinear coordinate system for a field containing any number of arbitrary two-dimensional bodies. J. Comp. Physics, Vol. 15.

Thwaites, B. (1949): Approximate calculation of the laminar boundary layer. Aero. Quarterly, Vol. 1, pp. 245.

Vatsa, V.N., Wedan, B.W. (1989): Development of an efficient multigrid code for 3D Navier-Stokes equations. AIAA-Paper 89-1791.

Veldman A.E.P. (1980): The calculation of incompressible boundary layers with strong viscous-inviscid interaction. AGARD CP 291, pp. 265.

Visbal, M.R., Shang, J.S. (1985): A comparative study between an implicit and explicit algorithm for transonic airfoils. AIAA-Paper 85-0480.

Weinerfelt, P.Å. (1992): A multigrid multiblock technique applied to the 2D Navier-Stokes equations. Department of Mathematics, Linköping University.

Wolfshtein, M. (1969): The velocity and temperature distribution in one-dimensional flow with turbulence augmentation and pressure gradient. Int. J. Heat and Mass Transfer, 12, pp. 301-312.

Xue, L., Thiele, F. (1989): An inverse boundary layer procedure with application to 3D-wing flow. 8th GAMM Conference, Delft.

Xue, L., Lesch, J., Thiele, F. (1989): Validierung und Dokumentation des dreidimensionalen Grenzschicht-Differenzenverfahrens. TUB-HFI Report.

Yee, H.C. (1987): Construction of explicit and implicit symmetric TVD schemes and their applications. Journal of Computational Physics, 68, 151-179.

Yee, H.C. (1989): A class of high resolution explicit and implicit shock-capturing methods. von Karman Institute Lecture Series 1989-04.

Yee, H.C., Warming, R.F. Harten, A. (1983): Implicit total variation diminishing schemes for steady state calculations. AIAA-Paper 83-1902.

Addresses of the Editors of the Series "Notes on Numerical Fluid Mechanics"

Prof. Dr. Ernst Heinrich Hirschel (General Editor)
Herzog-Heinrich-Weg 6
D-8011 Zorneding
Federal Republic of Germany

Prof. Dr. Kozo Fujii
High-Speed Aerodynamics Div.
The ISAS
Yoshinodai 3-1-1, Sagamihara
Kanagawa 229
Japan

Prof. Dr. Bram van Leer
Department of Aerospace Engineering
The University of Michigan
Ann Arbor, MI 48109-2140
USA

Prof. Dr. Keith William Morton
Oxford University Computing Laboratory
Numerical Analysis Group
8-11 Keble Road
Oxford OX1 3QD
Great Britain

Prof. Dr. Maurizio Pandolfi
Dipartimento di Ingegneria Aeronautica e Spaziale
Politecnico di Torino
Corso Duca Degli Abruzzi, 24
I-10129 Torino
Italy

Prof. Dr. Arthur Rizzi
FFA Stockholm
Box 11021
S-16111 Bromma II
Sweden

Dr. Bernard Roux
Institut de Mécanique des Fluides
Laboratoire Associé au C.R.N.S. LA 03
1, Rue Honnorat
F-13003 Marseille
France

Brief Instruction for Authors

Manuscripts should have well over 100 pages. As they will be reproduced photomechanically they should be typed with utmost care on special stationary which will be supplied on request.
In print, the size will be reduced linearly to approximately 75 per cent. Figures and diagrams should be lettered accordingly so as to produce letters not smaller than 2 mm in print. The same is valid for handwritten formulae. Manuscripts (in English) or proposals should be sent to the general editor, Prof. Dr. E. H. Hirschel, Herzog-Heinrich-Weg 6, D-8011 Zorneding.